ACOUSTICS RESEARCH AND TECHNOLOGY

SOUND WAVES

PROPAGATION, FREQUENCIES AND EFFECTS

ACOUSTICS RESEARCH AND TECHNOLOGY

Additional books in this series can be found on Nova's website
under the Series tab.

Additional E-books in this series can be found on Nova's website
under the E-book tab.

ACOUSTICS RESEARCH AND TECHNOLOGY

SOUND WAVES

PROPAGATION, FREQUENCIES AND EFFECTS

VITALE ABAGNALI
AND
GIAMPAOLO FABBRI
EDITORS

Nova Science Publishers, Inc.
New York

Library of Congress Cataloging-in-Publication Data

Sound waves : propagation, frequencies, and effects / editors, Vitale Abagnali and Giampaolo Fabbri.
p. cm.
Includes bibliographical references and index.
ISBN 978-1-61470-001-2 (hardcover : alk. paper) 1. Sound-waves. I. Abagnali, Vitale. II. Fabbri, Giampaolo.
QC243.S69 2011
534--dc23

2011020034

Published by Nova Science Publishers, Inc. † New York

CONTENTS

PREFACE

In this book, the authors present current research in the study of the propagation, frequency and effects of sound waves. Topics discussed include time resolved visualization and analysis on a single short acoustic wave generation; elastic vibrations of an isotropic plate with laser-induced atomic defects; sound velocity into turbulent medium; infrasound generation by turbulent convection; neutrons diffraction in a crystal under the influence of a sound wave and the transformation of sound waves in non-stationary media. (Imprint: Nova)

Chapter 1 - The generation and detection of short acoustic waves in liquid and solid media by short pulsed laser based techniques have been demonstrated to be efficient tools in many applied science and medical areas, including nondestructive evaluation and material characterization, laser cleaning of surface contaminants, laser tissue ablation, corneal sculpturing, and gall stone fragmentation. The laser energy interaction with the surface of an absorbing liquid or a transparent liquid in contact with an absorbing solid boundary, induces rapid heating, thermoelastic expansion or explosive phase change and finally emission of a strong ultrasonic wave or a shock wave, depending on the applied laser energy. Most of the laser induced acoustic wave experimental studies have aimed at the zero-dimensional point detection of the propagating pressure transient at a fixed location by utilizing capacitance transducers, piezoelectric transducers, or hydrophones. These approaches can only be applied to idealized zero- or one-dimensional geometries and fail to capture the detailed acoustic wave behavior inside the medium, especially for complex two- or three-dimensional configurations. The direct transient observation of two or three-dimensional laser induced acoustic wave interaction with solid structure is very important to understand the physics of short acoustic wave and to validate the numerical simulation studies. In this chapter, the state of the art laser based time resolved (100ns resolution) visualization technique of a single, short acoustic plane wave generation, propagation in various external and internal channels and the subsequent interaction with submerged solid structures are presented. The effect of liquid viscosity on the acoustic wave propagation velocity, pressure attenuation, and wave broadening was investigated. The fluid viscosity was varied by mixing glycerol with DI water to yield liquid of 1, 10, and 100 cp in viscosity at room temperature.

Chapter 2 - An analysis of the propagation of plane elastic waves in isotropic plates containing a distribution of non-equilibrium atomic point defects (interstitial atoms, vacancies, electron-hole pairs) is presented. The formation of atomic defects occurs as a result of the action of intense laser radiations on the solid plates. The study is based on coupled evolution equations for the elastic displacement of the medium and atomic defect density fields. The defect dynamics is governed by the strain-stimulated generation, transport, and annihilation processes. The frequency equations corresponding to the symmetric and antisymmetric elasto - concentration modes of vibration of the plate are obtained. Some limiting cases of the frequency equations are considered and a procedure for determining the phase velocity and the attenuation (or amplification) constants is discussed. Relevant results of previous investigations are deduced as special cases.

Chapter 3 - The development of techniques to study the dynamics of change of colloidal systems during processing or storage is critical, as these systems are often in a metastable state, and any disruption occurring during the analysis will affect the interpretation of the data. Acoustics and electroacoustics are a novel non-destructive technique that can simultaneously determine the ultrasonic properties and particle surface charge of emulsions. Since this method can be used in the analysis of highly concentrated and opaque samples, it has the potential to provide highly accurate data on the colloidal state of food emulsions. In addition, electroacoustics may be employed to study destabilization processes or interactions amongst ingredients. In this work, a model system of sodium caseinate emulsions is used to discuss the potential of acoustic spectroscopy and electroacoustics as means to follow the changes occurring to droplet interactions and emulsion destabilization. Sodium caseinate is one of the most frequently used emulsifiers in food emulsions, as it readily adsorbs at the interface during homogenization and stabilizes the oil droplets from coalescence due to electrostatic and steric repulsion. However, changes in environmental conditions affecting the polyelectrolyte layer surrounding the oil droplets (i.e., pH, ionic strength) as well as the presence of unadsorbed polymer can influence the stability of the emulsion droplets. The results demonstrate that ultrasonic techniques may assist in understanding the details of the destabilization with minimal sample disruption.

Chapter 4 - Recently, research on new techniques for single-chamber mufflers equipped with non-perforated intruding tube has been addressed; however, the research work on multi-chamber mufflers conjugated with open-ended perforated intruding inlet-tubes which may dramatically increase the acoustical performance has been neglected. Therefore, the main purpose of this paper is not only to analyze the sound transmission loss (STL) of a multi-chamber open-ended perforated inlet-tube muffler but also to optimize the best design shape within a limited space. In this paper, the four-pole system matrix for evaluating the acoustic performance — sound transmission loss (STL) — is derived by using a decoupled numerical method. Moreover, a genetic algorithm (GA), a robust scheme used to search for the global optimum by imitating a genetic evolutionary process, has been used during the optimization process. Before dealing with a broadband noise, the STL's maximization with respect to a one-tone noise is introduced for a reliability check on the GA method. Additionally, an accuracy check of the mathematical model is performed. To appreciate the acoustical ability of the open-ended perforated intruding inlet-tube and chambers inside a muffler, two kinds of traditional multi-chamber mufflers hybridized with simple expansion tubes and non-perforated intruding inlet-tubes have been assessed and compared. Moreover, the acoustical performance of the open-ended perforated intruding inlet-tube equipped with 1~3 chambers has also been analyzed. Results reveal that the maximal STL is precisely located at the desired tone. In addition, the acoustical performance of mufflers conjugated with perforated intruding inlet-tubes is superior to traditional mufflers. Also, it has been shown that the acoustic performance for both pure tone and broadband noise will increase if the muffler has more chambers. Consequently, the approach used for the optimal design of the noise elimination proposed in this study is easy and effective.

Chapter 5 - The sound velocity into the turbulent medium depends both on the sound velocity connecting with the usual molecular transport of the impulse and on the turbulent pulsation of the flow velocity too. Its value is higher than the usual sound velocity into the laminar medium as the turbulence transports the impulse additionally by its way. In this chapter, some generalization of the previous works touching with the problems of the sound waves propagation into turbulent medium is presented.

Chapter 6 - Variety of artificial and geophysical processes radiate infrasound, acoustic waves with frequencies below the normal limit of human hearing. Infrasonic signals are manifestations and/or precursors of extreme geophysical events or anthropogenic processes and, due to the property of infrasound to travel for great distances relatively undiminished, could be quite useful in advanced warning and monitoring purposes. Several universities and institutions around the world have research programs for studying acoustic generation and propagation mechanisms as well as for development of instrumentations and techniques for observation. Observations conducted over the last decades revealed that strong convective storms, such as supercells, that have cloud tops greater than 14 km or create a hook echo and are capable of producing strong tornadoes, generate significant infrasound in a passband from 0.5 to 2.5 Hz, with peak frequencies between 0.5 and 1 Hz, substantially before (0:5 ¡ 1 hrs) tornado appearance. Broad and smooth spectra of observed infrasound radiation indicates that turbulence is the most promising sources of the radiation. In this chapter, the authors review properties and characteristics of atmospheric infrasound waves. They study acoustic radiation from turbulent convection using Lighthill's acoustic analogy and taking into account the effects of stratification, temperature fluctuations and moisture in the air. It is shown that in saturated moist air turbulence in addition to the Lighthill's quadrupole and dipole sources of sound (related to stratification and temperature fluctuations), there exist monopole sources related to heat and mass production during the condensation of moisture. The authors determine the acoustic power of these monopole sources and show that radiation of a monopole source related to the nonstationary heat production during the condensation of moisture is dominant for typical parameters of strong convective storms. The results are in good qualitative agreement with the main observed infrasound characteristics e.g. total acoustic power and characteristic frequency. They perform spectral analysis of this source and give quantitative explanation of the high correlation between intensity of infrasound generated by supercell storms and later tornado formation. It is shown that low lifting condensation level (LCL) and high values of convective available potential energy (CAPE), which are known to favor significant tornadoes, also lead to a strong enhancement of supercells low frequency acoustic radiation. This qualitative analysis indicate the potential for infrasonic detection systems to determine potentially tornadic storms and improve tornado forecast.

Chapter 7 - In this chapter the authors consider neutron diffraction in a crystal under the influence of an external sound wave. They examine both the traveling and standing hypersonic waves' contribution in the neutron diffraction intensity, diffraction condition and the Debye–Waller factor (influence of thermal motion of atoms). The possibility of diffraction of both thermal and high-energy (short-wave) neutrons is shown in the process of multiphonon interaction of a neutron with a crystal and the field of a hypersonic wave. The difference between traveling and standing sound waves is discussed. The formation of a sublattice is illustrated in the process of neutron elastic scattering with respect to a standing hypersonic wave. The periodicity of a sublattice has the same order as the high-energy neutron wavelength, 0.01-0.1 Angstr¨oms. The analogy to the Kapitza–Dirac effect is considered for neutrons. The possible tuning of the Debye–Waller factor is examined as well. It is shown that the application of a hypersonic wave can either suppress or enhance the diffraction intensity. A realistic condition is obtained under which the negative influence of the thermal motion of atoms of which the matter is comprised is completely eliminated.

Chapter 8 - In this chapter the transformation of sound waves in the fluid dynamical approach is considered. In particular the reflection and transmission of sound waves for spatially homogeneous non-stationary media are considered. Time dependence of the properties of the medium (for example, mass density, sound velocity) is induced from the abruptness of its changes. Reflection and transmission coefficients for both sound wave amplitudes and energy fluxes are obtained. Quantitative relations between the reflection and transmission coefficients are adduced. It is shown that the sum of the energy flux reflection and transmission coefficients is greater than one and that the energy of a sound wave is not conserved, that is, exchange of the energy occurs between the wave and the medium. The non-conservation of the energy causes the sound wave to obtain a notable property: the transmitting wave carries an energy equal to the sum of the energies of the incident and reflected waves. The possibility of the amplification of sound waves is illustrated and a transformation of their frequencies is shown: the physical justification of these effects is also given. The problem is generalized for smooth variation of mass density and sound velocity and the transmission and reflection coefficients are obtained from the solution of generalized wave equation written for the potential of the velocity of the fluid.

Chapter 9 - Ocean Acoustic Tomography (OAT) is a remote-sensing technique for the collection of synoptic data from the ocean interior pertaining to density and meso-scale ocean flow fields that has been studied by many scientists. Dynamic ocean processes could be observed by measuring the change in the travel time of acoustic signals transmitted over a number of ocean paths. The travel time is subjected to change by thermal anomalies and inhomogeneity along the acoustic ray paths. An estimate of these quantities can be calculated by using Inversion Techniques. In this chapter the stochastic inverse method of estimating the sound velocity perturbations field from information on travel times of sound pulses through a real ocean has been discussed. A simulation experiment on forward and inverse problems for the observed sound velocity perturbation field has been discussed (in the vertical slice) keeping the single source- receiver configuration (at the channel axis depth) in the SOFAR channel, under peculiar characteristics viz, depth-limited environment and weak gradient waters of the northern Indian Ocean. For the formulation of the Stochastic Inverse, both vertical and horizontal structure of the ocean has been modeled using empirical orthogonal modes. In this chapter, the authors report (a) acoustic characteristics of a sub-surface cold core eddy observed (below the mixed layer between depths of 50 and 300m, with a diameter of about 200 km having temperature drop of 50 C at the center) in the Bay of Bengal during south-west monsoon period (21-29 July1984) and explore the possibility to reconstruct the acoustic profile of the eddy by Stochastic Inverse Technique and (b) results from a medium range, short duration acoustic tomography experiment for mapping of the synoptic ocean thermal field in the Arabian Sea during summer (2-12 May 1993). This OAT experiment has many components: preparation of available data on ocean parameters relevant to OAT studies, acoustic model simulation in reference ocean and relevant mathematics, optimal estimation of sound velocity perturbations prevailing at the time of observations and finally the interpretation of results for oceanographic applications.

In: Sound Waves: Propagation, Frequencies and Effects
Editors: V. Abagnali and G. Fabbri
ISBN 978-1-61470-001-2

Chapter 1

TIME RESOLVED VISUALIZATION AND ANALYSIS ON A SINGLE SHORT ACOUSTIC WAVE GENERATION, PROPAGATION AND INTERACTION

Seung Hwan Ko[1*] and Costas P. Grigoropoulos[2†]
[1] Department of Mechanical Engineering, KAIST
Daejeon, Korea
[2] Department of Mechanical Engineering
University of California, Berkeley
California, U. S.

Abstract

The generation and detection of short acoustic waves in liquid and solid media by short pulsed laser based techniques have been demonstrated to be efficient tools in many applied science and medical areas, including nondestructive evaluation and material characterization, laser cleaning of surface contaminants, laser tissue ablation, corneal sculpturing, and gall stone fragmentation. The laser energy interaction with the surface of an absorbing liquid or a transparent liquid in contact with an absorbing solid boundary, induces rapid heating, thermoelastic expansion or explosive phase change and finally emission of a strong ultrasonic wave or a shock wave, depending on the applied laser energy.

Most of the laser induced acoustic wave experimental studies have aimed at the zero-dimensional point detection of the propagating pressure transient at a fixed location by utilizing capacitance transducers, piezoelectric transducers, or hydrophones. These approaches can only be applied to idealized zero- or one-dimensional geometries and fail to capture the detailed acoustic wave behavior inside the medium, especially for complex two- or three-dimensional configurations. The direct transient observation of two or three-dimensional laser induced acoustic wave interaction with solid structure is very important to understand the physics of short acoustic wave and to validate the numerical simulation studies.

* E-mail address: maxko@kaist.ac.kr
† E-mail address: cgrigoro@me.berkeley.edu

In this chapter, the state of the art laser based time resolved (100ns resolution) visualization technique of a single, short acoustic plane wave generation, propagation in various external and internal channels and the subsequent interaction with submerged solid structures are presented. The effect of liquid viscosity on the acoustic wave propagation velocity, pressure attenuation, and wave broadening was investigated. The fluid viscosity was varied by mixing glycerol with DI water to yield liquid of 1, 10, and 100 cp in viscosity at room temperature.

1. Introduction

Photoacoustic pulses produced via the absorption of pulsed ruby laser radiation in solids and liquids were first observed in 1963 [1]. As the laser technology progressed, a number of efforts were reported on opto-acoustical effects, primarily of a pulsed nature.

The short pulsed laser induced acoustic wave generation and detection in liquid and solid media have been demonstrated to be efficient tools in many applied science and medical areas, including nondestructive evaluation and material characterization [2]-[4], laser cleaning of surface contaminants [5], laser tissue ablation, corneal sculpturing [6], and gall stone fragmentation [7]. The laser energy interaction with the surface of an absorbing liquid or a transparent liquid in contact with an absorbing solid boundary induces rapid heating, thermo-elastic expansion or explosive phase change and finally emission of a strong ultrasonic wave or a shock wave, depending on the applied laser energy [8], [9].

Most of the laser induced acoustic wave experimental studies have aimed at the zero-dimensional point detection of the propagating pressure transient at a fixed location by utilizing capacitance transducers [2], piezoelectric transducers [10], or hydrophones [11]. These approaches can only be applied to idealized zero- or one-dimensional geometries and fail to capture the detailed acoustic wave behavior inside the medium, especially for complex two- or three-dimensional configurations. Several numerical simulation studies [11], [12]-[15] were carried out to discuss multidimensional laser induced acoustic wave behavior. In addition, usually the laser pulse was tightly focused to generate a pressure point source for an expanding hemispherical wave induced by the material ablation and subsequent shock formation. Acoustic wave formation by surface ablation is not desirable in many technical applications due to the destructive nature and subsequent plasma formation in spite of the superior energy conversion efficiency [9]. Ko *et. al.* first reported the time resolved direct observation of two or three-dimensional laser-induced acoustic wave interaction with submerged solid structures such as single, double, 33° tilted single block, and concave cylindrical acoustic lens configurations [16], [17] and sudden expansion and contraction channels, bifurcating channels, gradual contraction wall, a cylinder [18]. They further demonstrated that these waves can be focused down to a domain of several tens of microns in size and acoustic focusing lens has potential applications to biomaterial processing and characterization.

In this chapter, the state of the art laser based time resolved (100 ns resolution) visualization technique of a single, short acoustic plane wave generation, propagation in various external and internal channels and the subsequent interaction with submerged solid structures are presented. The effect of liquid viscosity on the acoustic wave propagation velocity, pressure attenuation, and wave broadening was investigated. The fluid viscosity was varied by mixing glycerol with DI water to yield liquid of 1, 10, and 100 cp in viscosity at room temperature. The time resolved imaging of laser induced single, short acoustic plane wave propagation in various water filled internal channel and external channel configurations are be presented; internal channels: (a) sudden expansion and contraction channels, (b) bifurcating channels: T branched and Y branched (coronary artery bifurcation) channels, (c) gradual contraction wall channels: linear contraction and parabolic contraction (inkjet nozzle) wall channel, (d) a cylinder, and external channels: (a) a single block, (b) double blocks, (c) 33° tilted single block, and (d) concave cylindrical acoustic lens configurations. Experimental observations were compared with finite element method (FEM) based numerical simulations.

2. Experiment and Numerical Simulations

The generation/propagation/interaction of laser induced acoustic waves in internal channels were directly observed by two laser systems with different wavelength. The two laser systems were combined to achieve different functions; (1) neodymium-doped yttrium aluminum garnet (Nd:YAG) nanosecond pulsed laser with green wavelength (532nm) to generate the acoustic wave and (2) nitrogen laser pumped dye laser with blue wavelength (440nm) as an illumination source for the time resolved acoustic wave imaging. The two systems were synchronized via a delay generator to control the precise time delay between the two laser pulse firings [18]. All images went through background correction image processing to reduce background noise. Besides the experimental observations, numerical simulations were carried out to compare with experiment and also to further explain detailed physics.

A. Time Resolved Acoustic Wave Imaging: Laser flash Schlieren photography

Figure 1(a) shows how two laser systems with different wavelength were combined and synchronized to acquire time resolved images of the laser induced acoustic wave propagation in water and interaction with internal channel walls. A frequency doubled, homogenized Nd:YAG pulsed laser beam [New Wave Inc, wavelength = 532 nm, temporal pulse width = 5 ns full width half maximum (FWHM)], was applied normal to the vacuum deposited 100-nm-thick chromium (Cr) film on a 800 μm thick fused quartz substrate. The Nd:YAG laser is first absorbed by the Cr thin film directly attached to the de-ionized (DI) water filled channels. Upon laser induced heating, both the liquid and the Cr thin film experience thermo-elastic expansion and stress fields and thereby launches an acoustic wave into optically and acoustically transparent water. The Nd:YAG laser beam cross section was shaped to a 2 x 2 mm^2 flat-top beam profile of very good spatial uniformity by a microlens laser beam homogenizer [SUSS MicroOptics] combined with a 10× long working distance objective lens [Mitutoyo]. The Nd:YAG energy fluence applied to induce pressure wave was around 12-14 mJ/cm^2 which is substantially lower than the Cr melting threshold (120 mJ/cm^2) [19] and even smaller than bubble nucleation threshold (36.5 mJ/cm^2) [6]. For the laser energy levels considered in this work, transient laser induced pressure waves observed are attributed solely to the thermoelastic response of the metal thin film and the adjacent liquid. Rapid bubble growth, and collapse, as well as to plasma formation are not likely to contribute [5].

At a certain time delay after the firing of the Nd:YAG laser beam inducing acoustic wave, the nitrogen laser pumped dye laser beam [wavelength = 440 nm, pulse width = 10 ns FWHM, Laser Science Inc.] was expanded and applied parallel to the metal thin film surface as a synchronized illumination light source. A short pass filter was placed in front of the zoom lens to block the Nd:YAG laser beam and to accept only the nitrogen laser pumped dye laser beam. Upon encountering pressure-induced refractive index variations by Nd:YAG laser beam, the imaging laser beam (nitrogen laser pumped dye laser beam) bends from the original beam path. Time resolved frozen images of the acoustic waves were acquired via this mechanism.

The temporal interval between the processing Nd:YAG and the illuminating nitrogen laser pumped dye laser beams was controlled by a delay generator [Stanford Research Systems] as shown in Figure 1(b). Nd:YAG laser (indicated by a green colored line) was fired by two signals : lamp signal from channel 1 (Ch1) at time 1 (T_1) and Q-switch signal from channel 2 (Ch2) at time 2 (T_2). Nitrogen laser pumped dye laser illumination light source (indicated by a blue colored line) was fired by one trigger signal from channel 3 (Ch3) at time 3 (T_3). The time delay (ΔT) between T_2 and T_3 determines the flight time of the acoustic wave after it was launched from the Cr surface. The actual time delay was measured with a photo detector [Thorlab] and an oscilloscope [Agilent].

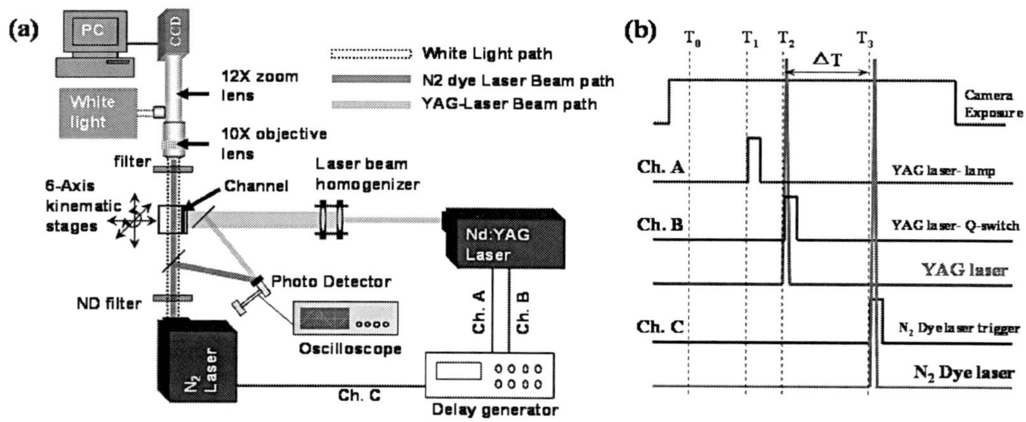

Figure 1. (a) Schematic diagram of experimental setup for a laser induced acoustic wave generation with a homogenized Nd:YAG laser (wavelength: 532 nm) and for the acoustic wave time resolved observation by a laser flash Schlieren photography with a delayed nitrogen laser pumped dye laser illumination (wavelength:440 nm). (b) Laser timing for time resolved Schlieren photography [18].

All images went through background correction image processing by subtracting a background image from the pressure wave images. Figure 2 exemplifies the image processing of an acoustic wave interaction with a 33° tilted single glass slide. First, a time resolved acoustic wave image (Figure 2(a)) was acquired with both Nd:YAG processing laser and nitrogen laser pumped dye laser illumination. The dark bar intercepting the acoustic wave is the cross section of a glass slide and the bright circle is the expanded nitrogen laser pumped dye laser illumination. A background image of the unperturbed medium (Figure 2(b)) was then acquired via applying only nitrogen laser pumped dye laser illumination. Finally, the corrected image (Figure 2(c)) enhancing the clarity of the acoustic wave was obtained by subtracting the background image (Figure 2(b)) from the original acoustic wave image (Figure 2(a)). This processing was done with image analysis software [ImagePro Plus].

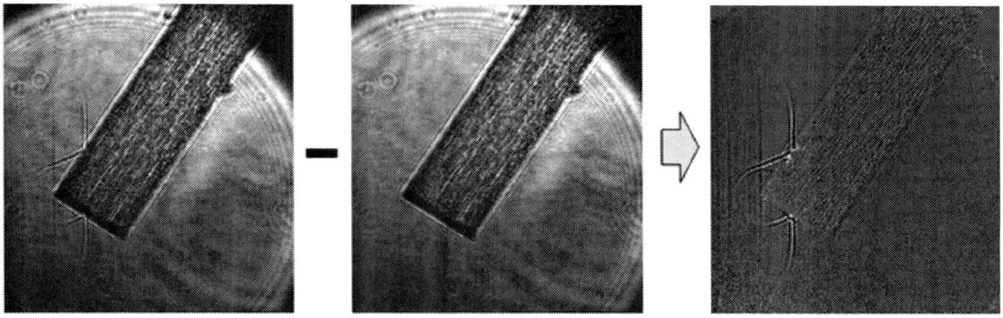

Figure 2. Image processing steps by background correction. (a) Time resolved acoustic wave image. (b) Background image without acoustic wave. (c) Background subtracted image [17].

B. Numerical simulations

With the aid of Comsol Multiphysics 3.4a acoustics module, finite element method (FEM) based numerical simulations were performed to compare with the experimental results. The acoustic pressure, $p(\mathbf{r},t)$ in a stationary lossless medium is governed by the following partial differential equation [20]:

$$\frac{1}{\rho_0 c_s^2}\frac{\partial^2 p}{\partial t^2}+\nabla\cdot\left(-\frac{1}{\rho_0}(\nabla p)\right)=\mathbf{Q}, \tag{1}$$

where ρ_0 is the density (kg/m^3), c_s is the speed of sound (m/s), Q is the source (1/s^2). The combination $\rho_0 c_s^2$ is called the adiabatic bulk modulus, commonly denoted by K. The density and speed of sound are assumed to be constant (ρ_0 = 998 kg/m^3, c_s = 1481 m/s) because they vary with time on scales much larger than the characteristic acoustic wave period [20]. Sound-hard boundary was employed to model the channel surfaces (Figure 3(b)-(i) & (iii)), where the normal component of the particle velocity vanishes. As there is no acoustic drift velocity, this condition is equivalent to the normal acceleration being zero. For constant fluid density and zero source (Q = 0 /s^2), this also means that the normal derivative of the pressure is zero at the boundary [20]:

$$\frac{\partial p}{\partial n}=0. \tag{2}$$

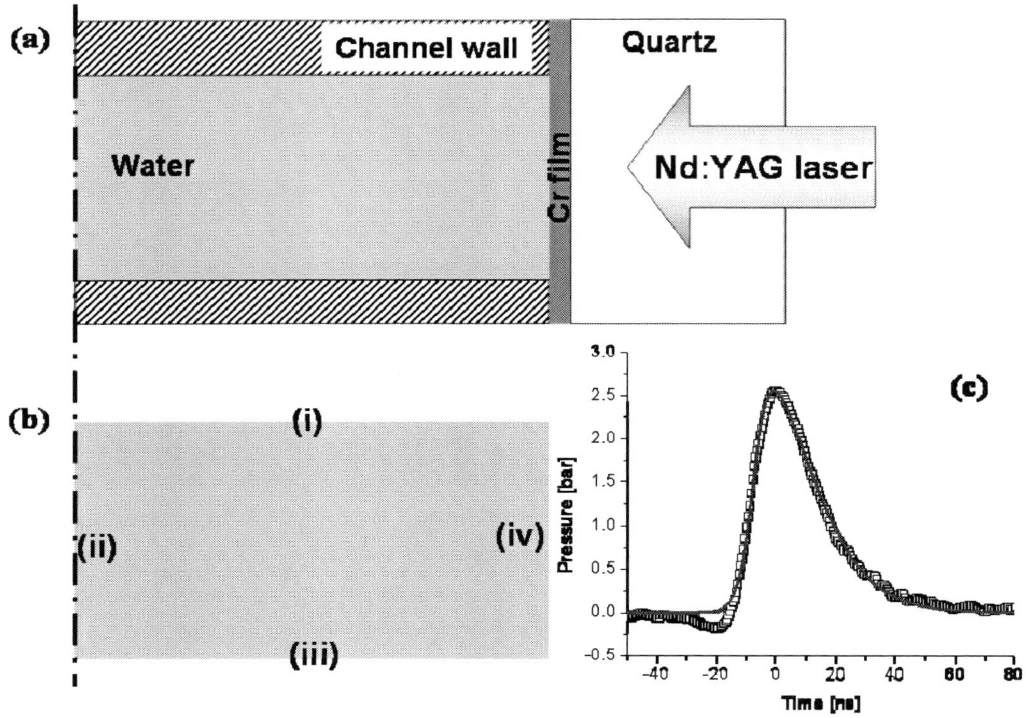

Figure 3. (a) Internal channel configuration for experiment, (b) computation domain and boundary condition, (c) measured and best fitted laser induced acoustic wave [18].

Radiation boundary condition was applied to model open boundaries (Figure 3(b)-(ii)). This condition allows the outgoing wave to leave the modeled domain with no or minimal reflection. The radiation boundary condition for transient analysis is the 1st order expression [20]:

$$\mathbf{n}\cdot\left(\frac{1}{\rho_0}(\nabla p)\right)+\frac{1}{\rho_0}\left(\frac{1}{c_s}\frac{\partial p}{\partial t}\right)=\frac{1}{\rho_0}\left(\frac{1}{c_s}\frac{\partial p_0}{\partial t}+\kappa(r)p_0+\mathbf{n}\cdot\nabla p_0\right), \tag{3}$$

where p_0 describes an incident wave, $p_0(\mathrm{r},t)$. Pressure boundary condition (Figure 3(b)-(iv)) was applied to model initial laser induced pressure field near the Cr thin film. The transient pressure source was assumed to follow the detected pressure (Figure 3(c) square symbol) and the following best fitted asymmetric double sigmoidal form (Figure 3(c) solid line) was assigned as the initial pressure boundary condition exerted by the Cr thin film.

$$P[bar]=\frac{4.12}{\left(1+\exp\left(-\frac{t+6.97}{2.51}\right)\right)}\left(1-\frac{1}{1+\exp\left(-\frac{t-6.97}{10.83}\right)}\right). \tag{5}$$

The time step and mesh size was carefully chosen to allow enough number of nodes per wavelength by setting the following CFL (Courant, Freidricks, Levy) number condition:

$$CFL = \frac{c \cdot \delta t_{\max}}{h} < 0.025\,, \qquad (4)$$

where c is the speed of sound, h is the mesh size, and $\delta t_{\max}$ is the maximum time step size. The constants for numerical simulation are density (ρ_0 = 998 kg/m3) [21], speed of sound in water at room temperature (c_s =1481 m/s) [21], source (Q = 0 /s^2).

3. Results and Discussions

3.1. Laser Induced Thermoelastic Acoustic Wave Generation in Water

The transient acoustic wave pressure amplitude was measured by a commercial lithium niobate ($LiNbO_3$) piezoelectric transducer [100 MHz, Science Brothers]. The pressure transducer was plugged into the cuvette facing the laser irradiated solid surface (4 mm away) at a normal angle. For the applied laser fluence (12-14 mJ/cm^2), the peak acoustic pulse amplitude released into the liquid was approximately 2.5 bar [Figure 3(c). square symbols].

The detected pressure transient of laser induced acoustic wave presented in Figure 3(c) was of 50 ns pulse width (corresponding frequency at about 20 MHz). Considering the acoustic attenuation coefficient (α_{ac} [m^{-1}]) in water that increases proportionally to the square of frequency [22] (i.e., $\alpha_{ac} = 2.5 \times 10^{-15} f^2$), the acoustic penetration depth at 20 MHz is calculated to be 10 cm. Laser induced acoustic wave generation is usually attributed to thermal expansion, rapid bubble growth, and collapse, as well as to plasma formation at higher applied laser energy densities. For the laser energy levels considered in this work, the latter two mechanisms are not likely to contribute [5]. Upon laser induced heating, both the liquid and the Cr thin film experience thermoelastic expansion and stress fields. The Nd:YAG energy fluence applied to induce pressure wave was around 12-14 mJ/cm^2 which is substantially lower than the Cr melting threshold (120 mJ/cm^2) [19], [23] and even smaller than bubble nucleation threshold (36.5 mJ/cm^2) [6]. Accordingly, no visual evidence of melting was found after repeated experiments. Therefore, the transient laser induced pressure waves observed are attributed solely to the thermoelastic response of the metal thin film and the adjacent liquid.

For numerical simulation, the initial pressure boundary condition [Figure 3(c). solid line] was assumed to follow the detected pressure [Figure 3(c). square symbols] by the piezoelectric transducer and was best fitted with the following asymmetric double sigmoidal form in equation (5).

3.2. Laser Induced Acoustic Waves in Various External Channels

Time resolved laser induced acoustic wave images of plane acoustic wave propagation and behavior in water inside various external channel configuration are presented in Figure

4~6, juxtaposed with the numerical simulation results. The external channel configuration of interest in this research include various channel configurations; (a) single block, (b) double block, (c) 33° tilted single block, and (d) concave cylindrical acoustic lens configurations in water.

A. Single Block

Time resolved laser flash Schlieren photography images of plane acoustic wave propagation and interaction with a single glass slide in water are presented in Figure 4(a) juxtaposed with the numerical simulation results in Figure 4(b) for 0–1900 ns range. The initial laser-induced acoustic wave just launched from the surface (0 ns) was very flat and of similar size (2×2 mm^2) to the incident homogenized Nd:YAG laser beam, except for slight smoothening at the edges due to acoustic wave diffraction. The primary acoustic wave was followed by several weaker secondary plane waves [indicated by arrows in Figure 4(a), 400 ns] that were generated by multiple reflections on the bounding surfaces of the quartz wafer bearing the laser absorbing Cr thin film. The plane acoustic wave traveled to the right side at sonic speed in water (1481 m/s) without significant damping within this time range. However, the acoustic waves were observed to decay after multiple reflections in water.

Upon arrival at the left surface of the glass slide (1.15 mm thick, borosilicate, indicated by hatched boxes), the upper half of the original flat acoustic wave interacts with the solid surface and bounces back to left side [Figure 4(a) 800 ns, indicated by an arrow] while the lower half is unobstructed and keeps moving forward. The upper half acoustic wave not only reflects on but also transmits through the solid structure. The transmitting acoustic wave re-emitted from the other side of the solid structure and moved ahead of the lower unobstructed wave [Figure 4(a) 1000ns, indicated by an arrow] due to the higher sonic velocity in solid than in water [21]. The reflected acoustic wave propagated toward the chrome surface and then bounced back to the original direction. Multiple reflections could be observed for 5 to 6 round trips between Cr/quartz and a glass slide. The transmitting, reflecting, and trespassing acoustic wave at the interface between the upper solid structure and water does not show a sharp cut but rather attains the form of an expanding cylindrical wave front due to diffraction by the glass slide edge.

Numerical simulation (Figure 4(b)) shows fairly good agreement with the experimental observation (Figure 4(a)). The reflection wave and the edge diffraction pattern are very similar to the laser flash Schlieren photography images. The only difference is the absence of wave transmission through the solid structure and the secondary waves. This is because the solid structures were not included in the simulation domain and the solid surface was considered as hard and perfectly reflecting without any losses or transmissions. A more realistic simulation can be carried out by including the solid structure in the calculation domain.

B. Double Block (A Channel)

Time resolved laser flash Schlieren photography images of plane acoustic wave propagation and interaction with double glass slides (indicated by hatched boxes) with 0.95 mm water channel are presented in Figure 4(c) juxtaposed with the numerical simulation results in Figure 4(d) for 0–1900 ns range. The overall behavior of the acoustic wave is very similar to that of the single block case except that now the acoustic wave passes through the liquid channel bounded by two solid walls. After the central 1/3 of the original acoustic wave passes through the channel, it forms an expanding cylindrical wave at both edges [Figure 4(c), 1800 ns]. Again, the numerical simulation (Figure 4(d)) is in agreement with the experimental observation (Figure 4(c)).

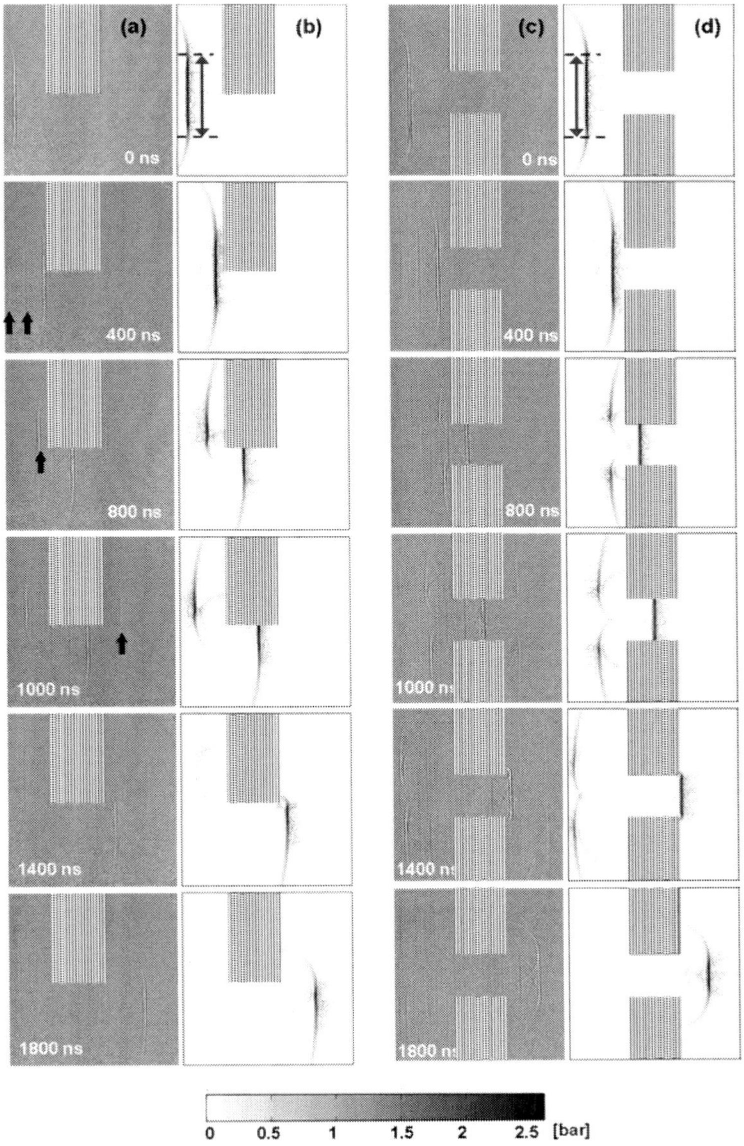

Figure 4. Time resolved laser flash Schlieren photography images of laser induced acoustic wave propagation and interaction with (a) a single glass block and (c) two parallel glass blocks in water. All images went through background correction image processing. Numerical simulation results for (b) a single glass block and (d) two parallel glass blocks in water. Blue arrow signifies the width of the initial flat acoustic wave and corresponds to 2 mm. Hatched boxes indicate glass slides (1.15 mm thick). The bottom legend presents the pressure in the numerical simulation [17].

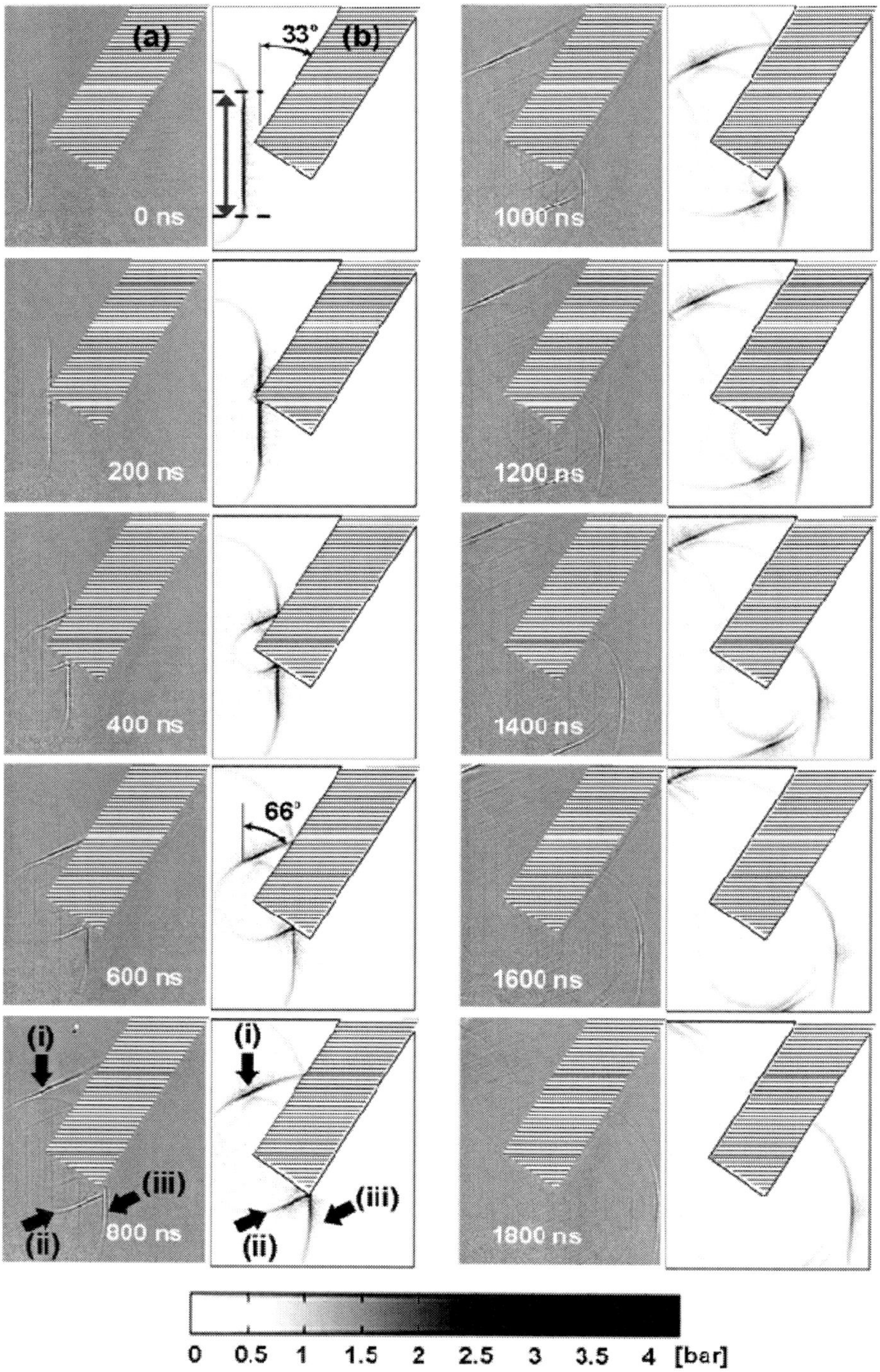

Figure 5. (a) Time resolved laser flash Schlieren photography images of laser induced acoustic wave propagation and interaction with a 33° tilted single glass block in water. All images went through background correction image processing. (b) Numerical simulation results. Blue arrow signifies the width of the initial flat acoustic wave and corresponds to 2 mm. Hatched boxes indicate glass slides (1.15 mm thick). The bottom legend presents the pressure in the numerical simulation [17].

D. 33° Tilted Single Block

Time resolved laser flash Schlieren photography images of plane acoustic wave propagation and interaction with a 33° tilted glass slide (indicated by hatched boxes) in water are presented in Figure 5(a) juxtaposed with the numerical simulation results in Figure 5(b) for 0–1900 ns range. The laser induced acoustic wave interaction with a solid structure in water is much more pronounced for acoustic wave reflection from an oblique flat structure as shown in Figure 5. The upper half of the original acoustic wave [Figure 5. 800 ns, (i)] intercepted the left long face of a glass slide that is tilted by 33° with respect to the wave front. The middle quarter of the wave [Figure 5. 800 ns, (ii)] met the bottom short face of the glass slide, and the lower quarter of the wave [Figure 5. 800 ns, (iii)] was transmitted without interaction with solid. The upper half acoustic wave [Figure 5. 800 ns, (i)] reflected on the glass slide surface and propagated at a deviation of 66° from the incidence direction. The acoustic wave reflected from the bottom short face of the glass slide [Figure 5. 800ns, (ii)] was exactly parallel to the upper half reflected acoustic wave [Figure 5. 800 ns, (i)]. The lower one quarter forward propagating acoustic wave [Figure 5. 800 ns, (iii)] did not have a sharp edge but attained the form of an expanding cylindrical wave front due to diffraction by the glass slide edge. Transmission through the glass slide and secondary waves were also observable.

E. Concave Cylindrical Lens (Acoustic Wave Focusing)

Time resolved laser flash Schlieren photography images of plane acoustic wave propagation, interaction and subsequent focusing by the fused silica concave cylindrical lens with 2.1 mm optical focal length (indicated by hatched boxes) in water are presented in Figure 6(a) and compared with the numerical simulation results in Figure 6(b) over the 0–5500 ns temporal range.

The propagating planar acoustic wave reflected [Figure 6. 600 ns] on the focusing lens and was transformed into a cylindrically shaped acoustic wave [Figure 6. 1000 ns] propagating to the left side. Finally the cylindrical wave focused into a point (if seen from side view, it appears as a line) at 1400 ns and then diverged into an expanding cylindrical acoustic wave whose images are depicted at 1800 ns. The expanding cylindrical wave reflected again on the Cr thin film surface where it was originally formed [Figure 6. 2500 ns] and kept bouncing between the two solid structures (the Cr film and cylindrical fused silica lens) inside the water. At 1400ns, the first acoustic wave focal point was at a distance of 1.04 mm from the cylindrical lens surface. This is very close to the geometric optical focal point (1.05 mm) calculated from the mirror formula [24] ($1/S_o + 1/S_i = -2/R$) where R is the radius of the mirror, S_0 is the distance between object and mirror surface, and S_i is the distance between image and mirror surface. Although there is an analogy between acoustic and light waves, obviously the specific wave characteristics such as propagation speed and wavelength are entirely different. While the first plane acoustic wave forms a focal point close to the predicted location, the second focal distance at 5000 ns is observed further away. This is because after the first focusing, the original flat acoustic wave turns into an expanding cylindrical wave and the incoming wave is not flat anymore. Consequently, the second focal spot changes depending on the curvature of the incident cylindrical wave.

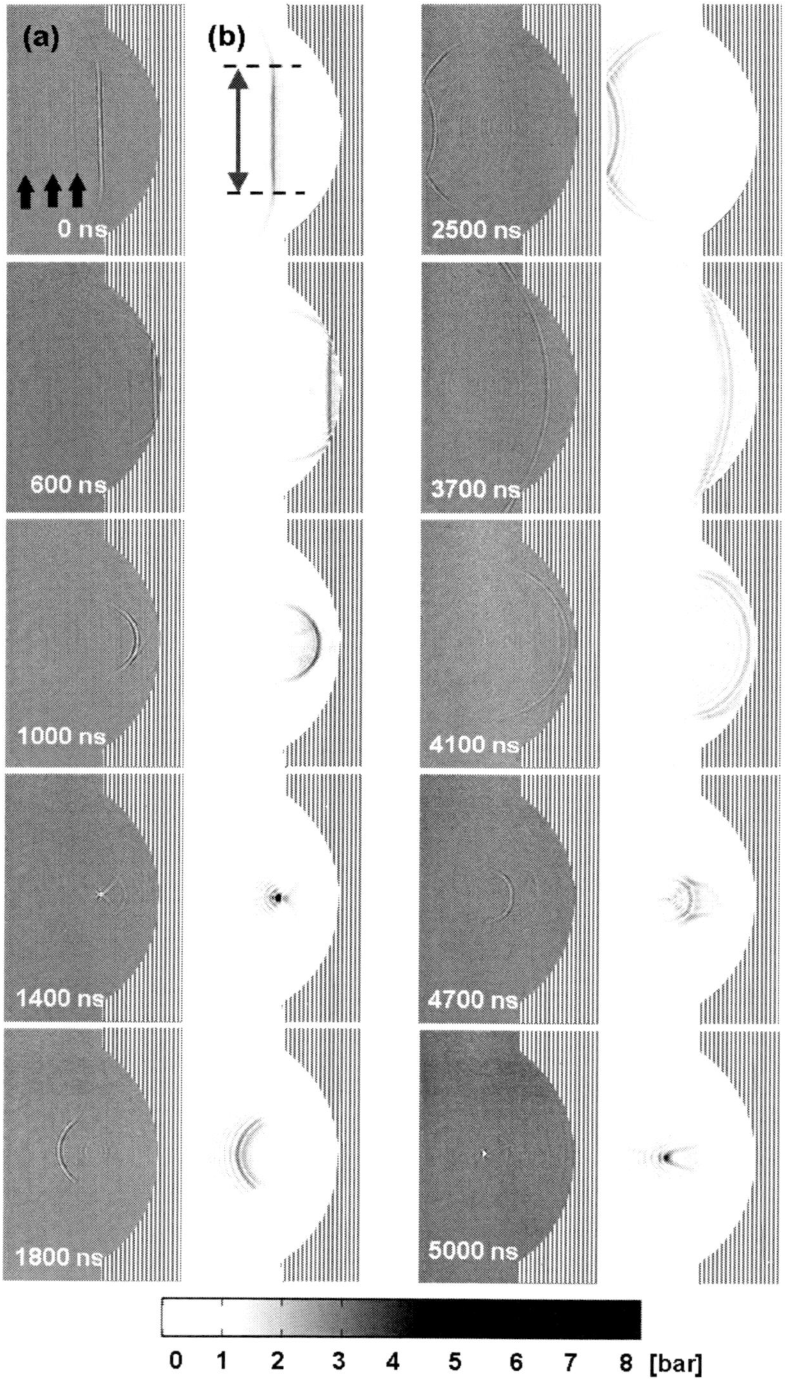

Figure 6. (a) Time resolved laser flash Schlieren photography images of laser induced acoustic wave propagation and focusing in water. All images went through background correction image processing. (b) Numerical simulation results. Blue arrow signifies the width of the initial flat acoustic wave and corresponds to 2 mm. Hatched boxes indicate the acoustical focusing lens with 2.1 mm focal length and made from fused quartz. The bottom legend presents the pressure in the numerical simulation [17].

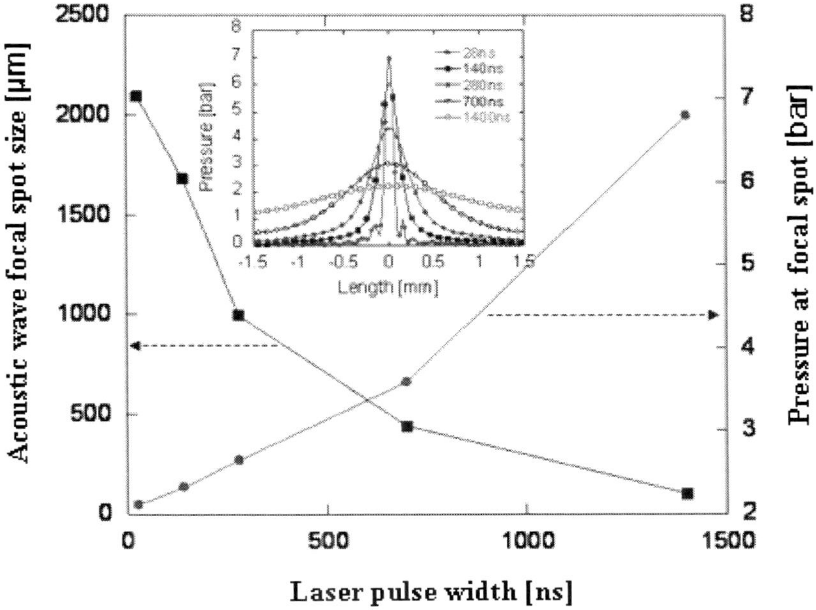

Figure 7. Simulated acoustic wave focal spot size (square symbol, FWHM) and focal pressure (circular symbol) for different laser pulse width. Inset graph presents the pressure distribution at focal point for different laser pulse width [17].

The acoustic wave size and pressure magnitude are of primary interest in this work. However, direct measurement of those values is difficult due to the small focus size. Our numerical simulations revealed that the focused acoustic wave has a 48 μm focal size (FWHM) and a 7 bar magnitude. The acoustic wave focal size and magnitude are greatly influenced by the laser pulse width that generates the initial thermoelastic acoustic wave. The numerical predictions of the acoustic wave focal spot size and pressure amplitude for different laser pulse widths (28, 140, 280, 700, and 1400 ns FWHM) are presented in Figure 7. As the laser pulse width shortens, the acoustic wave focusing becomes more notable. When a microsecond laser pulse was applied, almost no focusing effect could be observed. This result underlines the importance of utilizing short pulses to achieve efficient acoustic focusing and also suggests the possibility of focusing improvement with ultra short, pico- or femtosecond pulses. Picosecond laser ultrasonics (down to 100 fs) [25] have been extensively studied. However, further reduction of pulse width below 100 femtoseconds involves non-equilibrium interaction between carriers and phonons that may last over several picoseconds, as well as possible non-linear effects related to the mechanical material response at such time scales.

The calculated pressure at the focal spot of the focused acoustic wave seems to be somewhat low considering the ratio of the focused spot size (48 μm) and the initial laser spot size (2 mm). The computed spot size and pressure at focal point must have been broadened to exert lower pressure at a larger spot size due to numerical diffusion and unfaithful wave front tracking. These problems become much more severe for the numerical simulation of very short acoustic waves such as the one considered in this paper. Direct pressure measurement at the focal point will give us a more accurate estimate of the pressure value at the focal spot.

F. Viscosity Effect (Propagation Velocity and Attenuation)

The effect of liquid viscosity on the acoustic wave propagation velocity, pressure attenuation, and wave broadening was investigated. The fluid viscosity was varied by mixing glycerol with DI water to yield liquid of 1, 10, and 100 cp in viscosity at room temperature. The acoustic wave was observed to move faster in the higher viscosity liquid. Figure 8 shows the double exposure pictures of flat acoustic wave with 2 μs delay for 1, 10, and 100 cp from the top. 10 cp liquid showed a speed of sound of 1672 m/s and 100 cp liquid showed 1800 m/s that is close to speed of sound in glycerol (1900 m/s) at room temperature.

Viscosity also plays major role in acoustic wave attenuation and wave broadening. Figure 9 shows the initial acoustic waves (Figure 9(a), indicated by arrows) and acoustic waves after propagating a distance of 67 mm (Figure 9(b), indicated by arrows) in a glass cuvette [Starna Cell] with 2 mm height. Acoustic waves in a medium with a viscosity of 10 cp [Figure 9(b) middle picture] experience more noticeable attenuation and broadening compared to a medium with viscosity of 1 cp [Figure 9(b) top picture]. Furthermore, for media with viscosity of 100 cp, complete attenuation of the acoustic wave is observed after propagating a distance of 67 mm. These observations can be attributed to loss of acoustic energy resulting from three major mechanisms: viscous losses, heat conduction losses, and losses associated with internal molecular processes [21]. In polar liquids such as alcohols and water, the intermolecular forces are so strong that structural relaxation is more dominant than thermal relaxation and accounts for the observed excess absorption [26].

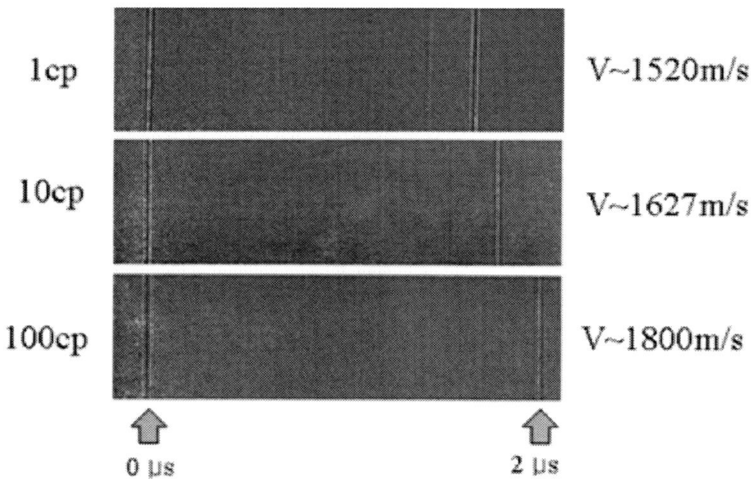

Figure 8. Acoustic wave propagation speed measurement for different viscosity fluid (1,10,100 cp from the top). Two pictures with 2 μs delay were taken and overlapped. All images went through background correction image processing [17].

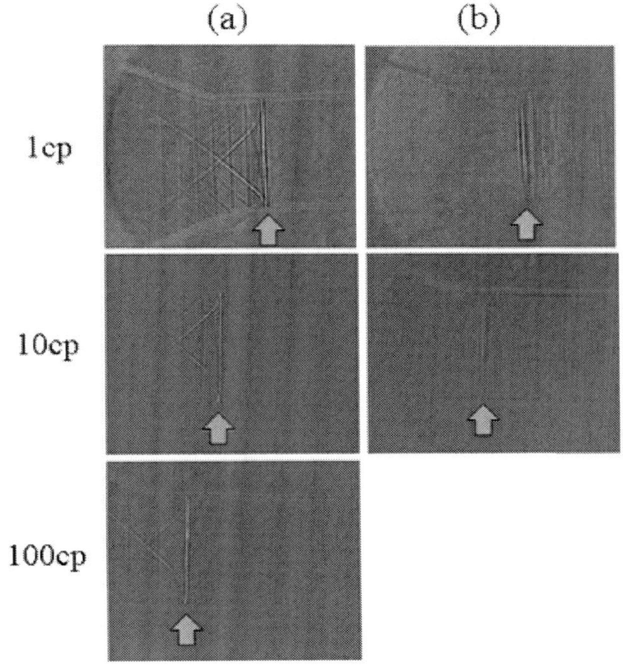

Figure 9. Acoustic wave attenuation for different viscosity fluid (1,10,100 cp from the top) after the initial acoustic wave (a) travels 67 mm (b) in a 2 mm high channel. All images went through background correction image processing [17].

3.3. Laser Induced Acoustic waves in Various Internal Channels

Time resolved laser induced acoustic wave images of plane acoustic wave propagation and behavior in water inside various channel configuration are presented in Figure 10~16, juxtaposed with the numerical simulation results. The internal channel configuration of interest in this research include various channel configurations; (a) sudden expansion and contraction channels, (b) bifurcating channels: T branched and Y branched channels, (c) gradual contraction wall channels: linear contraction and parabolic contraction (inkjet nozzle) wall channel, (d) a cylinder: trapped acoustic wave in a cylinder.

A. Sudden Expansion and Contraction Channels

In a sudden contraction channel (Figure 10(a)), the middle part of the flat acoustic wave kept proceeding along the narrow channel while the side parts reflected upstream. The middle proceeding part of the acoustic wave had sharp edges in the proximity of the narrow channel walls while the reflected acoustic wave induced a diverging cylindrical component due to diffraction at the sharp entrance of the contraction channel.

In the case of the sudden expansion channel (Figure 10(b)), the incoming acoustic wave in a narrow channel maintained a flat profile flanked by cylindrical waves caused by diffraction near the both sudden expansion edges. The cylindrical waves propagate both outwards and in the inwards. The outward expanding cylindrical acoustic waves attain a dove tail shaped acoustic wave followed by a secondary expanding dovetail acoustic waves from two inward expanding cylindrical acoustic waves. For both cases, numerical simulation showed fairly good agreement with experimental observation.

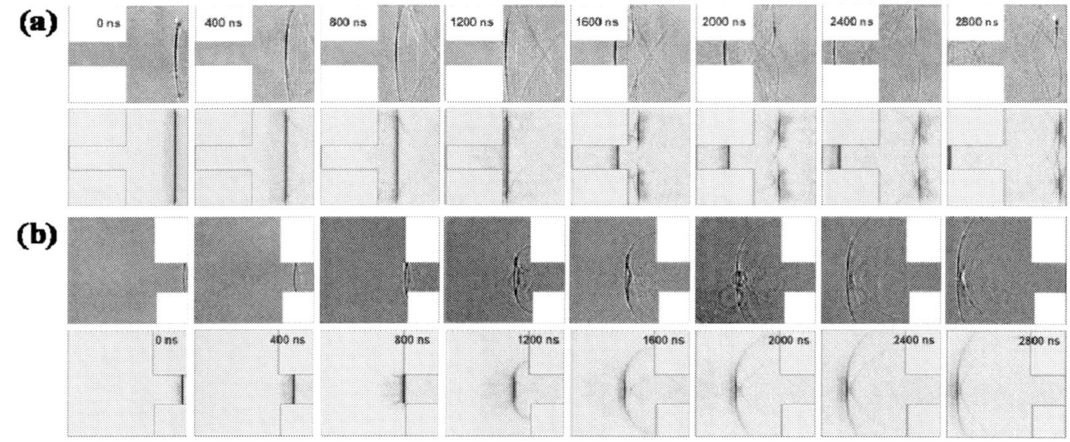

Figure 10. (a) Sudden contraction channel and (b) sudden expansion channel (top rows: experiment, bottom rows: simulation) [18].

B. Bifurcating Channels: T Branched and Y Branched Channels

Two bifurcating channel configurations were studied; a T-branched bifurcation wall channel and a Y-branched gradually bifurcating channel to simulate the pressure wave behavior at the coronary artery bifurcation especially for arteriosclerosis.

The T branched bifurcating channel case (Figure 11) is the inverse geometry of the sudden contraction channel. However, interestingly, after the middle acoustic wave part bounded back, the reflected wave behaved exactly as in the sudden expansion channel case (not sudden contraction channel case) forming a double expanding dovetail cylindrical acoustic wave. The side parts of the incoming flat acoustic wave propagated unobstructed with sharp edges across the wall interfaces.

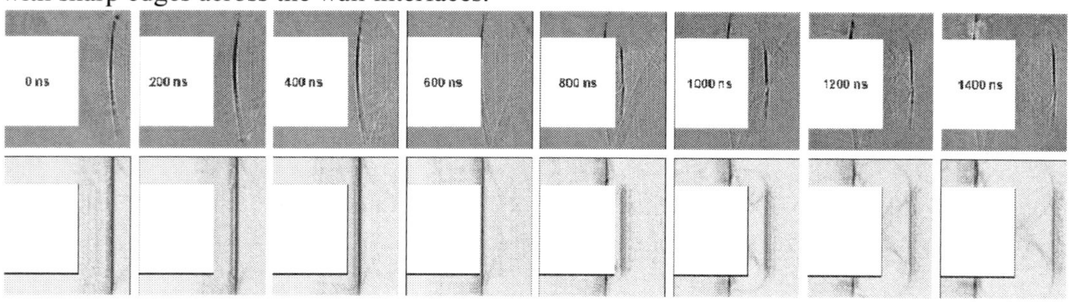

Figure 11. T-branched bifurcation channel (top row: experiment, bottom row: simulation) [18].

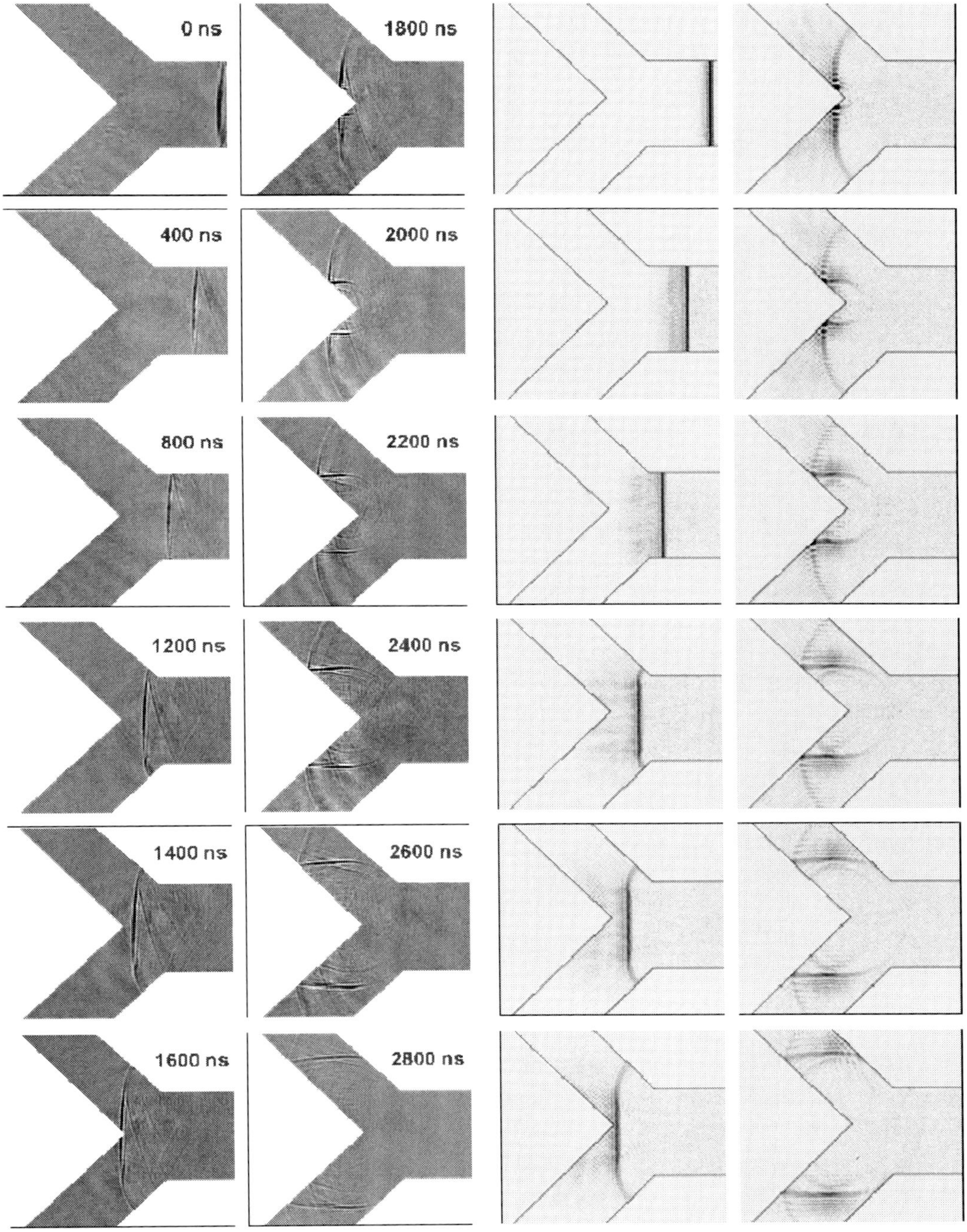

Figure 12. Y-branched bifurcation channel (left two columns: experiment, right two columns: simulation) [18].

In the Y branched gradually bifurcating channel case (Figure 12), the incoming flat acoustic wave was bisected upon meeting the sharp tip edge. The halved acoustic waves were first reflected specularly on the 45° tilted inner wall surface. The first reflected flat acoustic wave front moved upward (vertically) and then reflected back to left (horizontally) at the outer wall and kept propagating along the channel downstream via multiple reflections. The propagation behavior in this bifurcation channels can be exactly predicted from multiple reflections.

In both cases, numerical simulations showed fairly good agreement with the experimental observations.

C. Gradual Contraction Wall Channels: Linear Contraction and Parabolic Contraction Wall Channel

Gradual contraction channels with both linear and parabolic contraction walls channel were studied to simulate the pressure wave interaction and behavior inside the inkjet nozzle.

The jetting phenomenon in the piezoelectrically driven inkjet has been known to be strongly related to the pressure wave inside the printhead [27]. However, in many cases, it was difficult to observe the pressure wave directly and there have been scarcely any reports in the open literature [28]. Consequently, a CCD camera with a strobe LED illumination was typically employed to measure the droplet speed or volume, instead of directly measuring the pressure wave inside the dispenser. However, it is difficult to understand the pressure wave behavior inside the dispenser by just observing the droplet image [28]. Here, we conducted a direct observation of the pressure wave propagation and interaction inside nozzle geometry.

In the converging nozzle with straight wall (Figure 13(a)), the reflection was specular at an angle of 30°. As the central part of the original pressure wave approached the end of the nozzle, both ends of the reflected pressure wave extended to long wings. A considerable part of the reflected and extended pressure wave did not make it to the nozzle tip but reflected back. This implies that in this nozzle wall geometry, only a small portion of the pressure wave can reach the nozzle tip to contribute to the droplet formation while the rest of the pressure wave just reflects back and forth multiple times inside the capillary tube to finally dissipate. The slope of the nozzle wall may modulate the percentage and the amplitude of the pressure at the nozzle tip. Nozzles with larger slope may enhance the pressure amplitude via a slight focusing effect, but the back reflection will also increase. On the other hand, the nozzles with smaller slope do not impart focusing and experience little reflection.

In the converging nozzle with a parabolic wall (Figure 13(b)), the flat acoustic wave also reflected against the walls. However, the reflected pressure waves behaved totally differently compared to the nozzle with a linearly varying slope (Figure 13(a)). As the pressure wave entered the nozzle zone with small wall curvature, the reflection had a similar pattern that was previously examined in the nozzle with a linearly varying slope. As the pressure wave propagated toward the nozzle tip region where the curvature was higher, the pressure wave exhibited focusing near the nozzle wall. The focusing point was maintained near the nozzle wall and carried most of the concentrated pressure wave energy by suppressing the continuous expansion of the pressure wave after reflection from the nozzle wall that was previously observed in the linearly varying nozzle (Figure 14). Ko et al. [17] showed that the reflected pressure wave from a perfect cylindrical concave surface could be focused to a line R/2 distant from the cylinder wall after the reflection. A similar focusing effect is proposed to explain the local focusing effect of the reflected pressure wave on the curved nozzle wall. However, if the nozzle wall has a constant curvature like a perfect cylinder (Figure 15), the focus is not kept near the nozzle wall but will rather be fixed at R/2 and to the right of the nozzle wall. This means that a nozzle with one fixed curvature (such as cylinder) will produce a highly focused pressure point although this is just due to back reflected pressure wave. After the focusing, the pressure wave will reflect back to the incoming direction and will be dissipated without contributing to the droplet generation just like in the linearly varying nozzle case. More sophisticated nozzle wall shape is needed for the locally focused pressure wave to be maintained near the nozzle wall and to be carried to the nozzle tip not just fixed at one point. To keep the focused pressure wave near the nozzle wall, the nozzle must have an increasing curvature towards the nozzle tip instead of having one curvature.

Maximum pressure at the nozzle tip is implemented by carrying the concentrated pressure wave to the nozzle tip and minimizing the pressure wave back reflection loss. First, the nozzle wall should have a non-zero curvature (curved surface) imparting focusing effect to minimize the extension of the reflected pressure wave and backward reflection. Second, the nozzle must have varying (increasing) curvature to keep the reflected wave focused point near the nozzle wall and bring the locally focused pressure wave to the tip of the nozzle with minimum back reflection. The numerical simulations showed fairly good agreement with the experimental observations.

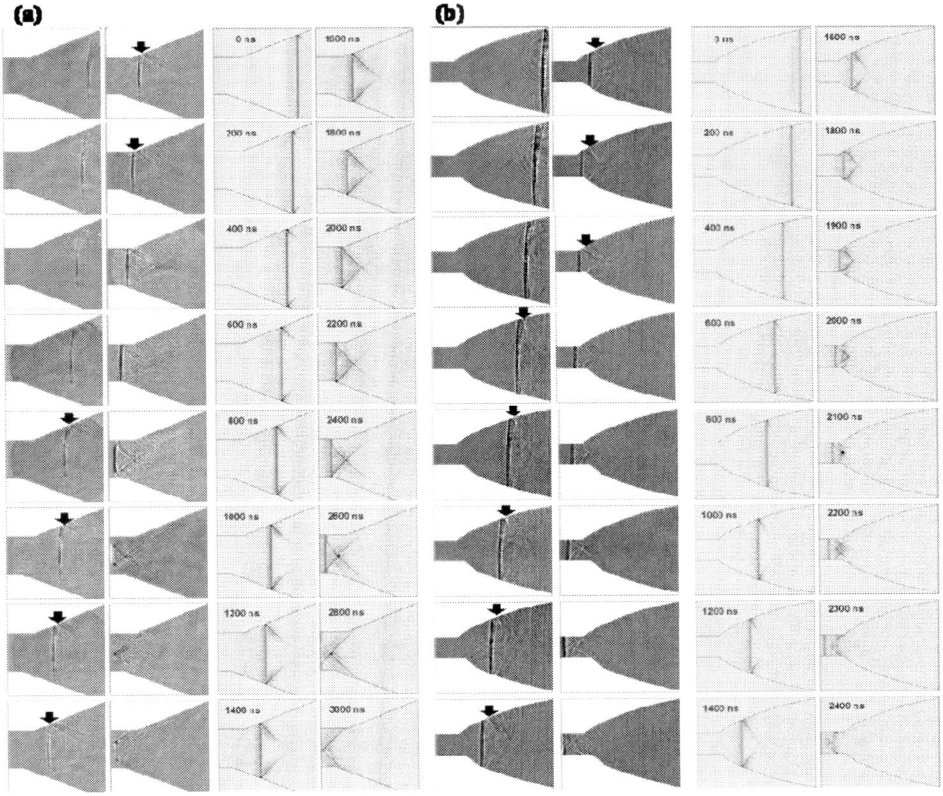

Figure 13. (a) Linearly varying contraction wall and (b) parabolic contraction wall (left two columns: experiment, right two columns: simulation) [18].

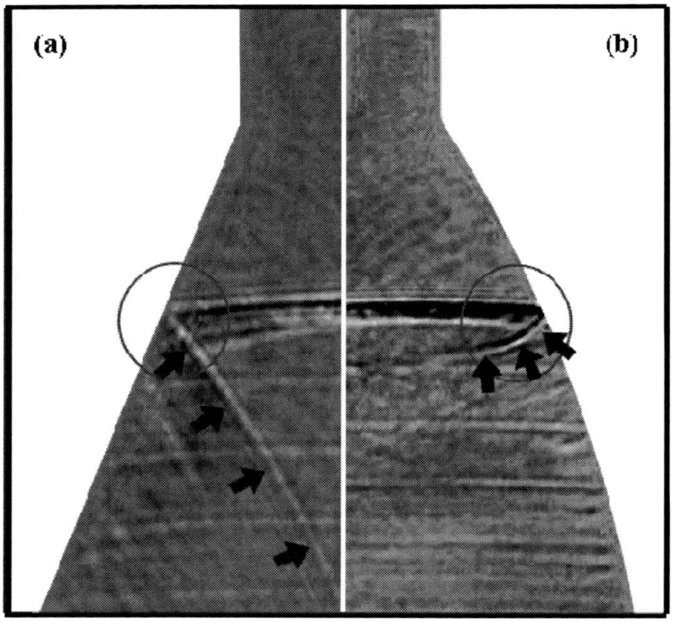

Figure 14. Magnified view from Figure 6 for reflected pressure wave at the nozzle wall for (a) linearly varying contraction wall and (b) parabolic contraction wall [18].

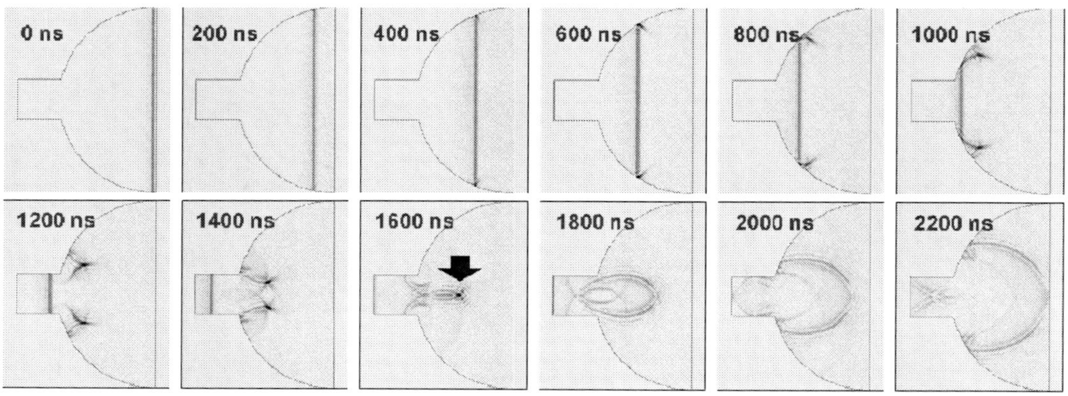

Figure 15. Pressure wave for perfect circular wall nozzle [18].

D. Trapped Acoustic Wave in a Cylinder

A plane acoustic wave impinged on a cylinder (Figure 16) in order to investigate trapped acoustic wave behavior within an enclosed geometry. First, the incoming plane acoustic wave was partially transmitted through glass cylinder and in part reflected toward the opposite direction diverging to a cylindrical wave. The transmitted acoustic wave propagated through the cylindrical enclosure maintaining a flat profile till it hit the opposite internal wall surface. The 1st reflected acoustic wave on the left inner cylinder wall focused in the left half of the cylinder (point "*1*" in Figure 16). This wave diverged after focusing and after reflection on the right inner wall was focused again in the right half of the cylinder (point "*2*" in Figure

16). These reflection-focusing events inside the cylinder kept repeating until the acoustic wave dissipates. During this reflection-focusing process, the focusing point kept alternating (left-right-left and so on) around and approaching the geometric center of the cylinder.

Acoustic wave inside the cylinder were studied to simulate the pressure wave interaction and behavior inside the human eyeball during the LASIK (laser-assisted in situ keratomileusis), a type of refractive surgery for correcting myopia, hyperopia, and astigmatism. The observation of the presence of the focusing point inside the cylinder tells us the possibility of the high pressure damage may occur inside the eye ball during the LASIK surgery. The numerical simulations showed fairly good agreement with the experimental observations.

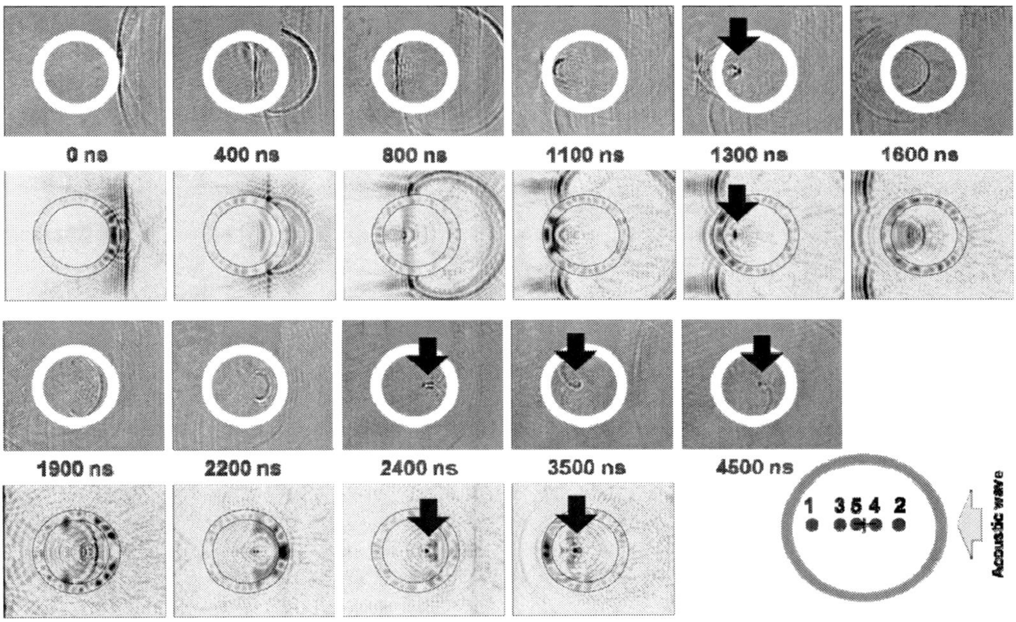

Figure 16. Trapped acoustic wave in a cylinder (first & third rows: experiment, second and fourth rows: simulation) [18].

Conclusion

In this chapter, the state of the art laser based time resolved (100ns resolution) visualization technique of a single, short acoustic plane wave generation, propagation in various external and internal channels and the subsequent interaction with submerged solid structures are presented. The effect of liquid viscosity on the acoustic wave propagation velocity, pressure attenuation, and wave broadening was investigated. The fluid viscosity was varied by mixing glycerol with DI water to yield liquid of 1, 10, and 100 cp in viscosity at room temperature.

The external channels include (a) single, (b) double, (c)33° tilted single block, and (d) concave cylindrical acoustic lens configurations, and internal channels include (a) sudden expansion and contraction channels, (b) bifurcating channels: T branched and Y branched channels, (c) gradual contraction wall channels : linear contraction and parabolic contraction wall channel, (d) a cylinder. Experimental observations were compared with finite element method (FEM) based numerical simulations. The generation/propagation/interaction of laser induced acoustic waves in internal and external channels were directly observed by two laser systems with different wavelength. The two laser systems were combined to achieve different functions; (1) neodymium-doped yttrium aluminum garnet (Nd:YAG) nanosecond pulsed laser with green wavelength (532nm) to generate the acoustic wave and (2) nitrogen laser pumped dye laser with blue wavelength (440nm) as an illumination source for the time resolved acoustic wave imaging. The two systems were synchronized via a delay generator to control the precise time delay between the two laser pulse firings. All images went through background correction image processing to reduce background noise. Among various submerged configurations, concave cylindrical acoustic focusing lens has potential applications to biomaterial processing and characterization. Effective control of the spatial and temporal structure of the generated sound pulse can be provided by the laser induced acoustic wave technique, especially in laser-based acoustic microscopy, biomedical ultrasound imaging, and biological material processing. Focusing of acoustic waves is important not only for achieving lateral resolution enhancement in ultrasound imaging but also in order to yield higher pressure amplitudes. Acoustic wave formation by thermoelastic expansion has poor laser energy conversion efficiency and needs to be amplified in most cases. Our simulation results demonstrate that the nanosecond laser induced acoustic wave can be focused to several tens of microns size with several bar pressure. Compared with submerged structures, acoustic wave propagation and interaction in internal channels are more frequently found in many practical applications such as in an inkjet head (gradual or sudden contraction wall channel) and coronary artery bifurcation (T branched and Y branched channels).

ACKNOWLEDGMENT

The authors would like gratefully acknowledge the financial supports to KAIST by the Industrial Strategic Technology Development Program from the Korea Ministry of Knowledge Economy (Grant No. 10032145), the National Research Foundation of Korea (Grant No. 2010-0003973), the cooperative R&D Program from the Korea Research Council Industrial Science and Technology (Grant No. B551179-10-01-00), and the support to the University of California, Berkeley by Berkeley by Xerox Wilson Center for Research and Technology, USA.

REFERENCES

[1] White, R. M. Generation of Elastic Waves by Transient Surface Heating, *J. Appl. Phys.* 1963, Vol. 34, 3559.

[2] Dewhurst, R. J.; Hutchins, D. A.; Palmer, S. B.; Scruby, C. B. Quantitative measurements of laser-generated acoustic waveforms, *J. Appl. Phys.* 1982, Vol. 53, 4064.

[3] McDonald, F. A. Quantitative Measurements of Laser-generated acoustic waveforms, *Appl. Phys. Lett.* 1989, Vol. 54, 1504.
[4] Mesaros, M.; Martı´nez, O.; Bilmes, G. M.; Tocho, J. O. Acoustic detection of laser induced melting of metals, *J. Appl. Phys.* 1997, Vol. 81, 1014.
[5] Park, H. K.; Grigoropoulos, C. P.; Leung, W. P.; Tam, A. C. A practical excimer laser-assisted cleaning tool for removal of surface contaminants, *IEEE Trans. Compon. Packag. Manuf. Technol.* 1994, Part A, Vol. 17A, 631.
[6] Vogel, A.; Schweiger, P.; Frieser, A.; Asiyo; M. N; Birngruber, R. Intraocular Nd: YAG laser surgery: light–tissue interaction, damage range, and reduction of collateral effects, *IEEE J. Quantum Electron*. 1990, Vol. 26, 2240.
[7] Teng, P.; Nishioka, N. S.; Anderson, R. R.; Deutsch, T. F. Optical studies of pulse laser fragmentation of biliary calculi, *Appl. Phys. B* 1987, Vol. 42, 73.
[8] Schilling, A.; Yavas, O.; Bischof, J.; Boneberg, J.; Leiderer, P. Absolute pressure measurements of a nanosecond time scale using surface plasmons, *Appl. Phys. Lett.* 1996, Vol. 69, 4159.
[9] Park, H. K.; Kim, D.; Grigoropoulos, C. P.; Tam, A. C. Quantitative Measurements of Laser-generated acoustic waveforms, *J. Appl. Phys.* 1996, Vol. 80, 4072.
[10] Bunkin, F. V.; Kolomensky, A. A.; Mikhalevich, V. G. *Lasers in Acoustics*, Switzerland, Harwood Academic, Chur, 1991.
[11] Varslot, T.; Måsøy, S. E. Forward propagation of acoustic pressure pulses in 3D soft biological tissue, *Model. Identif. Control.* 2006, Vol. 27, 181.
[12] Atalar, A.; Köymen, H. Use of conical axicon as a surface acoustic wave focusing device, *IEEE Trans. Ultrason. Ferroelectr. Freq. Control*. 1987, Vol. 34, 53.
[13] Köymen, H.; Atalar, A. Focusing surface waves using an axicon, *Appl. Phys. Lett.* 1985, Vol. 47, 1266.
[14] Yin, X.; Hynynen, K. A numerical study of transcranial focused ultrasound beam propagation at low frequency, *Phys. Med. Biol.* 2005, Vol. 50, 1831.
[15] Mal, A.; Feng, F.; Kabo, M.; Wang, J.; Bar-Cohen, Y. Interaction of Focused Ultrasound with Biological Materials, *Proc. SPIE* 2002, Vol. 4702, 339.
[16] Ko, S. H.; Ryu, S. G.; Misra, N.; Pan, H.; Grigoropoulos, C. P.; Kladias, N.; Panides, E.; Domoto, G. A. Laser induced short plane acoustic wave focusing in water, *Appl. Phys. Lett.* 2007, Vol. 91, 051128.
[17] Ko, S. H.; Ryu, S. G.; Misra, N.; Pan, H.; Grigoropoulos, C. P.; Kladias, N.; Panides, E.; Domoto, G. A. Laser induced plane acoustic wave generation, propagation and interaction with rigid structures in water, *J. Appl. Phys.* 2008, Vol. 104, 074103.
[18] Ko, S. H.; Lee, D.; Pan, H.; Ryu, S. G.; Grigoropoulos, C. P.; Kladias, N.; Panides, E.; Domoto, G. A. Laser induced acoustic wave generation/propagation/interaction in water in various internal channels, *Appl. Phys. A* 2010, Vol. 100, 391-400.
[19] Park, H. K.; Grigoropoulos, C. P.; Poon, C. C.; Tam, A. C. Modeling of pulsed laser heating of thin silicon films, *ASME Trans. J. Heat Transfer* 1996, Vol. 118.
[20] Acoustics module user's guide, COMSOL Multiphysics, 2006.
[21] Kinsler, L. E.; Frey, A. R.; Coppens, A. B.; Sanders, J. V. *Fundamentals of Acoustics*, Fourth Edition, New York, John Wiley & Sons, 2000.
[22] Markham, J. J.; Beyer, R. T.; Lindsay, R. B. Absorption of sound in fluids, *Rev. Mod. Phys.* 1951, Vol. 23, 353.

[23] Park, H. K.; Grigoropoulos, C. P.; Poon, C. C.; Tam, A. C. Quantitative Measurements of Laser-generated acoustic waveforms, *Appl. Phys. Lett.* 1996, Vol. 68, 596.

[24] Hecht, E. *Optics*, Fourth Edition, New York , Addison Wesley, 2002.

[25] Tas, G.; Maris, H. J. Picosecond ultrasonic study of phonon reflection from solid-liquid interfaces, *Phys. Rev. B* 1997, Vol. 55, 1852.

[26] Halpern, O.; Hall, H. The ionization loss of energy of fast charged particles in gases and condensed bodies, *Phys. Rev.* 1948, Vol. 73, 775.

[27] Bogy, D. B.; Talke, F. E. Experimental and theoretical study of wave propagation phenomena in drop-on-demand ink jet devices, *IBM J. Res. Dev.* 1984, Vol. 28(3), 314-321.

[28] Kwon, K.; Kim, W. A waveform design method for high-speed inkjet printing based on self-sensing measurement, *Sensors and Actuators A*, 2007, Vol. 140, 75.

In: Sound Waves: Propagation, Frequencies and Effects ISBN 978-1-61470-001-2
Editors: V. Abagnali and G. Fabbri

Chapter 2

ELASTIC VIBRATIONS OF AN ISOTROPIC PLATE WITH LASER-INDUCED ATOMIC DEFECTS

F. Kh. Mirzade*
Institute on Laser and Information Technologies
Russian Academy of Sciences
140700, Moscow, Russia

Abstract

An analysis of the propagation of plane elastic waves in isotropic plates containing a distribution of non-equilibrium atomic point defects (interstitial atoms, vacancies, electron-hole pairs) is presented. The formation of atomic defects occurs as a result of the action of intense laser radiations on the solid plates. The study is based on coupled evolution equations for the elastic displacement of the medium and atomic defect density fields. The defect dynamics is governed by the strain-stimulated generation, transport, and annihilation processes. The frequency equations corresponding to the symmetric and antisymmetric elasto - concentration modes of vibration of the plate are obtained. Some limiting cases of the frequency equations are considered and a procedure for determining the phase velocity and the attenuation (or amplification) constants is discussed. Relevant results of previous investigations are deduced as special cases.

Keywords: Propagation of elastic waves; laser-induced atomic defects; wave velocity equation; elastic-concentration instabilities; surface waves.

1. INTRODUCTION

Excitation of elasto – concentration waves by a pulsed laser in solid is of great interest due to extensive applications of pulsed laser technologies in material processing and nondestructive detecting and characterization, including the detection of defect concentration regions in them and the coating quality testing. When a solid is illuminated with a laser pulse,

* E-mail address: fmirzade@rambler.ru

absorption of the laser pulse results in a localized temperature increase, which in turn causes lattice defects and generates a elasto – concentration wave in the solid. Various structural imperfections in the crystal lattice, i.e., atomic point defects, which are produced from the lattice site atoms due to pulsed laser beam introduce a significant strain of the medium as a result of the difference between the radii of lattice atoms and defects [1], and play an important role in surface modification of solids exposed to laser radiation [2]-[5]. The formation of defects may occur also as a result of mechanical, thermal, and electric treatments of materials. Examples of such defects are interstitial atoms, vacancies, color centres and their clusters, electron-hole pairs, impurities atoms. Strains in an elastic wave cause the defects to move within a crystal cell (a strain-induced drift), whereas the strains and a variation in the temperature in the wave modulate the probabilities of generation and recombination of defects of the thermal-fluctuation origin (via variations in the energy parameters of the defect subsystem, i.e., the energies of the defect formation and migration) [2].

Under certain conditions, the nonlinearities related to these interactions may become essential for the propagation of elastic perturbations in condensed matter and result in a renormalization of lattice parameters (elastic modulus). The presence of point lattice defects with a finite recombination rate in a medium may induce the appearance of dissipative terms, which are absent in ordinary equations for elastic waves [3].

The study of the behavior of elastic waves in view of their interaction with structural defects is of certain theoretical and practical interest, in particular, when analyzing the mechanisms of anomalous mass transport observed in the cases of the laser-matter interaction and ion implantation into metallic materials and in the studies of mechanical activation of components in the case of solid-phase chemical reactions [6].

From the point of view of laser semiconductor technology, the phenomenon of elastic strain wave generation is of interest in connection with the radiant energy transfer by acoustic waves through distances far exceeding the size of the energy absorption region. If the transferred energy density is sufficiently high, the acoustic waves can be one of the sources of the so-called "long-range effect" observed in semiconductor structures exposed to laser radiation [7]. The strain wave generation is also one of the factors explaining the physical mechanisms of ion-beam gettering, which is widely used in modern microelectronics for improving the electrophysical characteristics of layered semiconductor structures [8].

The wave of elastic strains propagating in a condensed medium carries information about distortions of their form and energy and about the energy losses related to the defect structure; this information is needed for optical-acoustical diagnostics of various parameters and the structure of solids.

A large amount of work has been devoted to solving elastic wave propagation with the consideration of the coupling effect between defect density and strain fields. A model of the propagation of bending elastic waves in phononic crystal thin plates with a point defect has been considered in [9]. They sown, that the frequency and number of the defect modes are strongly dependent on the filling fraction of the system and the size of the point defect. The theory of elasticity concerning the solid elastic material consisting of a distribution of atomic defects [3], is receiving greater attention due to its theoretical and practical relevance. In the above theory the atomic defects have been included as an additional independent kinematic variable. This theory reduces to the classical theory of elasticity in the limiting case when the defect density field vanishes. This new theory can play an important role in practical problems of laser material treatment where the classical theory is inadequate. In Refs [10]-[12] non-linear dynamics of self-consistent longitudinal strain waves in laser-irradiated solids without taking into account the interaction with lattice defects has been theoretically investigated. In these studies, attention was mostly focused on the study of an influence of the strain-induced diffusion, generation, and recombination of defects on the propagation of one-dimensional (1D) elastic strain disturbances and their dispersion and dissipation properties. Self-organization of stationary elastic-concentration periodic structures on the surface of the laser-irradiated solids has been considered in [13].

Research has been proposed to solve 1D dispersive strain wave propagation in an isotropic solid with quadratic nonlinearity of elastic continuum by taking into account the interaction of the longitudinal displacements with the temperature field and the field of concentration of recombining atomic defects [14]. The governing nonlinear equation describing the evolution of the self-consistent strain fields has been derived and discussed. The influence of stress-induced decay of defect complexes on the evolution of strain waves has been also considered.

Our aim in the present paper is to investigate the propagation of elasto - concentration waves in a linear homogeneous isotropic elastic layer with atomic defect generation. The secular equation, that governs the propagation of elasto - concentration waves has been derived by solving a system of coupled partial differential equations. We obtain the frequency equations corresponding to the symmetric and antisymmetric modes of vibration of the plate. Some limiting cases of the frequency equations are considered and a procedure for determining the phase velocity and the amplification factor is discussed. Relevant results of previous investigations are deduced as special cases

2. Basic Equations

The basic equations governing the concentration and mechanical fields are given in this section. Let us assume that an external energy flux (e.g., laser radiation) creates point atomic defects in a surface layer. Let $n^{(j)}(x_i,t)$ ($i=1,2,3$) be the concentration of these defects of the j th - type ($j=v$ for vacancies (v - defects) and $j=i$ for interstitials (i - defects). In the context of thermal-fluctuation model of the point-defect production, the rate of defect generation from the lattice sites is governed by temperature (or intensity of laser beam) and stresses. Therefore, this rate may vary under the effect of elastic deformation fields, i.e., thermo-fluctuation-related defects may be generated and annihilate. Deformation field affects the characteristics of the defects. Thus, when the strain waves propagate, the formation energy of atomic defects changes in the compression and dilatation zones. If $e=u_{ll}=\nabla\cdot\vec{u}$ is the dilatation ($\vec{u}$ is the displacement vector of the medium) and $\vartheta_g^{(j)}$ is the deformation potential characterizing the variation of the formation activation energy of defects under the lattice deformation, the renormalized formation energy of atomic defects can be represented as $w_g^{(j)}=w_{g0}^{(j)}-\vartheta_g^{(j)}e$ ($w_{g0}^{(j)}$ is the formation energy for the jth type of defect in an unstrained crystal). If there is a deformation-related perturbation of the lattice, not only the formation energy of defects decreases, but also the activation energy for the defect migration

$w_m^{(j)} = w_{m0}^{(j)} - \vartheta_m^{(j)} e$ ($w_{m0}^{(j)}$ is the migration energy of the defects in the absence of deformation and $\vartheta_m^{(j)}$ is the deformation potential characterizing the variation of the migration activation energy of defects under the lattice deformation) decreases; this results in an increase in the diffusion coefficient of defects (D). Modulation of the formation brings about the corresponding modulations of the source function (g) and recombination rate (γ) of atomic defects

$$g_j = g\exp\left(\vartheta_g^{(j)}\nabla\cdot\vec{u}/k_B T\right),$$
$$\gamma_j = \tau_j^{-1}\exp\left(\vartheta_m^{(j)}\nabla\cdot\vec{u}/k_B T\right).$$

Here $g = g_0\exp\left(-w_{g0}/k_B T\right)$ is the rate of generation of atomic defects by an external source in the absence of the strain field, T is the absolute temperature, k_B is the Boltzmann constant, g_0 is the constant of defect generation; τ_j is the relaxation time of defects of the j th – type in the absence of the strain field).

The concentration of atomic defects is dependent on temperature. One thus needs to know how the laser irradiation affects the local temperature of the surface at laser spot. We will consider here situations where the laser only heats the material (the light energy absorbed by the medium is transformed into heat), and that an equilibrium between laser radiation and the temperature field (T) is reached on time scales much shorter than the characteristic time scale of defect density evolution. Typically, the time scale for equilibration between photon absorption and defect generation is on the order of picoseconds, while that for defect diffusion is of the order of microseconds. We also assume that the contribution of thermal strains to deformation fields is negligible compared to lattice dilatation due to atomic defects and the phase changes and chemical reactions in the medium are absent.

In this paper, we will analyze the problem of the wave propagation in elastic media irradiated over large area by CW or pulsed lasers. We will, furthermore, assume that the temperature profile has reached its equilibrium value. Its evolution is sufficiently slow compared to atomic defect generation, and can be considered as quasistationary. The solution of the heat conduction equation for this case is given by Duley [15].

The field equation in a linear solid with the generation of atomic defect in the absence of body forces, has the form in [3]

$$\rho\frac{d^2 u_i}{dt^2} = \nabla_j \sigma_{ij}^{el} + \nabla_j \sigma_{ij}^{d}, \tag{1}$$

$$\frac{dn_j}{dt} = -\nabla_i Q_i + g_j - \gamma_j n_j - \beta n_i n_v. \tag{2}$$

Here ρ is the density of the medium; $\sigma_{ik} = \sigma_{ik}^{el} + \sigma_{ik}^{d}$ is the symmetric stress tensor; Q_i is the defect flux.

The second term in the right-hand side of Eq. (1) takes into account the forces applied to the lattice because of the defect-deformation interaction. Eq. (1) represents the generalization of the known equation for the elastic waves in a solid [16] to the case of a system with concentration-related stresses [3], which are caused by the generation–recombination processes in the non-equilibrium atomic defect subsystem. In Eq. (2) the first term represents diffusion of the defects, the second term characterise generation of atomic defects stimulated due to presence of the elastic field, the last two term describe the strain-stimulated recombination of defects [2].

The constitutive equations for the stress tensor (σ_{ik}) and defect flux ($\vec{Q}$) in the media to the strain and the change in defect density.

Thus

$$\sigma_{ik} = c_{iklm} e_{lm} - \sum_j \vartheta_{ik}^{(j)} n_j = \sigma_{ik}^{el} + \sigma_{ik}^{d},$$

$$Q_i = -D_{ik} \nabla_k n_j - \vec{v}_j n_j .$$

Here $e_{ij} = \left(\nabla_j u_i + \nabla_i u_j\right)/2$, c_{iklm} is the elastic modulus, $\vartheta_{ik}^{(j)}$ is the deformational potential, and D_{ik} diffusion coefficients of the defects.

The defect flux Q_i, where the first term represents diffusion with a coefficient D_j, the second term corresponds to the drift of defects at the velocity $\vec{v}_j = \left(D_j / k_B T\right)\vec{F}_j$ under the influence of the force $F_j = -\nabla U_{\text{int}}$ resulting from the interaction of defects with an inhomogeneous strain field ($U_{\text{int}} = -K\Omega_d^{(j)} \nabla \cdot u$ is the energy of the interaction between one defect and the strain field).

If, the material is isotropic, then c_{iklm}, D_{ij} and ϑ_{ik} are given by

$$c_{iklm} = \lambda \delta_{ik} \delta_{lm} + \mu\left(\delta_{il} \delta_{km} + \delta_{im} \delta_{lk}\right)$$

$$D_{ik} = D\delta_{ik}, \ \vartheta_{ik}^{(j)} = \vartheta_d^{(j)} \delta_{ik},$$

and the constitutive equations becomes

$$\sigma_{ik} = \delta_{ik} \lambda e_{ll} + 2\mu e_{ik} - \delta_{ik} \sum_j \vartheta_d^{(j)} n_j , \tag{3}$$

$$Q_j = -D\nabla n_j - \vec{v} n_j , \tag{4}$$

where λ and μ are the Lame coefficients [16]; $\vartheta_d^{(j)} = K\Omega_d^{(j)}$, where K is the bulk modulus and $\Omega_d^{(j)}$ is the volume elastic strain caused by the relaxation of the j th-type defect volume. For *v*-defects, $\Omega_d^{(v)} = -\delta^{(v)}\Omega < 0$ (here, the coefficient is $\delta^{(v)} = 0.2 - 0.4$ and Ω is the atomic volume), whereas, for *i*-defects, $\Omega_d^{(i)} = \delta^{(i)}\Omega > 0$ (the coefficient is $\delta^{(i)} = 1.7 - 2.2$). v and i – defects are represented as a substitutional atom whose volume is smaller or greater than the volume of the matrix atoms, respectively.

The field equations governing the displacement field and the defect density fields are obtained by substituting the constitutive relations (3)–(4) into the equations of motion (1) and (2) as

$$\rho \frac{\partial^2 \vec{u}}{\partial t^2} - \mu \nabla^2 \vec{u} - (\lambda + \mu)\nabla(\nabla \cdot \vec{u}) = -\sum_{j=i,v} \vartheta_d^{(j)} \nabla n_j \,, \tag{5}$$

$$\frac{\partial n_j}{\partial t} = g_j - D\nabla Q_j - \beta n_i n_v - \tau_j^{-1} n_j \,, \tag{6}$$

Here $\nabla^2 = \partial^2 / \partial x_\alpha \partial x_\alpha$, $\alpha = 1, 2$. In the absence of defect generation, we have $n_j = 0$; and Eq. (1) reduces then to the Navier's equation of classical elasticity. From this point on, we limit our consideration to the case of only one type of atomic defects and drop the superscript j in Eqns. (5)-(6); i.e., we assume that $\vartheta_d^{(j)} = \vartheta_d$, $\Omega_d^{(j)} = \Omega_d$, $n_j \equiv n$, $\tau_j \equiv \tau$, $D_j = D$, etc.

3. Formulation of the Problem

We consider an isotropic elastic solid which occupies the Cartesian space $x_3 \in [-h, h]$, $x_1, x_2 \in [-\infty, \infty]$. The boundary $x_3 = \pm h$ are supposed to be free of stresses and strains. We choose x_1 - axis along the direction of wave propagation in such a way so that all the particles on a line parallel to x_2 - axis are equally displaced and hence all the field quantities are independent of x_2-coordinate. Thus the motion of the plate is supposed to take place in the $x_1 x_3$ plane and for the assumed motion of the plate the displacement vector $\vec{\mathrm{u}}$ has the component $(\mathrm{u}, 0, \mathrm{w})$ and all the other variables depend on x_1, x_3 and t only.

We seek solutions of (5) and (6) subject to the boundary conditions

$$\sigma_{33} = \sigma_{31} = 0 \text{ on } x_3 = \pm h. \tag{7}$$

Also, the boundary condition for n is given by

$$\frac{\partial n}{\partial x_3} = 0 \text{ for } x_3 = \pm h. \tag{8}$$

We can express the defect density field as $n = n_0 + n_1$ ($n_0 = g\tau$ is a spatially homogeneous solution; n_1 is a small non-homogeneous perturbations). Inserting in (6) and neglecting the nonlinear terms, we get the linearised equation as

$$\frac{\partial n_1}{\partial t} = ge - g_d \Delta e + D\nabla^2 n_1 - \gamma n_1, \tag{9}$$

where $g = g_0 (\vartheta_g - \vartheta_m)/k_B T$, $g_d = Dn_0 \vartheta_d / k_B T$.

The relation between the strain-induced generation and drift terms is given by $4\pi^2 (\vartheta_d/\vartheta_g)(D\tau/d_{latt}{}^2)$, where d_{latt} is the characteristic scale of the waves. Thus, the generation term dominates at formation of „large-scale" structures $\left(d_{latt} \geq 2\pi (\vartheta_d/\vartheta_g)^{1/2} \sqrt{D\tau}\right)$, and the drift term dominates at generation of "small-scale" structures $\left(d_{latt} < 2\pi (\vartheta_d/\vartheta_g)^{1/2} \sqrt{D\tau}\right)$. As $\sqrt{D\tau} = l$ (l is the average distance between sinks), and the scale of structures is always supposed to be greater than l, the choice of drift or generation terms depends on a ratio ϑ_d/ϑ_g. In this paper we believe that $\vartheta_d/\vartheta_g < 1$.

4. Basic Solutions

In order to solve the problem, we use Helmholtz decomposition theorem to express the displacement vector as $\vec{u} = \nabla\varphi + \nabla \times \vec{\psi}$, $\nabla \cdot \varphi = 0$ so that the displacement components are written as

$$u = \frac{\partial\varphi}{\partial x_1} + \frac{\partial\psi}{\partial x_3}, \quad w = \frac{\partial\varphi}{\partial x_3} - \frac{\partial\psi}{\partial x_1}, \tag{10}$$

where the vector point potential function is defined as where $\vec{\psi} = (0, -\psi, 0)$ and φ and ψ depend only x_1, x_3 and time t.

Upon introducing equations (10) in (5) we obtain

$$\frac{\partial^2\varphi}{\partial t^2} = c_L^2 \nabla^2 \varphi - \frac{\vartheta}{\rho} n_1, \tag{11}$$

$$\frac{\partial^2\psi}{\partial t^2} = c_T^2 \nabla^2 \psi. \tag{12}$$

where $c_L \left(= \sqrt{(\lambda + 2\mu)/\rho}\right)$ is the sound velocity for longitudinal acoustic waves and $c_T \left(= \sqrt{\mu/\rho}\right)$ is the sound velocity for transverse acoustic waves [16]. Equation (12) corresponds to purely transverse waves in the solid which get decoupled from rest of the motion and are not affected by the defect density fields.

Equations (9) becomes

$$\frac{\partial n_1}{\partial t} = g\nabla^2\varphi + D\nabla^2 n_1 - n_1 \tau^{-1}. \tag{13}$$

Now using (10) and (13) the boundary conditions (7) may be expressed in terms of φ and ψ as follows

$$2\frac{\partial^2\varphi}{\partial x_1\partial x_3}-\frac{\partial^2\psi}{\partial x_3^2}+\frac{\partial^2\psi}{\partial x_1^{\,2}}=0, \tag{14}$$

$$\frac{\partial^2\varphi}{\partial x_1\partial x_3}-\frac{\partial^2\psi}{\partial x_1^{\,2}}+\frac{1}{2c_T^2}\frac{\partial^2\psi}{\partial t^2}=0. \tag{15}$$

We consider the case of harmonic waves so that the solutions φ, ψ and n, of equations (11)-(12) take the forms of

$$\varphi(x_1,x_3,t)=\widetilde{\varphi}(x_3)\exp i(qx_1+\omega t),$$
$$\psi(x_1,x_3,t)=\widetilde{\psi}(x_3)\exp i(qx_1+\omega t),$$
$$n_1=\left(\rho c_L^2/\vartheta_d\right)\widetilde{n}\exp i(qx_1+\omega t), \tag{16}$$

where q and ω are wave number and complex angular frequency of the waves, respectively; $\widetilde{\varphi}(x_3)$, $\widetilde{\psi}(x_3)$ are functions of x_3. Evidently expressions (16) correspond to plane surface waves propagating along the positive x_1-direction with wave length $2\pi/q$. Here, the phase velocity is given by $c=\mathrm{Re}(\omega)/q$ and attenuation constant by $R=-\mathrm{Im}(\omega)$, where $\mathrm{Re}(\omega)$ and $\mathrm{Im}(\omega)$ denote the real and imaginary part of ω, respectively.

Substitution of (16) into (11) and (12) led to a coupled system of three equations in terms of ($\widetilde{\varphi}$, $\widetilde{\psi}$, $\widetilde{n}$)

$$\frac{d^2\widetilde{\varphi}}{dx_3^{\,2}}-\left(q^2-\frac{\omega^2}{c_L^2}\right)\widetilde{\varphi}-\widetilde{n}=0, \tag{17}$$

$$\frac{d^2\widetilde{\psi}}{dx_3^{\,2}}-\left(q^2-\frac{\omega^2}{c_T^2}\right)\widetilde{\psi}=0, \tag{18}$$

$$\frac{d^2\widetilde{n}}{dx_3^2}-\left(q^2+\frac{i\omega\tau+1}{l^2}\right)\widetilde{n}+\frac{\delta}{l^2}\left(\frac{d^2\widetilde{\varphi}}{dx_3^2}-q^2\widetilde{\varphi}\right), \tag{19}$$

where $\delta=g_0\vartheta_d\left(\vartheta_g-\vartheta_m\right)\tau/\rho c_L^2kT$ is the defect-strain coupling constant. The parameter δ may be of either sign. As signs of the deformational potentials ϑ_d,ϑ_g and ϑ_m are the same, we have: for v - defects $\delta>0$, if $|\vartheta_g|>|\vartheta_m|$; $\delta<0$, if $|\vartheta_g|<|\vartheta_m|$.

Eliminating $\widetilde{n}$ from (17) and (19), we obtain

$$\left\{\left[\frac{d^2}{dx_3^2}-q^2\left(1-\frac{\omega^2}{c_L^2k^2}\right)\right]\left[\frac{d^2}{dx_3^2}-q^2-\frac{1}{l^2}(1+i\omega\tau)+\frac{\delta}{l^2}\left(\frac{d^2}{dx_3^2}-q^2\right)\right]\right\}\tilde{\varphi}=0, \quad (20)$$

The requirement of the existence of non-trivial solution of equations (17) – (19) provides us a quartic polynomial characteristic equation in α^2, which give us three pairs of the characteristic roots $\pm\alpha_j$ ($j=1,2,3$). In general, the roots $\pm\alpha_j$ are complex and therefore, the solution is a superposition of the plane waves attenuating with depth. As we are considering surface waves only, we choose only that form of α_i which satisfies the radiation condition viz. $\mathrm{Re}(\alpha_j)>0$. Thus, we obtain the following formal solution for the functions ($\tilde{\varphi}$, $\tilde{\psi}$):

$$\tilde{\varphi}=\sum_{j=1}^{2}a_j\cosh\alpha_j x_3+b_j\sinh\alpha_j x_3,$$

$$\tilde{\psi}=a_3\sinh\alpha_3 x_3+b_3\cosh\alpha_3 x_3, \quad (21)$$

where a_j, b_j, $j=1,2,3$ are the arbitrary constants,

$$\alpha_1^2=q^2-\omega^2/c_T^2, \quad (22)$$

α_2^2 and α_3^2 are the roots of the equation:

$$\begin{aligned}&\alpha^4-\alpha^2\left[2q^2-\omega^2c_L^{-2}+(1+i\omega\tau-\delta)l^{-2}\right]+\\&\times\left\{\left(q^2-\omega^2c_L^{-2}\right)\left((i\omega\tau+1)l^{-2}+q^2\right)-\delta\, l^{-2}\right\}=0.\end{aligned} \quad (23)$$

Then α_2^2 and α_3^2 are defined from (23) as follows

$$\begin{aligned}&\alpha_2^2+\alpha_3^2=2q^2-\omega^2c_L^{-2}+(1+i\omega\tau-\delta)l^{-2},\\&\alpha_2^2\alpha_3^2=\left(q^2-\omega^2c_L^{-2}\right)\left(q^2+(i\omega\tau+1)l^{-2}\right)-\delta\, l^{-2}q^2.\end{aligned} \quad (24)$$

Upon inserting solutions for φ and ψ from (21) in Eq. (10), the displacement components are obtained as

$$u=\left(iq\tilde{\varphi}(x_3)+\alpha_3\left(a_3\cosh\alpha_3x_3+b_3\sinh\alpha_3x_3\right)\right)\exp i(qx_1+\omega t),$$

$$w=\left(-iqa_3\tilde{\psi}(x_3)+\sum_{j=1}^{2}\alpha_j\left(a_j\cosh\alpha_jx_3+b_j\sinh\alpha_jx_3\right)\right)\exp i(qx_1+\omega t).$$

Clearly, the displacements get modified due to the characteristic roots corresponding to diffusion field equations because of coupling among interacting fields. Similarly we can obtain expressions for the normal stresses.

5. Dispersion Equations if the Wave in an Infinitive Medium

In this section as a special case we consider the generation of plane harmonic structures in unbounded medium. Setting $\alpha_r = 0,\ \gamma = 1, 2, 3$ in (22) and (23) we obtain

$$\omega^2 = c_T^2 q^2, \tag{25a}$$

$$\left(q^2 - \omega^2 c_L^{-2}\right)\left[q^2 + (i\omega\tau + 1)l^{-2}\right] - \delta l^{-2} q^2 = 0. \tag{25b}$$

Equation (25a) corresponds to transverse waves which are independent of δ, having speed of propagation c_T. Eq. (25b) pertains to the coupled strain and concentration waves.

To explore and delineate the strain and defect generation effects, we shall seek solutions of (20b) for small values of δ. For $\delta = 0$, Eq. (25b) admits the following solutions $\omega_{1,2}^{(0)} = \pm c_L q$ (acoustical mode), $\omega_3^{(0)} = i\left(l^2 q^2 + 1\right)\tau^{-1}$ (diffusion mode). Now, for small $\delta << 1$, we may write

$$\omega_{1,2} = \omega_{1,2}^{(0)} + \varepsilon\omega_{1,2}^{(1)} + \varepsilon^2\omega_{1,2}^{(2)} + \cdots,$$

$$\omega_3 = \omega_3^{(0)} + \varepsilon\omega_3^{(1)} + \varepsilon^2\omega_3^{(2)} + \cdots,$$

substitute into (25b), and, equating the coefficients of like powers of δ, finally arrive at

$$\omega_{1,2} = \omega_{1,2}^{(0)} - \frac{\delta}{2}\frac{\omega_{1,2}^{(0)}\left(q^2 l^2 + 1 - i\omega_{1,2}^{(0)}\tau\right)}{\left(q^2 l^2 + 1\right)^2 + \left(\omega_{1,2}^{(0)}\tau\right)^2} + O(\delta^2), \tag{26}$$

$$\omega_3 = \omega_3^{(0)} - \frac{i\delta c_L^2 q^2 \tau}{c_L^2 q^2 \tau^2 + \left(q^2 l^2 + 1\right)^2} + O(\delta^2). \tag{27}$$

From the Eq. (26) we obtain

$$\mathrm{Re}\left(\omega_{1,2}\right) = \omega_{1,2}^{(0)}\left[1 - \frac{\delta}{2}\frac{q^2 l^2 + 1}{\left(q^2 l^2 + 1\right)^2 + \left(c_L q\tau\right)^2}\right], \tag{28}$$

$$\mathrm{Im}\left(\omega_{1,2}\right) = \frac{\delta\tau}{2}\frac{c_L^2 q^2}{\left(q^2 l^2 + 1\right)^2 + \left(c_L q\tau\right)^2}. \tag{29}$$

If $c_L q >> \tau^{-1}(l^2 q^2 + 1)$ and the viscosity is taken into account [by adding in Eq. (5) the terms $\eta_T \Delta\dot{\vec{u}}$ and $\eta_L \nabla(div\dot{\vec{u}})$, where $\eta_T = \eta$, $\eta_L = 4\eta/3 + \zeta$; η and ζ are the first and second viscosity coefficients] the dispersion Eq. (29) describes an instability of the amplitude of acoustic waves (i.e. laser generation of acoustic waves)

$$\mathrm{Im}(\omega_{1,2}) = \frac{1}{2\rho c_L^2}\eta q^2 + \frac{\delta}{2\tau}.$$

The excitation of acoustic waves as a result of the instability occurs, if $|\delta| > \eta q^2 \tau / \rho c_L^2$.

It follows that the frequency spectrum of acoustic wave is hardly changed $\left(\mathrm{Re}(\omega_{1,2}) \approx \omega_{1,2}^{(0)}\right)$, but the increment is renormalized.

If $c_L q << \tau^{-1}(l^2 q^2 + 1)$

$$\mathrm{Re}(\omega_{1,2}) = \omega_{1,2}^{(0)}\left(1 - \frac{\delta}{2(q^2 l^2 + 1)}\right), \tag{30}$$

$$\mathrm{Im}(\omega_{1,2}) = \frac{1}{2\rho c_L^2}\eta q^2 + \frac{\delta\tau}{2}\frac{c_L^2 q^2}{(q^2 l^2 + 1)^2}. \tag{31}$$

In this case there is a softening of frequencies of acoustic waves (instability of frequencies $\mathrm{Re}(\omega_{1,2}) \to 0, \mathrm{Im}(\omega_{1,2}) > 0$) and this is related to taking into account the generation of atomic defects. It is necessary to notice, that reduction of frequency occurs not up to zero, and up to value $\omega_\eta = \eta q^2 / 2\rho c_L^2 << c_R q$. The Eq. (31) describes attenuation of the amplitudes of acoustic waves.

At $\omega << c_L q$ from the (25b) we obtain

$$\omega = -i(\delta - l^2 q^2 - 1)\tau^{-1}. \tag{32}$$

It follows from the expression (32) that if

$\delta \geq \delta_{cr} = 1$ i.e.

$g \geq g_{cr} = \rho c_L^2 k_B T / \tau \vartheta_d (\vartheta_g - \vartheta_m)$,

there is a rang of wave numbers

$$0 < q < q_0 = \sqrt{(\delta - 1)/l^2},$$

such that $\operatorname{Im}\omega < 0$, i. e. the amplitudes of the Fourier harmonics of the small perturbations increase with time without limit. The result is an instability leading to the formation of coupled defect-strains structures. The period of these structures is

$$d_{lat} = \frac{2\pi}{q_0} = 2\pi l\sqrt{\frac{g_{cr}}{g - g_{cr}}}.$$

As temperature is increased, the period grows and this is related to an increase in the life-time of defects.

6. Dispersion Equation of the Plane Wave in Plates

Applying the boundary conditions (14) and (15) one obtains a set algebraic equations for the determination of the arbitrary constants d_j. Elimination of this constants, gives

$$|S_{ik}| \times |A_{ik}| = 0, \; i,k = 1,2,3, \tag{33}$$

where

$$S_{1l} = \left(2 - \omega^2/c_T^2 q^2\right)\coth\alpha_l h,$$
$$S_{13} = 2i\alpha_3 q^{-1}\coth\alpha_3 h,$$
$$S_{2l} = 2i\alpha_l q^{-1}, \; S_{23} = -(2 - \omega^2/q^2 c_T^2)$$
$$S_{3l} = \alpha_l q^{-1}\left(\alpha_l^2 + \omega^2 c_T^{-2} - q^2\right), \; S_{33} = 0, \; l = 1,2.$$

Expressions for A_{ik}'s can be obtained from S_{ik} by replacing $\coth(\alpha_j h)$ by $\tanh(\alpha_j h)$. Equations $|S_{ik}| = 0$ and $|A_{ik}| = 0$, respectively, are associated with the symmetric and antisymmetric motions of the plate relative to the middle surface. The frequency equations $|S_{ik}| = 0$ and $|A_{ik}| = 0$ can be combined and represented as

$$\left(2q^2 - \frac{\omega^2}{c_T^2}\right)^2 = \frac{4q^2\alpha_1\alpha_2\alpha_3\left(\alpha_2^2 - \alpha_1^2\right)}{\dfrac{\alpha_2\left(\alpha_2^2 + \omega^2 c_L^{-2} - 1\right)}{\gamma_1^{\pm 1}} - \dfrac{\alpha_1\left(\alpha_1^2 + \omega^2 c_L^{-2} - 1\right)}{\gamma_2^{\pm 1}}}, \tag{34}$$

where

$$\gamma_i = \tanh\alpha_3 h / \tanh\alpha_i h, \; i = 1,2.$$

In the equation (34), the positive power index is associated with the symmetrical motion and negative index with the antisymmetrical motion.

Equations (34) contains $c(= \operatorname{Re}(\omega)/q)$ as the unknown quantities and hence this equation may be used to determine c in terms of q. This equation is therefore the phase velocity equation for the waves considered. Since c depends on q in general, the waves are generally dispersive. This dispersive nature of the general waveform arises due to the presence of defect generation in the medium. The discussion of transcendental equation in general is difficult; we therefore, consider the results for some limiting cases.

7. LIMITING FORMS OF THE FREQUENCY EQUATIONS

We now discuss the limiting forms of the frequency equation, that are defined by the inequalities

i) $\alpha h << 1$ very long waves;
ii) $\alpha h >> 1$ very short waves.

In first case the hyperbolic tangents in (34) approaches unity and hence (34) for symmetrical motion becomes

$$\left(2q^2 - \omega^2 c_T^{-2}\right)^2 \left(\alpha_1^2 + \alpha_2^2 + \alpha_1\alpha_2 - \omega^2 c_L^{-2} + q^2\right) = 4q^2\alpha_1\alpha_2\alpha_3\left(\alpha_1 + \alpha_2\right). \quad (35)$$

Equation (35) determines the velocity of Rayleigh surface waves in a half - space solid with atomic defects. We shall now consider this in the following special cases.

a) In the particular case $\delta = 0$, which means that the defect - strain interaction is absent, we obtain

$$Ra\left(c^2\right)\left(i\omega\tau + l^2q^2 + 1\right) = 0,$$

$$Ra\left(c^2\right) = \left(1 - \frac{c^2}{2c_T^2}\right)^4 - \left(1 - \frac{c^2}{c_L^2}\right)\left(1 - \frac{c^2}{c_T^2}\right)$$

which yields

$$Ra\left(c^2\right) = 0, \; w = -i\left(Dq^2 + \tau^{-1}\right). \quad (36a,6)$$

Equation (36a) is equivalent to the classical Rayleigh equation, while (36b) defines the attenuation constant for the concentration waves. With $c_R^2 = \omega^2/k^2$ Eq. (36a) reduces to

$$\left(2 - c_R^2 c_T^{-2}\right)^4 = 16\left(1 - c_R^2 c_T^{-2}\right)\left(1 - c_R^2 c_L^{-2}\right). \quad (37)$$

b) $\delta \neq 0$. In this case it has solutions describing qualitatively different types of instability: 1) if the viscosity is taken into account the dispersion Eq. (35) describes laser generation of acoustical waves (SAWs); 2) generation of ordered static structures.

At $\delta << 1$ roots of Eqs (24) may by rewrite as

$$\alpha_2^2 = q^2 - \omega^2 c_L^{-2} + \eta_L i\omega^3 \left(\rho c_L^4\right)^{-1} - \delta \frac{\omega^2 c_L^{-2}}{i\omega\tau + 1 + \omega^2 c_L^{-2} l^2} + O(\delta^2),$$

$$\alpha_3^2 = q^2 + l^{-2}(i\omega\tau + 1) - \frac{\delta}{l^2} \frac{i\omega\tau + 1}{i\omega\tau + 1 + \omega^2 c_L^{-2} l^2} + O(\delta^2).$$

Then we have

$$\frac{\left(1 - \omega^2/2q^2 c_T^2 + i\omega^3 \eta_T / 2q^2 \rho c_T^4\right)^2}{\left(1 - \omega^2/q^2 c_T^2 + i\omega^3 \eta_T / q^2 \rho c_T^4\right)^{1/2}} = W^{1/2} \times$$

$$\frac{\left[2\sqrt{W} + 2q^2 - \omega^2 c_L^{-2} + (i\omega\tau + 1 - \delta)l^{-2} + \eta_L i\omega^3/\rho c_L^4\right]^{1/2}}{q\left[q^2 + (i\omega\tau + 1 - \delta)l^{-2} + \eta_L i\omega^3/\rho c_L^4 + \sqrt{W}\right]}, \tag{38}$$

where

$$W = \left(q^2 - \omega^2 c_L^{-2} + \eta_L i\omega^3 \left(\rho c_L^4\right)^{-1}\right)\left(q^2 + (i\omega\tau + 1)l^{-2}\right) - q^2 \delta\, l^{-2}.$$

Assuming that $\delta << 1$ is sufficiently small, set

$$\omega^2/k^2 = c_R^2\left(1 + \varepsilon\xi_1 + \varepsilon^2\xi_2 + \ldots\right), \tag{39}$$

where $c_R = \omega_R/q$ is the classical Rayleigh wave speed which is a root of (37).

Substituting expression (39) into (38), and retaining only terms of $O(\varepsilon)$, we receive the following dispersion relation

$$\omega^2 = q^2 c_R^2 \left(1 - \delta \frac{(1 - \beta\varsigma) f(c_R, q)}{l^2 c_R^2 Ra'(c_R)}\right), \tag{40}$$

where

$$f(c_R, q) = \frac{1}{b}\left(1 - \frac{a}{q^2} \frac{b + 2\sqrt{ab} + q^2}{a + b + 2\sqrt{ab}}\right),$$

$$a = q^2\left(1 - c_R^2 c_L^{-2}\right),\ b = q^2 + (ic_R q\tau + 1)l^{-2} + \eta_L i q^3 c_R^3 \left(\rho c_L^4\right)^{-1},$$

$$\beta = c_R^2/c_L^2,\ \varsigma = c_L^2/c_T^2 \quad Ra'(c_R) = \left.\frac{dRa}{d(\omega^2/q^2)}\right|_{\omega^2/k^2 = c_R^2}.$$

If ${\omega_R}^2 \eta D << \rho c_T^4$ and $c_R q >> Dq^2 + \tau^{-1}$ from the Eq. (40) we have

$$\omega^2 = q^2 c_R^2 - iq|\delta| \frac{c_R(1 - \beta\varsigma)}{\tau c_L^2 Ra'(c_R)} + iq^3 \eta \frac{c_R \beta\varsigma}{\rho}, \tag{41}$$

After decomposition of Eq. (41) into real and imaginary parts, for the frequency of SAWs we have the familiar expression $\omega_R = c_R q$. The increment $\Gamma = \mathrm{Im}(\omega)$ of SAWs obtained from Eq. (41) is

$$\Gamma = \eta q^2 \frac{\beta\varsigma}{2\rho} - |\delta| \frac{(1-\beta\varsigma)}{2\tau c_L^2 Ra'(c_R)}. \tag{42}$$

The dispersion Eq. (42) describes an instability of the amplitude of acoustic waves.

In the absence of defect-strain interaction ($\delta = 0$) waves decay due to viscosity of the medium. The excitation of SAWs as a result of the instability occurs if

$$|\delta| > \delta_{cr} = \eta\tau q^2 \frac{\beta\varsigma c_L^2 Ra'(c_R)}{\rho(1-\beta\varsigma)}.$$

It follows that the SAW frequency is hardly changed, but the increment is renormalized. If $c_L q << \tau^{-1}(l^2 q^2 + 1)$ from the Eq. (40) we have

$$\mathrm{Re}(\omega_{1,2}) = q c_R \left(1 - \delta \frac{1-(1-\beta)/s(\beta)}{(q^2 l^2 + 1)c_R^2 R'(c_R)}\right)^{1/2},$$

$$s(\beta) = \frac{2\left(1+\sqrt{1-\beta}\right)-\beta}{2+\sqrt{1-\beta}}.$$

It is seen that in this case there is a softening (as like for the infinite volume) of frequencies of acoustic waves (instability of frequencies $\mathrm{Re}(\omega_{1,2}) \to 0, \mathrm{Im}(\omega_{1,2}) > 0$) and this is related to taking into account the defect-strain interaction. This reduction of frequency occurs not up to zero, and up to value $\omega_\eta = \sigma\eta q^2 / 2\rho c_L^2 << c_R q$, where $0.87 < \sigma < 0.95$ [16].

Equation (38) leads to the dispersion equation for static structures [on condition that $\Gamma^2 / c_{T,L}^2 << q^2$, $(\Gamma + \tau^{-1})/D$, $\delta > 0$, $\mathrm{Re}(\omega) = 0$] in the form:

$$\Gamma = \left(\frac{\delta}{1 - c_T^2 c_L^{-2}} - l^2 q^2 - 1\right)\tau^{-1}. \tag{42}$$

From the expression (42) follows that if

$$\delta \geq \delta_{cr} = \left(1 - c_T^2 c_L^{-2}\right)$$

i.e.

$$g \geq g_{cr}^{(s)} = \rho c_L^2 \left(1 - c_T^2 c_L^{-2}\right) k_B T / \tau \vartheta_d \left(\vartheta_g - \vartheta_m\right)$$

there is a rang of wave numbers

$$0 < q < q_0 = l^{-1}\sqrt{\delta\left(1 - c_T^2 c_L^{-2}\right)^{-1} - 1},$$

in which $\Gamma > 0$, i. e. the amplitudes of the Fourier harmonics of the perturbations increase with time without limit. This gives rise to an instability and its growth results in the formation of coupled periodic structures of defect density and strain on surface. The period of surface static structures is

$$d_{latt}^{(s)} = 2\pi l \sqrt{\frac{g_{cr}^{(s)}}{g - g_{cr}^{(s)}}}. \tag{43}$$

The value of the maximum increment is:

$$\Gamma_{\max} = \delta\left(1 - c_T^2 c_L^{-2}\right)^{-1} - \tau^{-1}.$$

Then the critical value of the pump parameter is found from the condition $\Gamma_{\max} = 0$:

$$\delta_{cr} = \left(1 - c_T^2 c_L^{-2}\right).$$

The formation time of a surface concentration-elastic structure is given by

$$\tau_s \approx \Gamma_{\max}^{-1} = \tau\left(\delta/\delta_{cr} - 1\right)^{-1},$$

which tends to infinity near an instability threshold ($\delta \to \delta_{cr}$) (critical slowing down occurs at a phase transition). When the threshold is exceeded significantly ($\delta >> \delta_{cr}$), we find that $\tau_s = \tau_d \delta_{cr} \delta^{-1}$.

For antisymmetical motion, the ratio of hyperbolic tangents in (34) may by approximated to unity and (34) reduces to (35).

From the results presented we observe that the influence of coupling between the strain and defect fields affects the phase velocity of the wave motion and the amplification factor as well.

Consider now the case of very long waves, by making use of the approximation $\tanh x \approx x - x^3/3$. So (28) for symmetrical motion becomes

$$4\alpha_1^2\alpha_2^2 - \left(2 - c^2 c_T^{-2}\right)^2 \left[\alpha_1^2 + \alpha_2^2 - q^2(1 - c^2 c_L^{-2})\right] q^2 = 0. \tag{44}$$

Equation (44) determines the vibration of thin plates with defect generation. When the medium is free from defects we have $\alpha_1^2 = q^2\left(1 - c^2/c_L^2\right)$ and we get the classical results of Rayleigh:

$$\left(2 - c^2 c_T^{-2}\right)^2 - 4\left(1 - c^2 c_L^{-2}\right) = 0 .$$

For small frequency waves we ignore higher degree terms in. In view of this approximation Eq. (44) becomes

$$\left(2 - \frac{c^2}{c_T^2}\right)^2 = 4\left(1 - \frac{c^2}{\tilde{c}_L^2}\right)^2, \tag{45}$$

where

$$\tilde{c}_L^2 = c_L^2(1-\theta),\ \theta = \delta\tau/\rho\, c_L^2 .$$

For antisymmetical motion equation (34) reduces to

$$\begin{aligned}\left(2-c^2c_T^{-2}\right)^2\Big[\alpha_2\left(\alpha_2^2+q^2\left(c^2c_L^{-2}-1\right)\right)\left(1-\tfrac{1}{3}h^2\alpha_1^2\right)\\ -\alpha_1\left(\alpha_1^2+q^2\left(c^2c_L^{-2}-1\right)\right)\left(1-\tfrac{1}{3}h^2\alpha_2^2\right)\Big]\\ =4q\left(1-c^2c_T^{-2}\right)\left(\alpha_2^2-\alpha_1^2\right)\left(1-\tfrac{1}{3}h^2\alpha_3^2\right).\end{aligned} \tag{46}$$

Equation (46) can be considered as modified version of the classical result obtained by Rayleigh in an elastic plate with defect generation. If the plate is free from defects, (46) transforms to

$$c^2/c_T^2 = \left(\tfrac{4}{3}\right)q^2h^2\left(1 - c_T^2/c_L^2\right),$$

which is the classical result of Rayleigh [16].

CONCLUSION

In the present work, study has been made to study the problem of elastic wave propagation in an isotropic plate containing a distribution of non-equilibrium atomic defects (interstitial atoms, vacancies, electron-hole pairs). The formation of atomic defects occurs as a result of the action of intense external energy fluxes (laser radiations) on the solid plates. The analysis is based on coupled evolution equations for the elastic displacement of the medium and defect density fields. The defect dynamics is governed by the strain-stimulated generation, transport, and annihilation processes. We have provided an exact formal solution for the displacements and defect density fields in an infinite plate of finite thickness. We have obtained dispersion equations corresponding to the symmetric and antisymmetric modes of vibration of the plate assuming that the boundaries of the plate are stress-free. We used Lame's potentials to derive the dispersion equations. A procedure for determining the phase velocity and the amplification factor is discussed. The proposed analysis is applied for the special cases of very short and very long waves. A procedure for determining the phase velocity and the amplification factor has been discussed. The phase velocity and attenuation constants of the waves get modified due to the generation-recombination effects in a defect subsystem. Relevant results of previous investigations are deduced as special cases.

We also sowed, that at certain conditions concentration – elastic instabilities in a system of atomic point defects on the surface of a semi-infinite condensed medium can be developed. The underlying idea of the model of an instability considered in this paper is related to a reduction in the activation (formation and migration) energy of atomic defects under the influence of elastic fields. This influence modulates the rates of generation (or recombination) processes. At small concentration of atomic defects, their return influence on strain fields a little also can be ignored. However under strongly non-equilibrium conditions, when the pump parameter is $g > g_{cr}$ (or the density of atomic defects $n > n_{cr}$), it is necessary to take into account that the processes of recombination or generation react on the stress fields. The equation for the kinetics of point defects is then supplemented by an equation from the theory of elasticity.

We have observed that, if the pump parameter is above a critical value, due to the strain-defect instability a coupled elastic-defect periodic structure on the surfaces of plates arises. A mechanism on the development of the instability is due to the coupling between defect dynamics and the elastic field of the solids. Laser radiation (or, in general, a flux of particles) generates high concentrations of atomic defects in the surface layer of the irradiated material. When a fluctuation harmonic of the elastic deformation field appears in a medium because of the generation of atomic defects, the activation energies of formation and migration of the defects are modulated and a strain-induced drift of atomic defects occurs. This is a consequence of defect–strain interaction. The associated modulation of the rates of defect generation (recombination) and strain-induced flux of defects gives rise to periodic spatial– temporal fields of the defect concentration. The redistribution of defects creates forces proportional to their gradients. These forces lead in turn to additional growth of strain fluctuations. When the defect density or a critical rate of defect generation exceeds the critical value, concentration–elastic instabilities develop as a result of positive feedback, which result in the formation of ordered concentration–strain structures.

The limiting case of these instabilities, when the spatial inhomogeneity of the defect distribution is the result solely of the strain-induced fluxes, can be called the diffusion– elastic instability and the case in which only modulation of the defect generation rate is important can be considered as the generation-elastic instability. It follows that the system of atomic defects with the elastic interaction is internally unstable against a transition to a spatially inhomogeneous state.

We sown that, the increment of the elastic-concentration structures is proportional to pump parameter. The criterion of self - organization of atomic defects and the dependence of a lattice period on temperature of medium are determined analytically. The growth in period with increase in temperature is related to a decrease in the life-time of atomic defects. For a fixed value of the temperature field the period of the lattice decreases with increase in pump parameter. The critical value of the pump parameter for self-organization of periodic structures is governed by the dilatation volume of defects, by the potential energy of elastic interaction, and the elastic parameters of the medium.

As an example let us consider the formation of concentration-strain structures in laser-irradiated semiconductors (in particular, *Si*). To evaluate the concentration of generated lattice defects (n_0), we consider here conditions when the duration of a laser pulse (τ_{Las}) exceeds the defect-relaxation time (τ_d). In this case the density of defects on the surface of the solid reaches a steady-state value

$$n_0 = g_0\tau = g^{(0)}\tau^{(0)} \exp\left(-w_{f0}/k_B T_0\right),$$

where

$$w_{f0} = w_{f0} - w_{m0},\ g_0 = g(T_0)$$

is the defect generation rate, T_0 is the steady-state value of the temperature field on the surface, $G^{(0)}$ and $\tau_d^{(0)}$ are constants. If $I_{Las} = const$ (uniform irradiation) and the optical absorption length ($\alpha_{Las}{}^{-1}$) sufficiently less than heat diffusion length $l_T = (\chi\tau_{Las})^{1/2}$, the maximum temperature rise cat the substrate surface owing to the laser pulse action may be evaluated [15] as

$$T_0 = \frac{2(1-R)I_0}{\lambda_T}\sqrt{\frac{\chi\tau_{Las}}{\pi}}, \tag{47}$$

where R is the reflectivity coefficient, λ_T the thermal conductivity coefficient; χ the thermal diffusion coefficient. So putting values into (47) for *Si* ($\lambda_T = 0.8\,\mathrm{W/(cm\,^0K)}$, $\chi = 0.4\,\mathrm{cm^2/s}$, $I_0 = 10^5\,\mathrm{W/cm^2}$, pulse duration $2\times10^{-3}\,\mathrm{s}$ and $R = 0.4$ one can get $T_0 = 1.6\times10^3\ {}^0\mathrm{K}$. Then, taking $\Omega = 2\times10^{-22}\,\mathrm{cm^3}$, $w_{d0} = 1\,\mathrm{eV}$, a value of $10^{19}\,1/\mathrm{cm^3}$ may be estimated for the critical density of defect concentration ($n_{cr} = g_{cr}\tau$), which is several orders of magnitude less than the concentration of the host atoms and shows that this mechanism of the formation of ordered structures may be realized on practice. For the period (d_{latt}) of the resultant surface structure we have an estimate $0.9\,\mu\mathrm{m}$, which follows from expression (43) (for typical values of parameters $\rho = 2.3\,\mathrm{g/cm^3}$, $\lambda = 6.4\times10^{10}\,\mathrm{Pa}$, $\mu = 7.9\times10^{10}\,\mathrm{Pa}$, $|\vartheta_g| = 15\mathrm{eV}$, $|\vartheta_m| = 10\mathrm{eV}$, $|\vartheta_d| = 10^2\mathrm{eV}$, $g = 1.1 g_{cr}^s$, $D = 10^{-5}\,\mathrm{cm^2/s}$ and $\tau_d = 3\times10^{-4}\,\mathrm{s}$). The maximum growth increment of the instability is $\Gamma_{max} = 10^4\,\mathrm{s^{-1}}$; the instability increment exceeds the reciprocal of the duration of a laser pulse acting on the surface of a solid $\left(\Gamma_{max}\tau_{Las} = 10 >> 1\right)$.

In addition to one-dimensional (or two-dimensional) lattices, the growth of the concentration elastic instability on the surface can also create structures of different types. If the laser field is radially symmetric relative to the z axis, then surface structures in the form of radial rays and rings are formed. Then, the substitution

$$\exp(iqx_1) \to J_m(qr)\cos(m\varphi)$$

is made in solutions (16); here, J_m is a Bessel function of the first kind and of order m (m is an integer). The specific case of $m = 0$ corresponds to structures which are concentric rings; r and φ are the polar coordinates in the x_1x_2 - plane. An analysis similar to that given in Sec. 4 leads to a dispersion equation which is identical; with Eq. (34) and the increment is independent of the number of a harmonic.

The degeneracy in respect of m is lifted if the Gaussian dependence of the laser radiation intensity on the radial coordinate, $I = I_0 \exp\left(-r^2/r_0^2\right)$, is taken into account. This leads to the appearance of structures in the form of radial rays and it is then found that

$$\exp(ikx_1) \to (r/r_0)^m \cos(m\varphi)\exp\left(-r^2/r_0^2\right).$$

The dispersion dependences for these structures are again described by Eq. (34), but with q^2 replaced by $\tilde{q}^2 = 4m/r_0^2$. The maximum number of rays in a structure deduced from the condition $q_m^2 = \tilde{q}^2$ is $m \approx \pi^2 r_0^2/d_{latt}^2$.

The appearance of ordered surface-relief structures in the form of radial and ring rays was detected experimentally [17] when the surface of nickel was irradiated in air by pulses from a solid-state laser ($\lambda_{Las} = 1.06\,\mu$m, duration $\tau_{Las} = 1.6\times10^{-3}$ s) with a Gaussian distribution and with the maximum intensity $I = 5\times10^5$ W/cm^2. For $r_0 = 3\times10^{-2}$ cm and $d_{latt} = 30\,\mu$m, the maximum number of rays is 10^3, roughly in agreement with the experimental results [17]. Radial ring periodic structures of the precipitate have been also observed after laser precipitation of ions from the liquid phase on the surfaces of metals in Ref. [18].

We considered the self-organization of atomic defects of same j-th type due to the interaction of these defects through the self-consistent strain field. In an irradiated medium, defects are generated as pairs of interstitial atoms ($j = i$) and vacancies ($j = v$). Usually the inequality $\vartheta_i >> \vartheta_v$ is met for such processes. Therefore, we kept only the term with $\vartheta_d^{(i)} \equiv \vartheta_d > 0$ in equations (1) and (2) for self consistent displacements. Interstitial atoms form self –consistent periodic structures with defect concentration maxima within the regions where strain field $divu_{ll} = \zeta > 0$. Vacancy defects are accumulated within the regions where $\zeta < 0$. Thus, double interstitial atom – vacancy periodic structures are formed.

Clearly, the linear theory considered here describes the early stage of the development of an instability only. However, the nature of generated ordered structures (due to an instability) and the amplitudes of these structures as functions on material and irradiation conditions can only be determined by considering the influence of nonlinear effects in the model. In this connection, computer simulation of fluctuation development is needed.

It follows that, for the concentration elastic instabilities ordered structures of different type may appear depending on the parameters of the radiation and the sample itself. Formation of one-dimensional lattices, and of concentric-ring and radial structures, is governed by the spatial characteristics of the laser radiation (uniform irradiation of the whole sample, a laser spot with a uniform intensity distribution, a circular spot with a radial intensity distribution). More complex structures can also be obtained. Variation of the spatial characteristics of the laser beam provides an effective means for the control of the formation of specific (specified in advance) surface structures.

REFERENCES

[1] A. M. Kosevich. *Physical Mechanics of Real Crystals*, Moscow: Nauka, 1986.
[2] F. Kh. Mirzade, V. Ya. Panchenko, L. A. Shelepin. *Phys. Usp.* 39 1 (1996).
[3] F. Kh. Mirzade, in: V. Ya. Panchenko (Eds), *Laser technologies of materials treatment*, Moscow: Fizmatlit, 2009, p. 220 (in Russian).
[4] D. Walgraef, N. Ghoniem, M. Lauzeral, *Phys. Rev. B* 56 (1997) 15361.
[5] V. A.Gnatyuk, A. I. Vlasenko, P. O. Mozol, O. S. Gorodnychenko. *Semicond. Sci. Technol.* 13 1298 (1998).
[6] Yu. A. Bykovskii, V. N. Nevolin, Yu. V. Fominskii. *Ionic and Laser Implantation of Metallic Materials*, Energoatomizdat, Moscow, 1991 (in Russian).
[7] A. I. Vlasenko, A. Baidullaeva, E. I. Kuznetsov, et al. *Semiconductors*. 35, 924 (2001).
[8] V. D. Skupov and S. V. Obolenskii. *Tech. Phys. Lett.* 26, 645 (2000).
[9] Z. Yao, G. Yu, Y. Wang, Z. Shi. *Int. J. of Solids and Structures* 46 2571 (2009).
[10] F. Kh. Mirzade. *J. Appl. Phys.* 97 084915 (2005).
[11] F. Kh. Mirzade. *Phys. Stat. Sol. (b)* 244 529 (2007).
[12] F. Kh. Mirzade *J. Phys.: Condens. Matter* 20 275202 (2008).
[13] F. Kh. Mirzade. *Physica B* 406 119-124 (2011).
[14] F. Kh. Mirzade. *J. Appl. Phys*. 103 044904 (2008).
[15] W. W. Duley. *Laser processing and analysis of materials*, New York: Plenum Press, 1983.
[16] L. D. Landau, E. M. Lifshitz. *Theory of Elasticity*, 3rd edition, Oxford: Pergamon Press, 1986.
[17] E. M. Kreutz, M. Krosch., G. Herziger, S. Wagner. *Proc. SPIE*, 650, 202 (1986).
[18] V. E. Gusev, E. K. Kozlova, A. I. Portnyagin. *Sov. J Quantum Electron.* 17 195 (1987).

In: Sound Waves: Propagation, Frequencies and Effects ISBN 978-1-61470-001-2
Editors: V. Abagnali and G. Fabbri

Chapter 3

ELECTROACOUSTIC MONITORING OF COLLOIDAL STATE CHANGES IN SODIUM CASEINATE STABILIZED OIL IN WATER EMULSIONS

İbrahim Gülseren and Milena Corredig
Department of Food Science, University of Guelph
Guelph,Ontario, CANADA

Abstract

The development of techniques to study the dynamics of change of colloidal systems during processing or storage is critical, as these systems are often in a metastable state, and any disruption occurring during the analysis will affect the interpretation of the data. Acoustics and electroacoustics are a novel non-destructive technique that can simultaneously determine the ultrasonic properties and particle surface charge of emulsions. Since this method can be used in the analysis of highly concentrated and opaque samples, it has the potential to provide highly accurate data on the colloidal state of food emulsions.

In addition, electroacoustics may be employed to study destabilization processes or interactions amongst ingredients. In this work, a model system of sodium caseinate emulsions is used to discuss the potential of acoustic spectroscopy and electroacoustics as means to follow the changes occurring to droplet interactions and emulsion destabilization. Sodium caseinate is one of the most frequently used emulsifiers in food emulsions, as it readily adsorbs at the interface during homogenization and stabilizes the oil droplets from coalescence due to electrostatic and steric repulsion. However, changes in environmental conditions affecting the polyelectrolyte layer surrounding the oil droplets (i.e., pH, ionic strength) as well as the presence of unadsorbed polymer can influence the stability of the emulsion droplets.

The results demonstrate that ultrasonic techniques may assist in understanding the details of the destabilization with minimal sample disruption.

1. Introduction

1.1. Fundamentals of electroacoustics

The stability of suspensions is one of the most important aspects of colloid science due to its importance for many industrial products. The majority of techniques available to date to observe the destabilization of colloidal systems such food emulsions are somewhat disruptive, and the particle-particle aggregation mechanisms cannot be studied in detail. Although the principles of ultrasonics and electroacoustics have been known for a considerably long period of time, their potential applications in the analysis of emulsions and colloids are underexploited. Ultrasonic and electroacoustic sensors have the capabilities to accurately analyze concentrated systems *in situ*. The changes occurring to colloidal systems can be of physical or chemical nature and can be followed measuring parameters such as surface charge (i.e., ζ-potential) of the colloidal particles or velocity and attenuation of the ultrasonic wave traveling through the sample. These measurements allow for the study of real-time changes in oil-in-water emulsions.

In colloidal systems, the changes in ultrasonic attenuation have been used to derive information on aggregation (Chanamai et al. 1998). Ultrasonic attenuation is the dissipation of the ultrasonic energy of the sound wave while propagating through the sample. The attenuation can detect changes in a large range of scales (from nanometers to micrometers), and it depends on the measurement frequency. The four basic mechanisms contributing to the energy loss of the ultrasonic wave in colloidal dispersions include: i) intrinsic attenuation, which is a function of the material properties, ii and iii) viscoinertial and thermal losses due to the density and temperature gradient, respectively, between the continuous and dispersed phases, and iv) scattering losses due to the refraction, diffraction and reflection of ultrasound by the particles (McClements & Coupland, 1996; Richter et al. 2007). Visco-inertial and thermal effects are the main contributors in the long wavelength regime, where the wave number (ka) <<1. The wave number is a function of frequency, *f*, diameter of the colloid, d, and the fluid sound velocity (c_{fl}): [$ka = (\pi f d) / c_{fl}$]. In particular, for aqueous systems and food emulsions, where the density gradient between the particles is low, thermal attenuation is generally the dominant loss mechanism, at frequencies < 100 MHz. Temperature gradient is a consequence of the difference in thermophysical properties of two phases (i.e. heat conductance, thermal expansion, heat capacity). The ultrasonic compression and decompression cycles will generate a temperature gradient at the boundary layer between the dispersed particle and the solvent. Since all four loss mechanisms are dependent on the thermophysical properties of the components present in the system, the size distribution of the colloidal particles can be extracted from its attenuation spectrum (McClements & Coupland, 1996; (Dukhin & Goetz, 2002).

The dependence of ultrasonic attenuation with frequency and particle diameter is affected by of the wavenumber, ka (Serabian & Williams, 1978). At the long wavelength regime, as in the case of small particles, attenuation is proportional (to the 4th power) to frequency and (to the 3rd power) diameter. In the intermediate (phase) regime, attenuation is proportional to particle diameter and frequency (squared), while in the diffusion (short wavelength) regime attenuation is proportional to the inverse diameter (Serabian & Williams, 1978). Therefore, as the particle diameter increase to ka values ≥ 1, attenuation is less likely to increase rapidly with particle diameter, as its frequency dependence is weak. Figure 1 shows the frequency dependence of ka assuming an approximated ultrasonic velocity of 1500 m.s^{-1}. It is clear that for small oil droplets, the analysis lies within the confines of Rayleigh (long wavelength) regime.

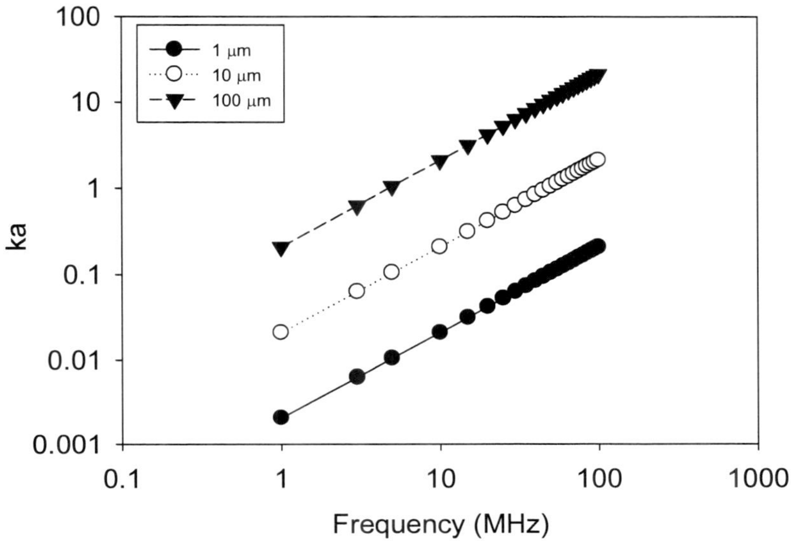

Figure 1. Frequency dependence of wavenumber (ka) for oil in water emulsions, depending on average diameter. The ultrasonic velocity was assumed to be 1500 m.s^{-1}.

Electroacoustic analysis is based on the coupling of acoustic and electric fields. One approach to electroacoustics is the colloid vibration current method (CVI), where an ultrasonic input can be utilized to induce an electrical response (colloid vibration current) in the medium. In the CVI method, while the ultrasound wave propagates, it causes vibrations in the particle by means of planar wave compressions. As a consequence, the motion of the particles relative to the continuous phase affects their charged double layers, and the electroacoustic signal is then measured (Dukhin & Goetz, 2002). Once the CVI is established, using the density contrast between the dispersed and the continuous phase, the electrodynamic mobility and ζ-potential can be calculated.

In this work, a model system of sodium caseinate emulsions is employed to discuss the potential of ultrasonic spectroscopy and electroacoustics as means to follow the changes in the colloidal properties of oil droplets during various destabilization mechanisms, namely, acidification or interactions with adsorbing or unadsorbing polymers.

1.2. Sodium Caseinate Stabilized Emulsions and Challenges in Formulation

Emulsions are commonly utilized in the food, pharmaceutical and cosmetic industries and are formed by the dispersion of one or more liquid phases in a continuous liquid phase. In most cases, emulsion formulations include an aqueous phase and a natural or synthetic oil phase and surface active agents, and both water-continuous and oil-continuous emulsions are produced as industrial products. Due to the density contrast between the continuous and dispersed phases and the unfavorable interactions between the polar and non-polar molecules, emulsions are thermodynamically unstable systems that can be kinetically stabilized for a defined period of time, which, in the case of industrial products, has to be significantly longer than the shelf-life of the product. In the production of dairy food emulsions, it is a common practice to use food-grade small-molecule surfactants and dairy proteins as emulsifiers, to aid in the stabilization of the oil droplets.

Sodium caseinate is produced by the isoelectric precipitation (pH 4.6) and re-dispersion at pH near neutral using sodium hydroxide, followed by spray drying. Sodium caseinate has very good emulsifying properties, because of the amphiphilic character of the constituent proteins, mainly α_s- and β-caseins. When properly used, sodium caseinate can impart long-term stability to dairy emulsions primarily through steric stabilization, whereas electrostatic repulsion might also enhance the stability under favorable conditions (Dickinson & Golding, 1997). Since the interfacial conformation and emulsion stabilizing capacity of sodium caseinate are highly dependent on environmental factors, the formulators of sodium caseinate emulsions must consider the possible agents and mechanisms of destabilization that could alter the microstructural properties of emulsions (Dickinson, 2010). Sodium caseinate emulsions can show destabilization through a number of mechanisms, including, but not limited to, the formation of bridges during emulsification (when not enough protein is present), the reduction of surface charge density and the thickness of the double layer (with changes in pH or ionic strength), depletion effects by non adsorbed polymers, or the formation of bridges between droplets by polymers or proteins (Dickinson, 2010).

The employment of electroacoustic techniques could enable the investigators to analyze highly concentrated emulsions without dilution and provide highly accurate data on the colloidal state of emulsions including the evolution of surface charge of droplets, ultrasonic properties and particle size distribution over time. Ultrasonic spectroscopy has been previously employed to monitor the changes in food colloids *in situ* (see for example, Liu et al., 2008; Gülseren et al. 2010). The present work aims at illustrating how it is possible to follow the colloidal interactions involved in the destabilization of sodium caseinate using ultrasonic spectroscopy and electroacoustics. In particular, the changes that occur in the emulsions during acid-induced aggregation, depletion flocculation or bridging with high methoxyl pectin are described, to show the potential of acoustics in describing a number of different destabilization mechanisms.

2. Materials and Methods

2.1. Emulsion Preparation

Solutions of 1% sodium caseinate (Alanate 180, NZMP, Lamoyne, PA, USA) were prepared in 5 mM imidazole (03196-500, Fisher Scientific, Fair Lawn, NJ, USA) buffer (pH 6.7) and kept refrigerated overnight prior to use. Appropriate amounts of soy oil (S7381, Sigma-Aldrich Corp., St. Louis, MO, USA) were mixed with sodium caseinate solutions to prepare 20% oil-in-water emulsions. The mixtures were pre-homogenized using a shear mixer and subjected to high pressure homogenization at 750 bar for five passes using Emulsiflex

C5, Avestin (Ottawa, Ontario, Canada). Immediately after homogenization, the samples were analyzed for particle size distribution.

When analyzing the effect of volume fraction, the stock emulsions (20% oil in water based on volume) were diluted using a buffered sodium caseinate solution (1% in 5 mM imidazole buffer, pH 6.7), to have a comparable amount of unadsorbed protein in all the samples. During acidification experiments, the emulsions were acidified adding varying amounts of glucono-δ-lactone (GDL) (0.2 to 0.4%) (G4750, Sigma-Aldrich Corp., St. Louis, MO, USA). The gradual hydrolysis of GDL causes a slow decrease in pH. In the experiments with high-methoxyl pectin (HMP) (DE 72.8, CP Kelco, Lille Skensved, Denmark) a stock solution was prepared 5 mM imidazole buffer (pH 6.7) and various amounts were added to the sodium caseinate emulsion. All solutions prepared in this study contained 0.02% sodium azide as bacteriostatic (S227I, Fisher Scientific, Fair Lawn, NJ, USA). All the measurements were carried out in duplicate.

2.2. Determination of Size Distribution Using Light Scattering

Using a dynamic light scattering (DLS) technique (Zetasizer Nano, Malvern Instruments, Worcestershire, UK), volume frequency based particle size distribution of the emulsions were determined. When necessary, the measurements were compared to a static light scattering method (Mastersizer 2000, Malvern Instruments, Worcestershire, UK). The reference samples used here for size measurements were obtained from Duke Scientific , Palo Alto, CA, USA (Catalog Numbers: 3550A and 5009A). For the DLS measurements, emulsion samples were appropriately diluted using pre-filtered imidazole buffer. For Mastersizer measurements, refractive index of oil and water were taken as 1.47 and 1.33, respectively. A small quantity of emulsion sample was diluted in a large volume of water in the sample analysis chamber of the Mastersizer.

2.3. Acoustic and Electroacoustic Spectrometer

The acoustic and electroacoustic spectrometer used in this study (DT-1200, Dispersion Technology, Bedford Hills, NY, USA) was based on the colloidal vibration current (CVI) method for the determination of ζ-potential and swept-distance interferometry for the determination of ultrasonic attenuation. Ultrasonic determination of particle size distribution of the emulsions was based on the frequency dependence of ultrasonic attenuation and the thermophysical properties of the continuous and dispersed phases of the emulsion (see below) and calculated using an in-built algorithm (Dispersion Technology). The instrument enabled to monitor the electroacoustic properties of soy oil emulsions as a function of time, pH and changes in colloidal state at relatively high dispersed phase concentration (mostly 15-20%).

The instrument has two separate sensor systems for the measurement of ζ-potential and ultrasonic properties. The measurement of ζ-potential of emulsions is based on the CVI method. In the setup employed here, a piezoelectric transducer transmits ultrasonic pulses at a frequency of 3 MHz, and the difference in electric potentials in the medium is registered as CVI. An electroacoustic analog of the Smoluchowski equation is employed to derive the ζ-potential of emulsion droplets. This high accuracy and precision method has been shown to measure the ζ-potential of monodisperse rutile and latex particles up to 50% volume fraction (Dukhin & Goetz, 2001; 2002). Previously the electroacoustic properties of casein micelles have been measured up to a volume fraction of 40% (Gülseren et al. 2010). The Smoluchowski theory is known to be valid regardless of concentration and particle shape, if a thin double layer ($\kappa a >> 1$, where κ is the inverse Debye length and a is the radius of the particle) exists, and the surface conductivity of the particles is negligible. In the present work, both conditions were met, since the conductance of droplets is negligible and inverse Debye length is in the order of a few nm, considerably smaller than the size of emulsion droplets studied.

A 10% w/w suspension of silica Ludox particles (AS-40, Sigma-Aldrich Corp, St. Louis, MO) in 10 mM KCl was used as a daily ζ-calibration. ζ-potential of Ludox under these conditions is approximately -38.5 mV. In the electroacoustic determination of ζ-potential, soy oil droplets were considered as liquid soft particles, and four thermophysical properties were assumed to calculate ζ-potential and size (density = 0.917 $g.ml^{-1}$ [manufacturer data]; viscosity, C_p= 1.917 J/kg/K, thermal conductivity k = 0.17 mW/cm/K, ultrasonic velocity c = 1469.8 $m.s^{-1}$ and thermal expansion coefficient β = 7.24 x 10-4 K^{-1} (Coupland & McClements, 1997)). Based on this assumption, oil droplets have negligible viscoinertial attenuation, whereas a relatively high thermal attenuation will be observed at low frequencies (< 10 MHz) and will increase linearly with frequency (Dukhin & Goetz, 2001). Experiments were performed in the long wavelength regime, where the contribution of scattering to attenuation was low even at the high frequencies (10-100 MHz).

The acoustic sensor has identical transmitting and receiving piezoelectric broadband L-wave transducers separated by a variable motor controlled gap. The frequency range used was from 3 to 99.5 MHz. Usually 18 different frequencies were used in this range. When the changes in the colloidal state of the emulsions were rapid, only 5 frequencies were selected in order to accelerate data acquisition.

3. Results and Discussion

3.1. Electroacoustic Properties of Sodium Caseinate Emulsions as a Function of Concentration

The electroacoustic properties of oil-in-water emulsions as a function of volume concentration are shown in Figure 2. These data are representative of a 20% oil-in-water emulsion stabilized with 1% sodium caseinate, and diluted in buffer at pH 6.7 containing 1% sodium caseinate at various volume fractions. The presence of sodium caseinate in the soluble phase ensured a comparable amount of unadsorbed protein in all cases. These emulsions were stable to creaming. For all the volume fractions < 20%, the ultrasonic attenuation increased with volume concentration and frequency at frequencies > 10 MHz, whereas below 10 MHz, attenuation remained fairly constant. On the other hand, the sample containing 20% oil showed a decrease in thermal attenuation at the low frequencies, and an increase in attenuation at frequencies > 50 MHz. The experiments were carried out in duplicate and the measurements agreed with each other within 0.7%.

From the attenuation spectra, using appropriate equations taking into consideration the various contributions to the loss in amplitude of the sound wave (McClements & Coupland, 1996), it is possible to derive a particle size distribution. In Figure 3 the particle size distribution determined by ultrasonic spectroscopy is compared to that measured using dynamic light scattering. In all cases, the emulsions showed a monomodal distribution of sizes, independently on the volume fraction, and a comparable average size with a peak at about 0.1-0.2 μm. The light scattering data also showed a monomodal distribution of droplets with an average size larger than that measured with ultrasonic spectroscopy. The discrepancy between the dynamic light scattering and the ultrasonic method may derive to the fact that light scattering signal is proportional (to the 6th power) to the particle diameter, whereas acoustics is based on weight concentration (i.e., size to the 3rd power) (Dukhin et al. 2010). The results clearly demonstrated that an accurate particle size distribution can be obtained using ultrasonic spectroscopy for a large range of volume fractions (from 1 to 20%).

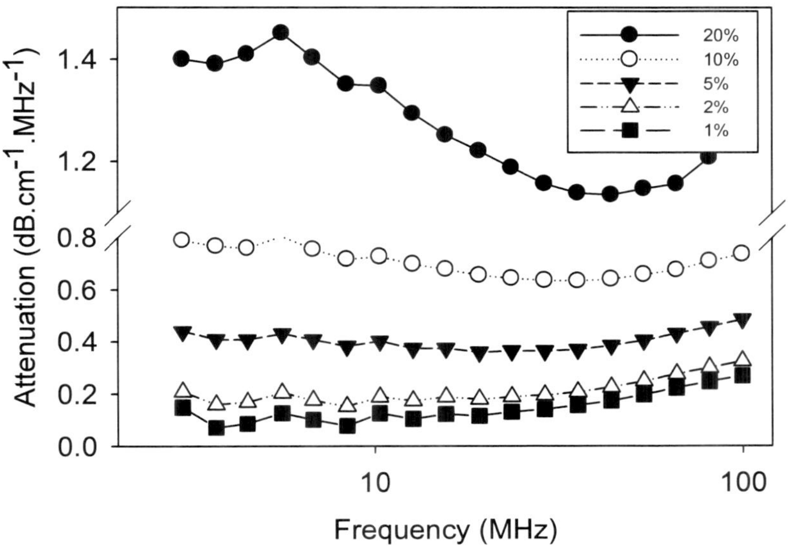

Figure 2. Ultrasonic attenuation spectra of soy oil emulsions stabilized with 1% sodium caseinate, pH 6.7, measured at various volume fractions.

The dynamic mobility and ζ-potential of sodium caseinate emulsions analyzed at different volume fractions are shown in Figure 4. As the interparticle distance decreases with increasing number of oil droplets per volume, the measured dynamic mobility decreased with the volume fraction of oil, because of hydrodynamic effects. Once the volume fraction effect was taken into account, the ζ-potential of the oil droplets did not vary significantly with dilution, as the particle size, ionic strength and medium composition remained constant. The results in Figure 4 clearly show that the ζ-potential of the oil droplets can be measured accurately using electroacoustics, once the thermal properties and the density of the samples are known. Unlike the measurements carried out using light scattering, the ζ-potential measured by electroacoustics can be derived *in situ* without dilution of the emulsions.

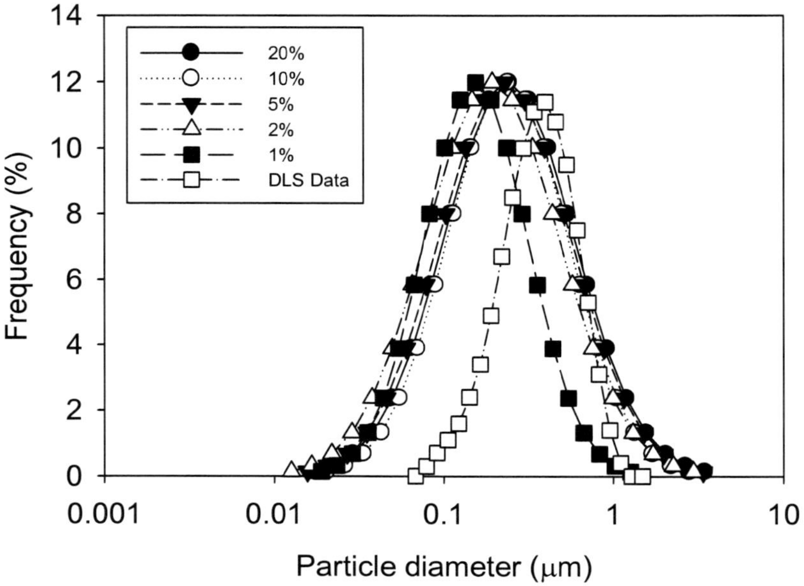

Figure 3. Particle size distribution of soy oil emulsions stabilized by 1% sodium caseinate (pH 6.7) as determined by ultrasonic spectroscopy (at various volume fractions, using Figure 1 data) and dynamic light scattering (DLS) method.

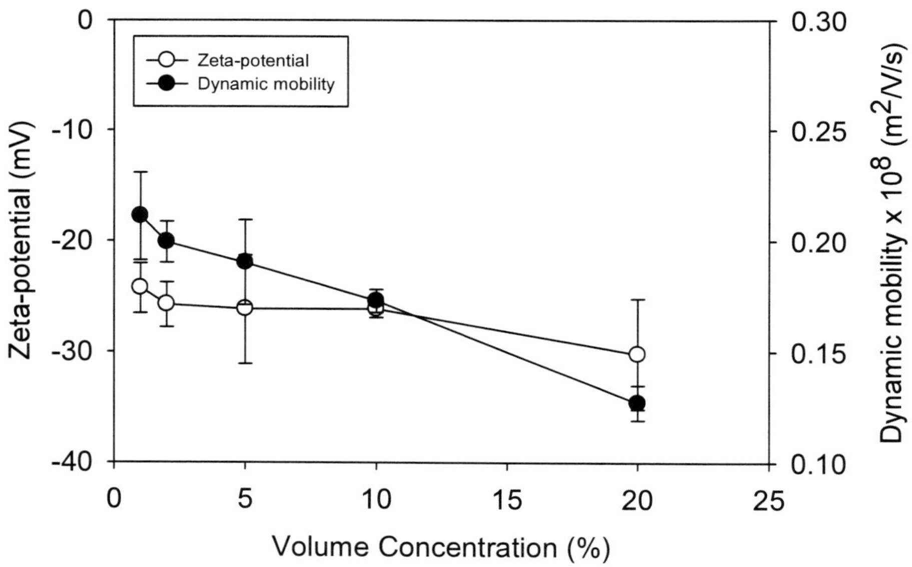

Figure 4. Dynamic mobility and ζ-potential of a soy oil emulsion emulsion stabilized by 1% sodium caseinate (pH 6.7), as a function of different oil volume fractions.

3.2. Destabilization of emulsion droplets: acidification

The ability of acoustic and electroacoustic spectroscopy to follow the physico-chemical changes of colloidal systems *in situ* is of great importance in the study of dynamics of change, as for example, during pH induced destabilization of emulsions. In sodium caseinate-stabilized emulsions, with acidification, the surface charge density progressively decreases, and with this, the ability of the sodium caseinate molecules to stabilize the emulsion droplets. Hence, at a pH close to the isoelectric point of the proteins, there is a net attraction between the droplets leading to aggregation, because of the loss of both electrostatic and steric repulsion forces. To be able to follow the details of such sol-gel transition, the emulsions need to be analyzed without disruption during the testing (Liu et al. 2007).

The changes occurring to the electroacoustic properties of soy oil in water emulsions stabilized with sodium caseinate are shown in Figures 5-7. Acidification was carried out by addition of glucono-delta-lactone (GDL), an internal ester of D-gluconic acid. As GDL slowly hydrolyzes, it is possible to carry out a detailed analysis of the changes occurring during acidification. Figure 5 illustrates the changes occurring in the attenuation spectra for a 15% soy oil emulsion during acidification with 0.4% GDL. Final pH after 4 h was 4.57. While no changes in attenuation were seen between pH 6.7 and 6.1, there was a decrease in attenuation at pH 5.9, and after about 50 min, at pH 5.7, rapid changes in attenuation took place. After pH 5.1, there was no further decrease in the attenuation (Figure 5A). The changes in attenuation were also accompanied by the changes in the derived particle size distribution (Figure 5B). At pH 5.5 there was an obvious shift in particle size to larger values, and with further decrease in pH there was an increase in average size distribution, and the quality of fit also rapidly deteriorated (i.e., error % > 10%).

Since the changes in attenuation were quite rapid during acidification, to better depict the changes in charge and size with pH, slower acidifications were carried out with both 15 and 20% oil in water emulsions, as shown in Figure 6. Initially, the sodium caseinate emulsions near neutral pH carried a negative charge, with a ζ-potential of about – 26 mV. The decrease in the surface charge of the oil droplets can be clearly noted in Figure 6A, where the overall extent of the charge gradually decreased, and the samples shifted to a positive charge approximately at pH 5.02. It was clear that during acidification, there was an instability region between pH 5.5 and 5.0, where charge can not be measured accurately. It is important to note that in the case of 0.2% GDL, the pH reached a plateau at pH 5.5. With further decrease in pH, the ζ-potential changed to positive values, reaching +8.88 mV at pH 4.6.

When looking at the aggregation of the sodium caseinate emulsions with acidification, it is clearly shown that when the surface charge reached a minimum, at pH 5.5, aggregation of the oil droplets occurred (Figure 6B). The model system used in this study, clearly demonstrates that the acid-induced sol-gel transitions of oil-in-water emulsions can be followed well using electroacoustics, measuring both the state of aggregation as well as the extent of surface charge of the oil droplets. Furthermore, the attenuation of the sound wave clearly described the changes occurring to the sodium caseinate emulsions during acidification (Figure 7). The value of attenuation remained constant down to pH 5.7, when a sudden drop in pH was noted. The value of attenuation dropped at pH 5.5, where aggregation occurred.

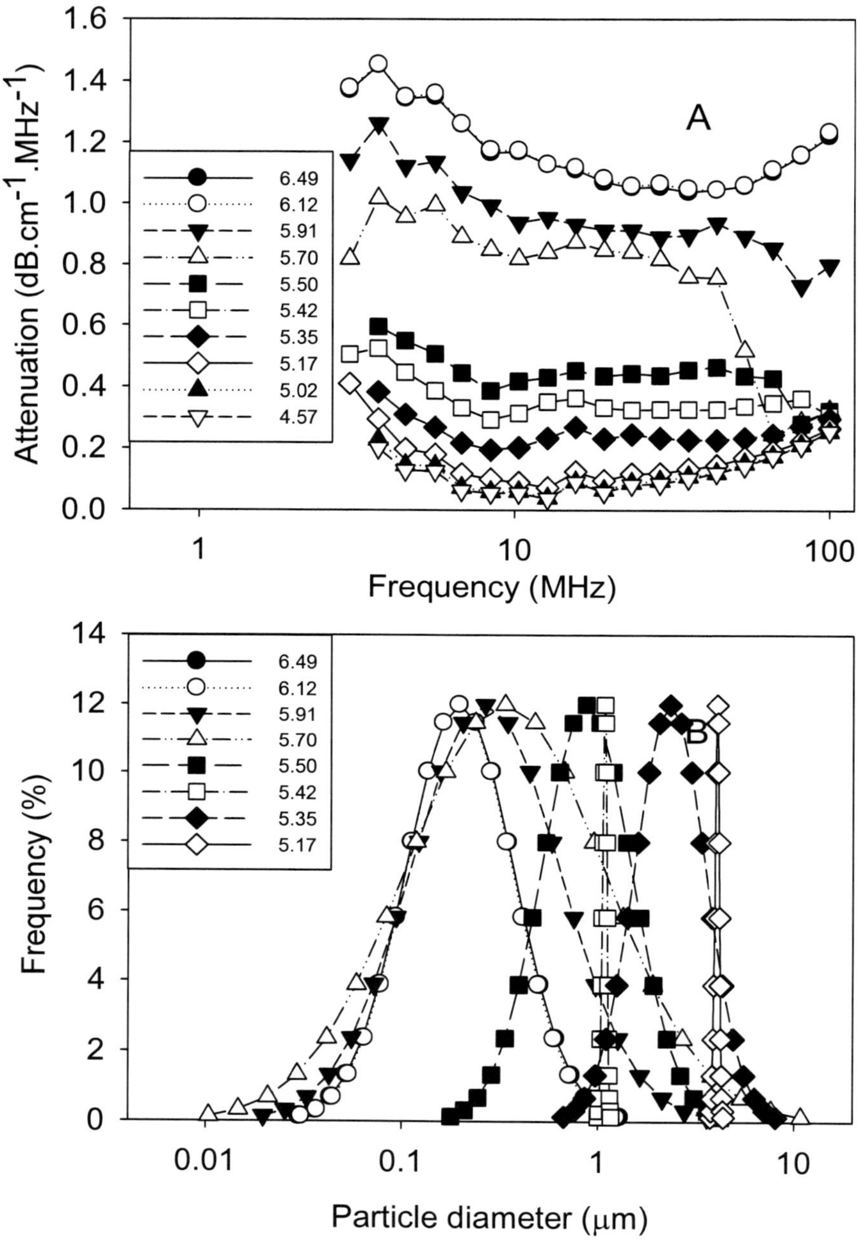

Figure 5. Effect of pH on the ultrasonic attenuation spectra (3-100 MHz) (A) and particle size distribution (B) of a soy oil in water emulsion (15%) stabilized with 1% sodium caseinate (pH 6.7). Emulsions were acidified with addition of 0.4% GDL.

Previously, Bryant & McClements (1999) have demonstrated that both theoretical and experimental attenuation spectra of thermally induced protein aggregates are considerably different than those collected for the corresponding native protein solutions, due to the changes in thermal attenuation characteristics which are a strong function of particle size. Since the formation of aggregated oil droplets would lead to an overlap of the thermal waves, the consequence of this process is reduced thermal losses and attenuation (Liu et al. 2007). That is, the increase in the number of thermal skin depths in the effective medium will reduce the extent of thermal attenuation (Allegra & Hawley, 1972).

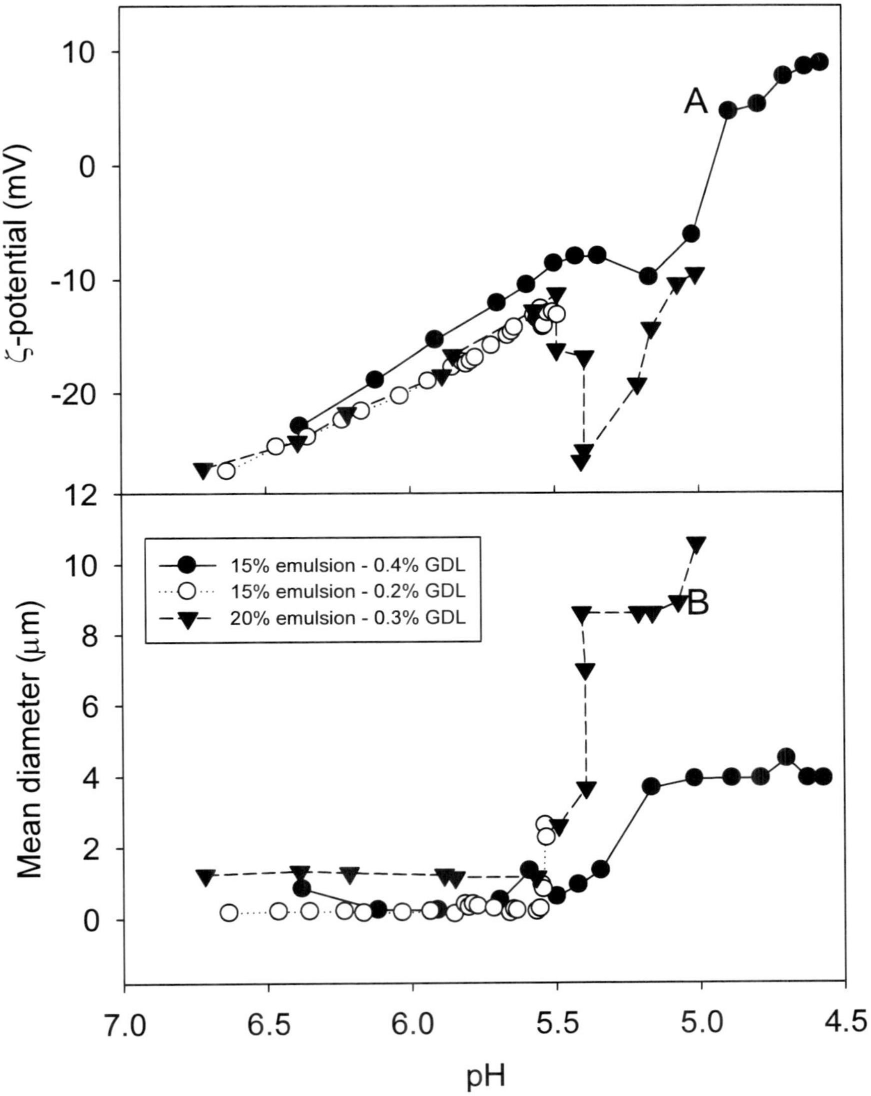

Figure 6. Effect of pH on the overall charge (ζ-potential) (A) and average size (B) of soy oil emulsions (15-20% by vol) stabilized by 1% sodium caseinate (pH 6.7) during acidification with 0.2-0.4% GDL.

3.3. Interactions between the Oil Droplets and a Charged Polymer: High Methoxyl Pectin (HMP)

High methoxyl pectin (HMP) is a negatively charged polysaccharide at neutral pH, with an isoelectric point of about 3.5. It mainly consists of a backbone of galacturonic acid (a portion of which methylesterified) and branches containing arabinose, galactose and xylose (Schols & Voragen, 2002). High methoxyl pectin is commonly used in the stabilization of acid yogurt drinks because of its association with the casein micelles at low pH (Liu et al. 2006). It is important to note that the influence of added polysaccharides in protein stabilized emulsions is highly dependent on environmental conditions such as pH, temperature and ionic strength.

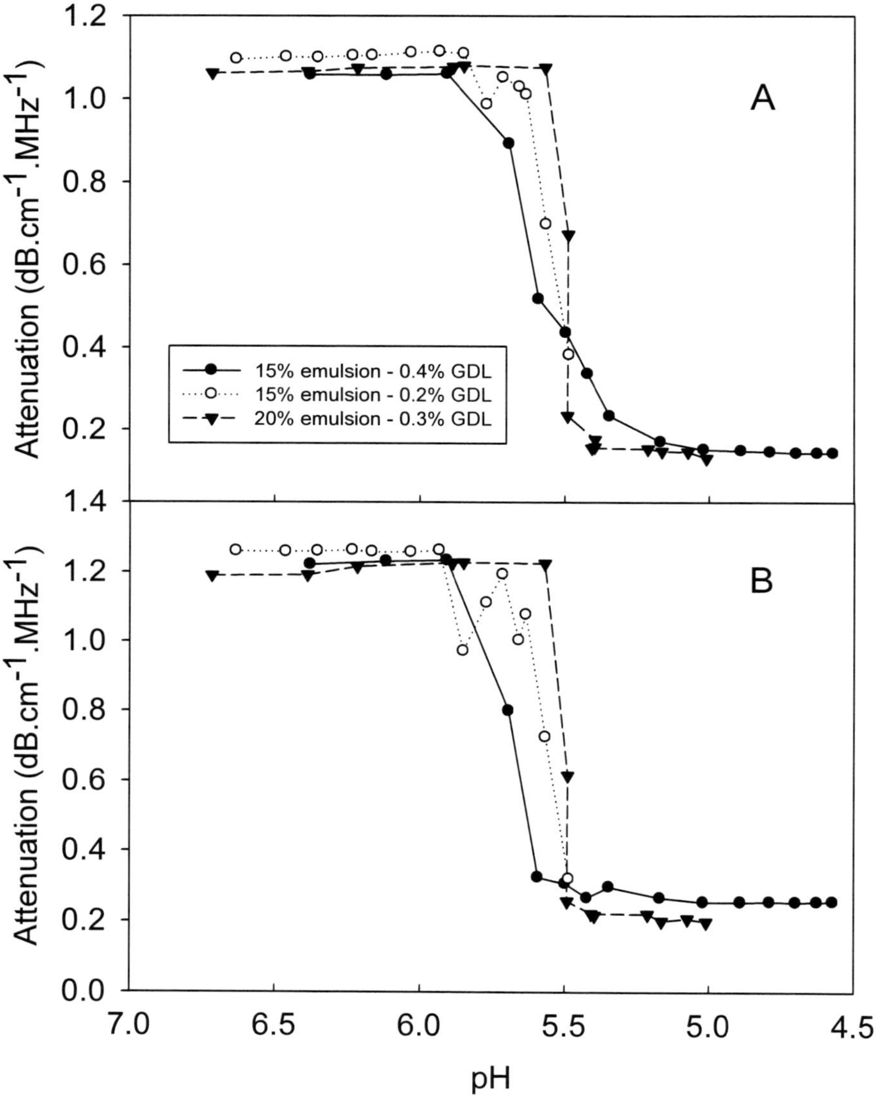

Figure 7. Effect of pH on ultrasonic attenuation of soy oil-in-water emulsions prepared with 1% sodium caseinate and acidified with GDL. Measurements at 53 (A) or 99 (B) MHz.

Polysaccharides can increase the viscosity of emulsion, facilitate the formation of an emulsion gel or interact with the interfacial or unadsorbed protein to form molecular complexes (Liu et al. 2008). During the acidification of casein stabilized emulsions, HMP adsorbs onto casein molecules by electrosorption, most likely below pH < 5 (i.e., around the pI of caseins) and it inhibits the aggregation of the oil droplets. The extent of the protection is both of electrostatic and steric nature and highly dependent on the concentration of pectin molecules, degree of methylation, and pH of the medium (Liu et al. 2007). The charge, thickness, and structure of the interfacial layer will be affected by the adsorption of HMP (Liu et al. 2007).

At concentrations below those necessary to cover all the available interface, pectin can induce bridging flocculation. For example, low concentrations of HMP were observed to induce bridging flocculation in whey protein stabilized emulsions (Gancz et al. 2005). It is also important to note that, at high pH, the presence of HMP can cause depletion flocculation because of the like charge of the polysaccharide with the protein covered oil droplets. It has been recently shown that partially depleted HMP-containing emulsions at pH near neutral will be stabilized because of the HMP adsorption on casein molecules during acidification (Liu et al. 2008).

During acidification, the complex formation between HMP and dairy proteins (sodium caseinate, in this study) becomes more likely, as the both protein and pectin molecules are progressively less negatively charged and beyond the pI of proteins, the electrostatic attraction between HMP and sodium caseinate becomes favorable. Therefore, the presence of HMP in a dairy emulsion prevents the aggregation of the protein-stabilized oil droplets through steric stabilization.

The ultrasonic attenuation spectra of sodium caseinate emulsions containing 0.3% HMP at different incubation times at pH 6.7 are shown in Figure 8. The changes in pH with HMP addition were < ± 0.05 pH units at all cases. The particle size distribution derived from the ultrasonic attenuation data is also shown in Figure 8.

The emulsions containing 0.3% HMP had a stable ultrasonic attenuation spectrum over time, and a monomodal size distribution with an average diameter of about 150 nm. The value of ζ-potential of these emulsions was -17 mV and the change over 5 h of incubation was < 2 mV. The stability of this system at pH near neutral is in agreement with previous data with whey protein stabilized emulsions, also showing stability at comparable concentration levels as those in Figure 8 (Gancz et al. 2005).

Using electroacoustics, it is possible to follow, *in situ*, without dilution, the adsorption of HMP on the oil droplets covered with sodium caseinate during acidification.

Figure 9 describes the changes in the value of ζ-potential for the original emulsion (same as Figure 6A) during acidification, with a final value at low pH of +8.88 mV. On the other hand, acidification of the oil droplets in the presence of 0.3% HMP showed a different behaviour. The first pH measurement, at 6.4 showed a ζ-potential of -14 mV, and with decreasing pH, the absolute value of the charge continued to decrease, reaching a plateau at about pH 5.4 with a ζ-potential of -5.47 mV. This clearly shows that it is possible to follow the binding of the HMP to the casein proteins on the surface of the oil droplets, without dilution of the emulsion.

The ultrasonic attenuation of the emulsion containing 0.3% HMP remained constant up to pH 5.2, when it gradually decreased, reaching a plateau at about 0.4 $dB.cm^{-1}.MHz^{-1}$ (Figure 10A), a value significantly higher than that of the acidified emulsions without pectin (Figure

5B). For example, the final attenuation value after 5 h at 99.5 MHz was 0.5 and 0.2 dB.cm^{-1}.MHz^{-1}, respectively, for the HMP containing and no HMP samples.

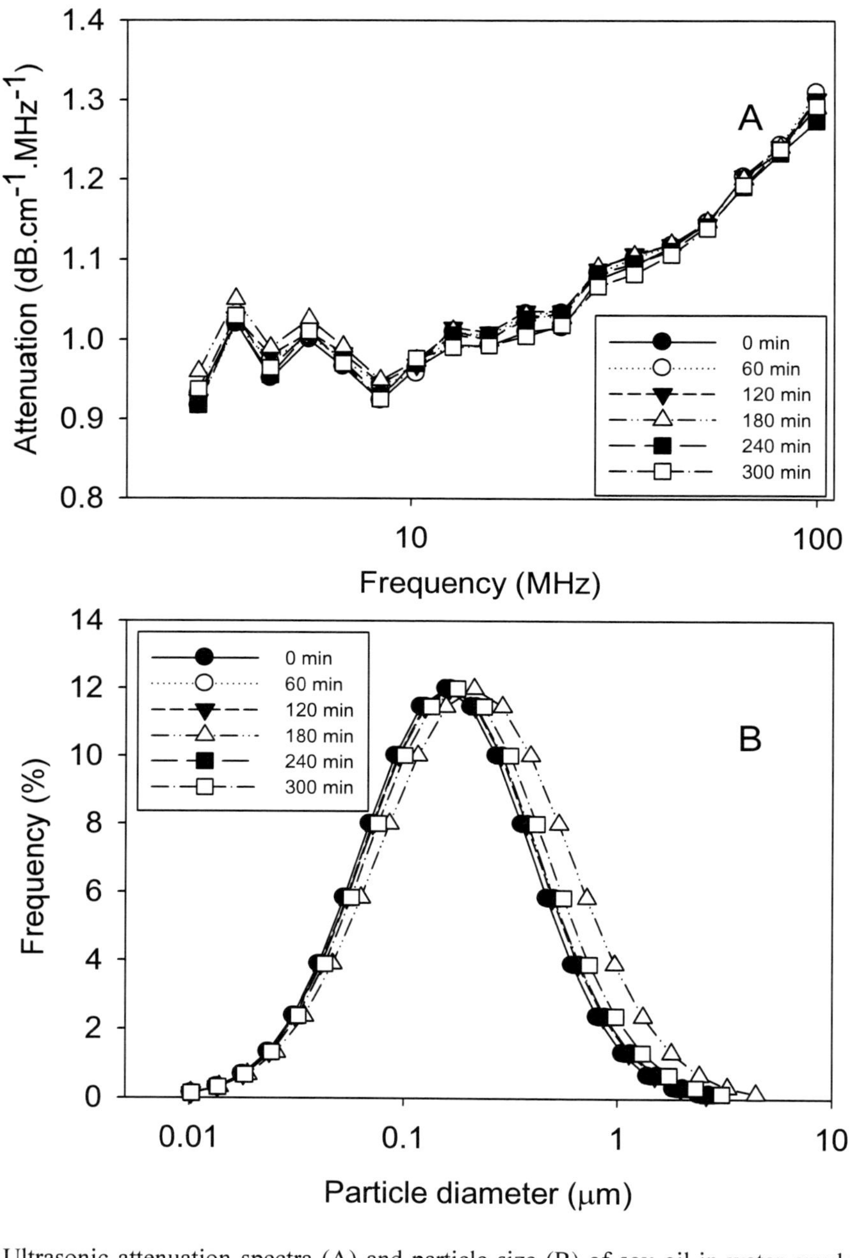

Figure 8. Ultrasonic attenuation spectra (A) and particle size (B) of soy oil-in-water emulsions (15% v/v) stabilized with 1% sodium caseinate with 0.3% HMP, at pH 6.7. Data collected at various times at 25°C.

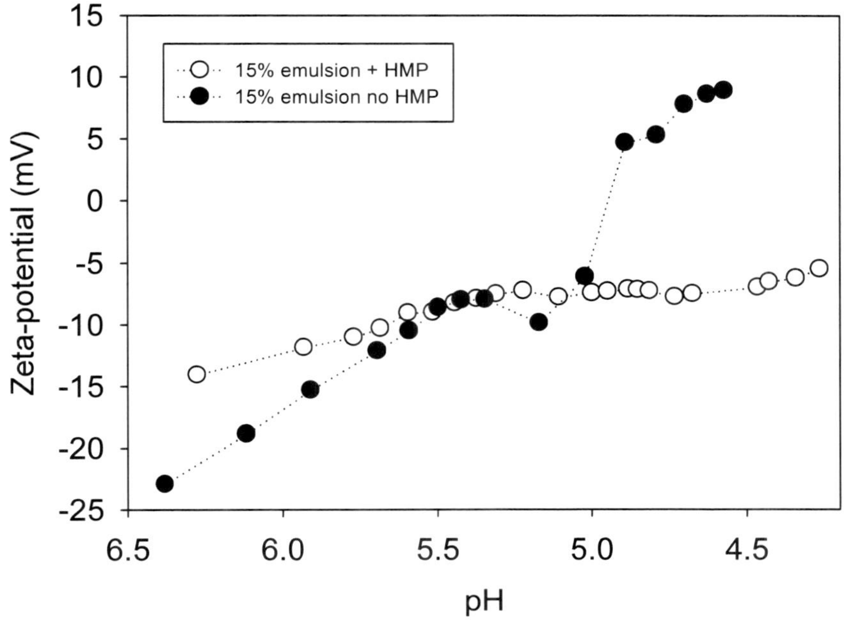

Figure 9. Values of ζ-potential of soy-oil in water emulsions (15% by vol) prepared with 1% sodium caseinate, during acidification with 0.4% GDL. Comparison between a control emulsion, not containing HMP, and the same emulsion containing 0.3% HMP.

These attenuation values indicated that attenuation of sound still decreased with the presence of HMP, most likely because of a limited aggregation of the oil droplets and an increase in their size. The particle size distribution of these emulsions is shown in Figure 10B. The increase in size was measured in the concentrated systems, and it may reflect the formation of bridging between the oil droplets. In previous work using light scattering, it was shown also a limited aggregation for these oil droplets once acidified and in the presence of HMP (Bonnet et al. 2005).

3.4. Destabilization of the oil droplets due to depletion flocculation

The presence of unadsorbed polymers in the emulsions can cause depletion flocculation, especially when their size is only a few times smaller than the emulsion droplets (Dickinson et al. 1997; Dickinson, 2010). The polymers generate a concentration gradient at the distance of separation of two droplets which is accompanied by an osmotic pressure gradient. In order to reduce the gradients, the droplets have to come into close contact and reversibly flocculate. It has been previously shown that unbound sodium caseinate destabilizes oil-in-water emulsions due to depletion interactions, the extent of which is proportional to the concentration of unbound polymers (Dickinson & Golding, 1997).

At relatively low sodium caseinate concentrations (for example, 1% sodium caseinate for 35% *n*-tetradecane emulsions), incomplete surface coverage will cause bridging flocculation due to the adsorption of protein molecules on multiple droplets, whereas weak depletion is observed at high concentrations (i.e., especially beyond 4%) (Dickinson et al. 1997).

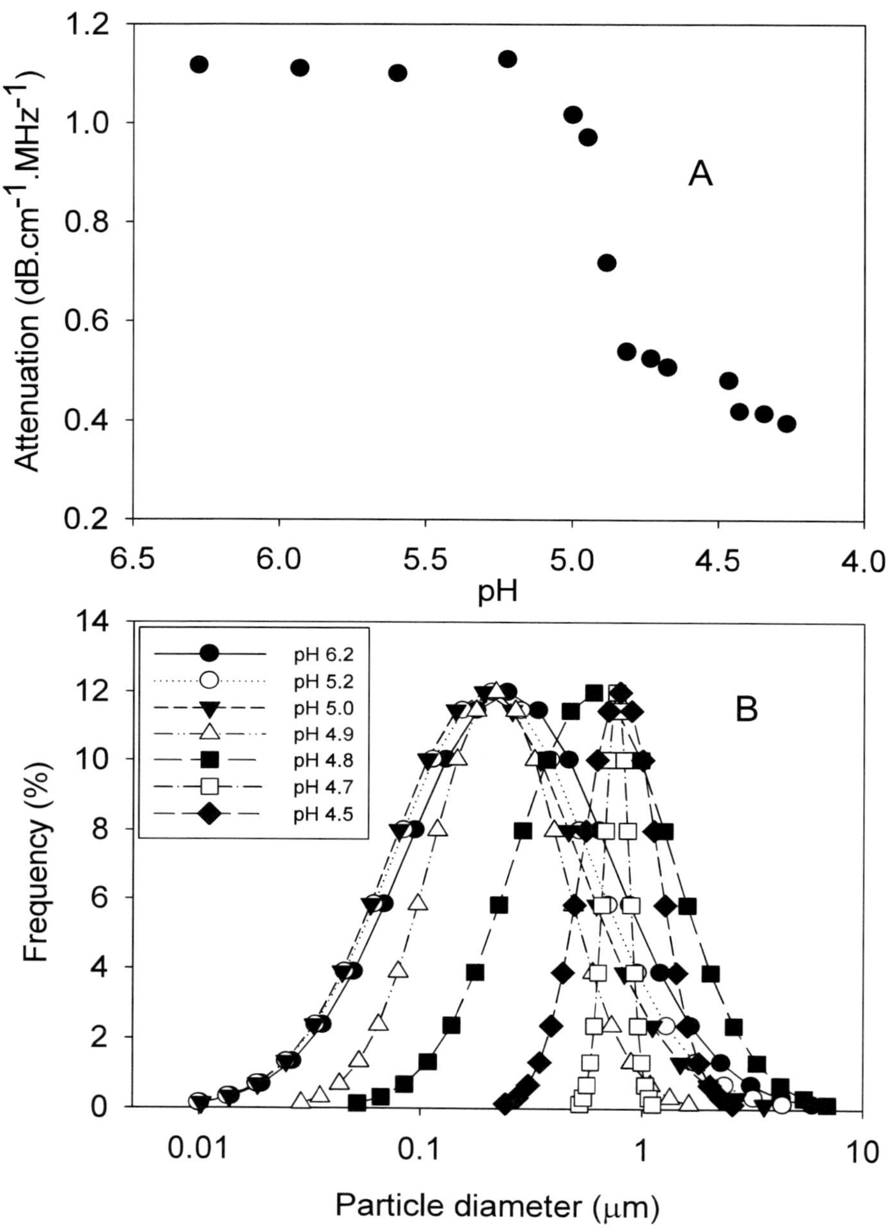

Figure 10. Effect of pH on ultrasonic attenuation (A) measured at 53 MHz, and particle size distribution (B) of a soy oil emulsion (15% v/v) stabilized with 1% sodium caseinate, during acidification with 0.4% GDL, in the presence of 0.3% HMP.

Once again, the use of an *in situ* technique to be able to follow the changes occurring to the system while destabilizing would be of great importance to better understand the details of the depletion flocculation processes. The ultrasonic attenuation and the ζ-potential measured by electroacoustics for a sodium caseinate emulsion containing 4% protein are shown in Figures 11 and 12.

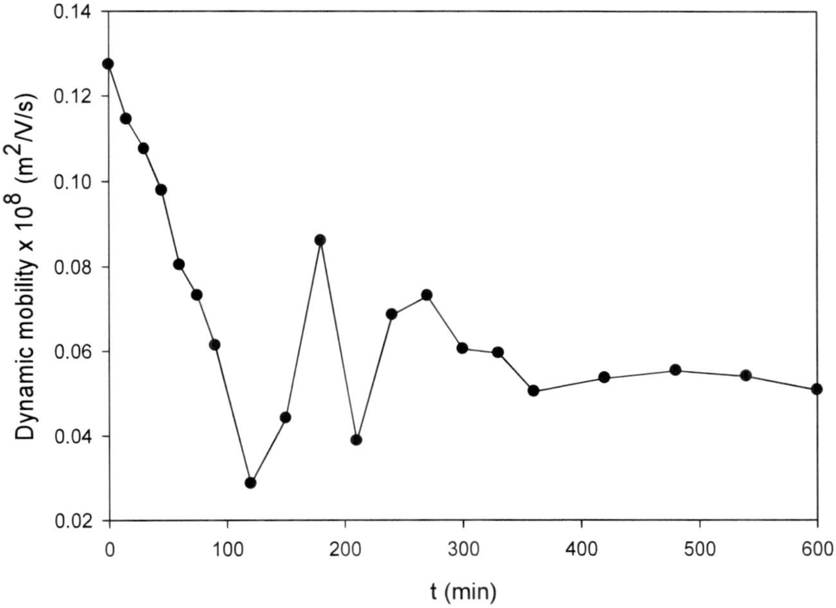

Figure 11. Changes in dynamic mobility of a soy oil emulsion (15% by vol) containing 4% sodium caseinate (pH 6.7) over time. Experiments were conducted quiescently at 25°C.

Visible destabilization of the 15% soy oil-in-water emulsion with unadsorbed sodium caseinate occurs over 10 h. Figure 11 illustrates the changes occurring over time to the value of dynamic mobility. It is clear that the electrophoretic mobility decreases significantly after about 100 min of storage under quiescent conditions. In addition to the changes noted very clearly using electroacoustics, changes were observed in the ultrasonic attenuation spectra over time (Figure 12A).

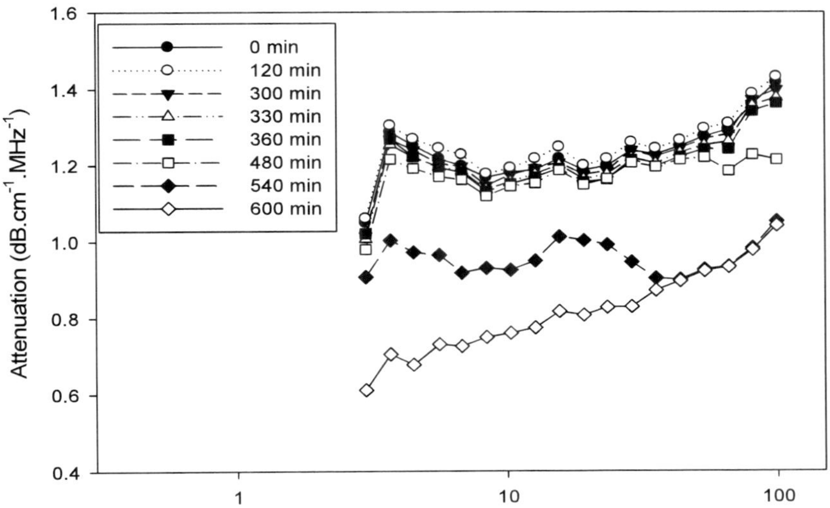

Figure 12. continued on next page.

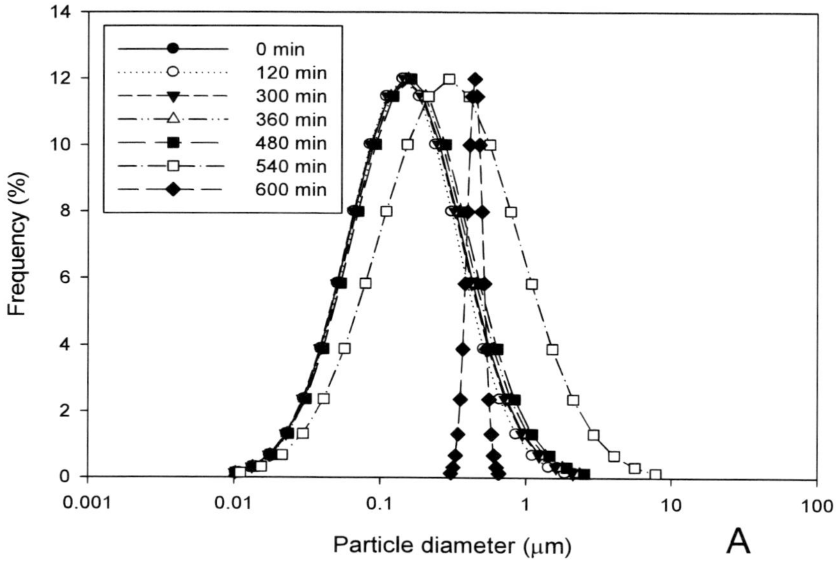

Figure 12. Changes in ultrasonic attenuation spectra (A) and particle size distribution (B) of a soy oil emulsion (15% v/v) containing 4% sodium caseinate (pH 6.7), over storage time.

However, although there seemed to be a gradual decrease in the ultrasonic attenuation since the beginning of the experiments, significant differences can be noted using ultrasonic spectroscopy only after longer quiescent storage, when flocs start to be large enough to cause a change in the ultrasonic spectra. This is also reflected in the particle size distribution, where aggregation is noted after 540 min from mixing (Figure 12B). This example of the use of acoustic and electroacoustic spectroscopy clearly suggests that electroacoustics is very sensitive to physical changes such as flocculation, as those change the spatial correlation between the particles and their motion relative to the continuous phase, which in turn, will affect their electrical double layer.

Conclusion

Electroacoustics is a novel technology which can be utilized in the characterization of the colloidal dispersions *in situ*. Due to their industrial applications, the stability of colloidal dispersions is of paramount importance. Using acoustics and electroacoustics, the colloidal state changes of oil-in-water emulsions can be followed in great detail. Although this exemplifies the use of acoustic and electroacoustic spectroscopy to study depletion, bridging or acid induced aggregation, this technique may be of high interest in the study of any type of colloidal stabilization *in situ*.

References

[1] Allegra, J. R.; Hawley, S. A. *J Acoust Soc Am.* 1972, 51, 1545-64.

[2] Bonnet, C.; Corredig, M.; Alexander, M. *J Agric Food Chem*. 2005, 53, 8600-6.

[3] Bryant, C. M.; McClements, D. J. *J Sci Food Agric*. 1999, 79, 1754-60.

[4] Chanamai, R.; Herrmann, N.; McClements, D. J. *J Colloid Interface Sci.* 1998, 204, 268-276.

[5] Coupland, J. N.; McClements, D. J. *J Am Oil Chem Soc*. 1997, 74, 1559-64.

[6] Dickinson, E. *Colloid Surface B.* 2010, 81, 130-140.

[7] Dickinson, E.; Golding, M.; Povey, M. J. W. *J Colloid Interface Sci.* 1997,185, 515-529.

[8] Dickinson, E.; Golding, M. *Food Hydrocolloid.* 1997, 11, 13-18.

[9] Dukhin, A. S.; Goetz, P. J. *Adv Colloid Interface Sci*. 2001, 92, 73-132.

[10] Dukhin, A. S.; Goetz, P. J. *Ultrasound for characterizing colloids: particle sizing, zeta potential, rheology*. Elsevier: Amsterdam, The Netherlands, 2002.

[11] Dukhin, A. S.; Goetz, P. J.; Fang, X.; Somasundaran, P. *J Colloid Interface Sci.* 2010, 342, 18-25.

[12] Gancz, K.; Alexander, M.; Corredig, M. *J Agric Food Chem*, 2005, 53, 2236-41.

[13] Gülseren, İ.; Alexander, M.; Corredig, M. *J Colloid Interface Sci*, 2010, 351, 493-500.

[14] Liu, J.; Nakamura, A.; Corredig, M. *J Agric Food Chem*, 2006, 54, 6241-46.

[15] Liu, J.; Alexander, M.; Verespej, E.; Corredig, M. *Food Biophysics*. 2007, 2, 67-75.

[16] Liu, J.; Verespej, E.; Corredig, M.; Alexander, M. *Food Hydrocolloids*, 2008, 22, 47-55.

[17] McClements, D. J.; Coupland, J. N. *Colloids Surface A,* 1996, 117, 161-170.

[18] Richter, A.; Voigt, T.; Ripperger, S. *J Colloid Interface Sci*. 2007, 315, 482-492.

[19] Schols, H. A.; Voragen, A. G. J. In *Pectins and their manipulation*. Seymour, G.B.; Knox, J. P.; Ed.s.; CRC: USA, 2002.

[20] Serabian, S.; Williams, R. S. *Mater Eval.* 1978, 36, 55-62.

In: Sound Waves: Propagation, Frequencies and Effects ISBN 978-1-61470-001-2
Editors: V. Abagnali and G. Fabbri

Chapter 4

NUMERICAL ASSESSMENT OF MULTI-CHAMBER MUFFLERS HYBRIDIZED WITH MULTIPLE PERFORATED INTRUDING TUBES USING GA METHOD

Min-Chie Chiu
Department of Mechanical and Automation Engineering
Chung Chou University of Science and Technology
6, Lane 2, Sec.3, Shanchiao Rd., Yuanlin
Changhua 51003, Taiwan, R.O.C.

Abstract

Recently, research on new techniques for single-chamber mufflers equipped with non-perforated intruding tube has been addressed; however, the research work on multi-chamber mufflers conjugated with open-ended perforated intruding inlet-tubes which may dramatically increase the acoustical performance has been neglected. Therefore, the main purpose of this paper is not only to analyze the sound transmission loss (*STL*) of a multi-chamber open-ended perforated inlet-tube muffler but also to optimize the best design shape within a limited space.

In this paper, the four-pole system matrix for evaluating the acoustic performance — sound transmission loss (*STL*) — is derived by using a decoupled numerical method. Moreover, a genetic algorithm (*GA*), a robust scheme used to search for the global optimum by imitating a genetic evolutionary process, has been used during the optimization process. Before dealing with a broadband noise, the *STL*'s maximization with respect to a one-tone noise is introduced for a reliability check on the *GA* method. Additionally, an accuracy check of the mathematical model is performed. To appreciate the acoustical ability of the open-ended perforated intruding inlet-tube and chambers inside a muffler, two kinds of traditional multi-chamber mufflers hybridized with simple expansion tubes and non-perforated intruding inlet-tubes have been assessed and compared. Moreover, the acoustical performance of the open-ended perforated intruding inlet-tube equipped with 1~3 chambers has also been analyzed. Results reveal that the maximal *STL* is precisely located at the desired tone. In addition, the acoustical performance of mufflers conjugated with perforated intruding inlet-tubes is superior to traditional mufflers. Also, it has been shown that the acoustic performance for both pure tone and broadband noise will increase if the muffler has more chambers.

Consequently, the approach used for the optimal design of the noise elimination proposed in this study is easy and effective.

Keywords: open-ended perforated tube; decoupled numerical method; space constraints; genetic algorithm.

1. Nomenclature

This paper is constructed on the basis of the following notations:

chrmlength: bit length;

c_o: sound speed (m s^{-1});

dh_i : the diameter of a perforated hole on the i-th inner tube (m);

D_i: diameter of the i-th perforated tubes (m);

D_o: diameter of the outer tube (m);

elt: elitism (1 for yes, 0 for no);

f: cyclic frequency (Hz);

j : imaginary unit;

k : wave number $\left(=\frac{\omega}{c_o}\right)$;

ff_1, ff_2, ff_3, ff_4: coefficients in function $ff_i e^{\beta\beta_i x}$;

$iter_{max}$: maximum iteration;

L_{C1}, L_{C2} : lengths of perforate straight ducts (m);

L_o :total length of the muffler (m);

M_i : mean flow Mach number at the i-th node;

OBJ_i : objective function (dB);

p : acoustic pressure (Pa);

pc: crossover ratio;

$\overline{p}_i$: acoustic pressure at the i-th node (Pa);

pm: mutation ratio;

popsize: number of population;

Q : volume flow rate of venting gas (m^3 s^{-1});

S_i : section area at the i-th node(m^2);

STL : sound transmission loss (dB);

$SWLO$: unsilenced sound power level inside the muffler's inlet (dB);

SWL_T : overall sound power level inside the muffler's output (dB);

t_i: the thickness of the i-th inner perforated tube (m);

$TS1_{ij}$, $TS2_{ij}$, $TS3_{ij}$, $TS4_{ij}$, $TS5_{ij}$: components of four-pole transfer matrices for an acoustical mechanism with straight ducts;

$TPOE_{ij}1$: components of a four-pole transfer matrix for an acoustical mechanism with an expanded perforated intruding tube;

$TPOC_{ij}1$: components of a four-pole transfer matrix for an acoustical mechanism with a contracted perforated intruding tube;

T_{ij}^* : components of a four-pole transfer system matrix;

T : current temperature;

T_o : initial temperature;

u : acoustic particle velocity (m s^{-1});

$\bar{u}_i$: acoustic particle velocity at the i-th node (m s^{-1});

u_{ij} : acoustical particle velocity passing through a perforated hole from the i-th node to the j-th node (m s^{-1});

V_1 : mean flow velocity at the inner perforated tube (m s^{-1});

V_2 : mean flow velocity at the outer tube (m s^{-1});

ρ_o: air density (kg m^{-3});

ρ_i : acoustical density at the i-th node;

ς_i : specific acoustical impedance of the i-th inner perforated tube;

η_i : the porosity of the i-th inner perforated tube;

$\beta\beta_i$: i-th eigen value of $[NN]_{4x4}$;

$[\Omega\Omega]_{4x4}$: the model matrix formed by four sets of eigen vectors $\Omega\Omega_{4x1}$ of $[NN]_{4x4}$.

2. Introduction

Research on mufflers was started by Davis *et al.* in 1954 [1]. On the basis of the plane wave theory, studies of simple expansion mufflers without perforated holes have been made [2], [3]. To increase a muffler's acoustical performance, an internal perforated tube which is an essential acoustical element used to depress low frequency sound energy, was introduced and discussed by Sullivan and Crocker in 1978 [4]. Based on the coupled equations, a series of theories and numerical techniques in decoupling the acoustical problems have been proposed [5]-[7].

In 1981, Jayaraman and Yam [8] developed a method for finding an analytical solution; however, a presumption of the velocity equality within the inner and outer duct, which is not reasonable in the real world, is required. To overcome this drawback, Munjal *et al.* [9] provided a generalized de-coupling method. Regarding the flowing effect, Peat [10] publicized the numerical decoupling method by finding the eigen value in transfer matrices.

In order to maintain a steady volume-flow-rate in a venting system, limiting a muffler's back pressure is necessary. Therefore, Wang [11] developed a perforated intruding-tube muffler (a low back-pressure muffler with perforated open-ended tubes inside the cavity) using the BEM (Boundary Element Method).

However, the investigation of the optimal muffler design within certain space constraints has rarely been addressed. In previous work [12], [13], the shape optimization of a low back-pressure muffler (a two-chamber simple expansion muffler and a two-chamber muffler conjugated with non-perforated intruding tubes) using a GA (genetic algorithm) within a space-constrained milieu has been addressed; yet, the acoustical performance of above mufflers is still insufficient. In order to effectively improve the performance of the noise control device and maintain a steady volume-flow-rate within a space-constrained environment, an optimal design on a two-chamber muffler equipped with perforated intruding tubes is presented. In this paper, the numerical decoupling methods used in forming a four-pole system matrix are in line with the genetic algorithm.

3. Theoretical Background

In this paper, a three-chamber muffler with perforated intruding tubes was adopted for noise elimination in the fan room shown in Figure 1. Before the acoustical fields of the mufflers were analyzed, the acoustical elements had been distinguished. As shown in Figure 2, five kinds of muffler components, including five straight tubes, one expanded perforated intruding tube, one contracted perforated intruding tube, one sudden expanded tube, and one sudden contracted tube, are identified and symbolized as I, II, III, IV, and V.

Additionally, the acoustical field within the muffler is represented by ten points. The outline dimension of the three-chamber muffler with perforated intruding tubes is shown in Figure 3. As derived in previous work [12], [13] and appendices A ~ B, individual transfer matrices with respect to straight ducts, expanded/contracted perforated intruding ducts, and sudden expanded/contracted ducts are described as follows:

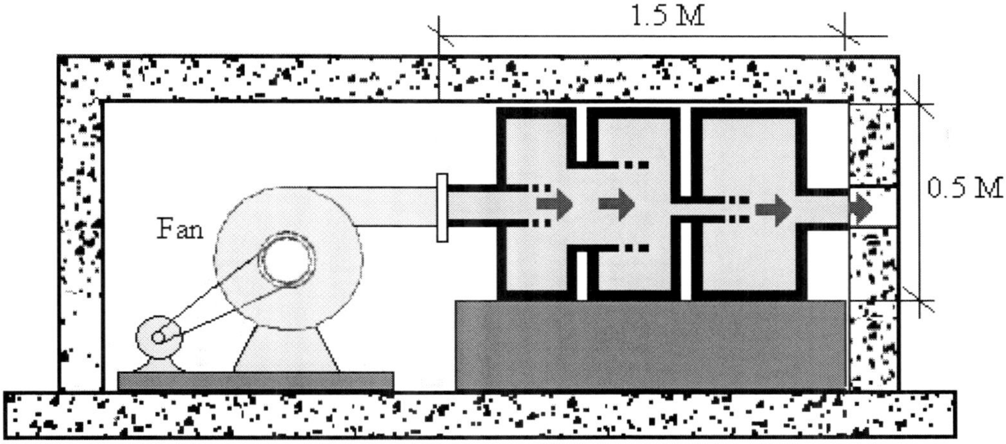

Figure 1. Noise elimination on a fan system within a limited space.

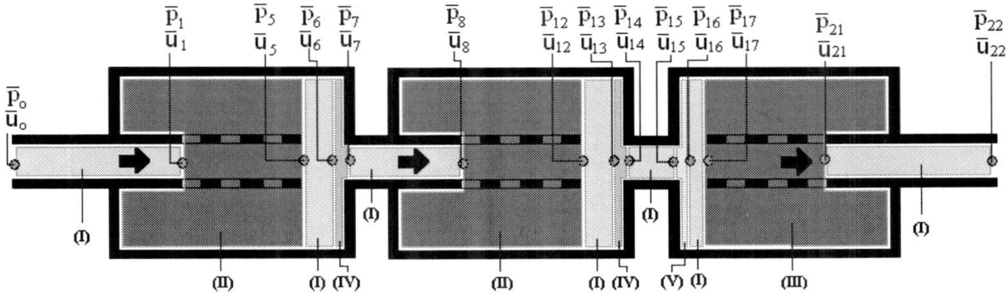

Figure 2. Acoustical elements in a three-chamber muffler hybridized with perforated intruding tubes (muffler A).

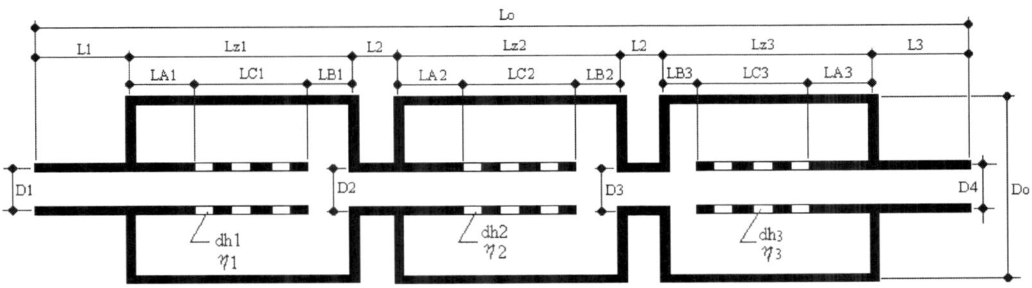

Figure 3. The outline dimension of a three-chamber muffler hybridized with perforated intruding tubes (muffler A).

3.1. Four-pole Transfer Matrices

$$\begin{pmatrix} \bar{p}_o \\ \rho_o c_o \bar{u}_o \end{pmatrix} = e^{-j\frac{M_o k(L_1+L_{A1})}{1-M_o^2}} \begin{bmatrix} TS1_{1,1} & TS1_{1,2} \\ TS1_{2,1} & TS1_{2,2} \end{bmatrix} \begin{pmatrix} \bar{p}_1 \\ \rho_o c_o \bar{u}_1 \end{pmatrix}, \tag{1a}$$

where

$$TS1_{1,1} = \cos\left[\frac{k(L_1+L_{A1})}{1-M_o^2}\right], \quad TS1_{1,2} = j\sin\left[\frac{k(L_1+L_{A1})}{1-M_o^2}\right],$$
$$TS1_{2,1} = j\sin\left[\frac{k(L_1+L_{A1})}{1-M_o^2}\right], \quad TS1_{2,2} = \cos\left[\frac{k(L_1+L_{A1})}{1-M_o^2}\right], \tag{1b}$$

$$\begin{pmatrix} \bar{p}_1 \\ \rho_o c_o \bar{u}_1 \end{pmatrix} = \begin{bmatrix} TPOE1_{1,1} & TPOE1_{1,2} \\ TPOE1_{2,1} & TPOE1_{2,2} \end{bmatrix} \begin{pmatrix} \bar{p}_5 \\ \rho_o c_o \bar{u}_5 \end{pmatrix}, \tag{2}$$

$$\begin{pmatrix} \bar{p}_5 \\ \rho_o c_o \bar{u}_5 \end{pmatrix} = e^{-j\frac{M_5 k L_{B1}}{1-M_5^2}} \begin{bmatrix} TS2_{1,1} & TS2_{1,2} \\ TS2_{2,1} & TS2_{2,2} \end{bmatrix} \begin{pmatrix} \bar{p}_6 \\ \rho_o c_o \bar{u}_6 \end{pmatrix}, \tag{3a}$$

where

$$TS2_{1,1} = \cos\left[\frac{kL_{B1}}{1-M_5^2}\right], \quad TS2_{1,2} = j\sin\left[\frac{kL_{B1}}{1-M_5^2}\right],$$
$$TS2_{2,1} = j\sin\left[\frac{kL_{B1}}{1-M_5^2}\right], \quad TS2_{2,2} = \cos\left[\frac{kL_{B1}}{1-M_5^2}\right], \tag{3b}$$

$$\begin{pmatrix} \bar{p}_6 \\ \rho_o c_o \bar{u}_6 \end{pmatrix} = \begin{bmatrix} 1 & 0 \\ 0 & S_7 / S_6 \end{bmatrix} \begin{pmatrix} \bar{p}_7 \\ \rho_o c_o \bar{u}_7 \end{pmatrix}, \tag{4}$$

$$\begin{pmatrix} \bar{p}_7 \\ \rho_o c_o \bar{u}_7 \end{pmatrix} = e^{-j\frac{M_7 k(L_2+L_{A2})}{1-M_7^2}} \begin{bmatrix} TS3_{1,1} & TS3_{1,2} \\ TS3_{2,1} & TS3_{2,2} \end{bmatrix} \begin{pmatrix} \bar{p}_8 \\ \rho_o c_o \bar{u}_8 \end{pmatrix}, \tag{5a}$$

where

$$TS3_{1,1} = \cos\left[\frac{k(L_2+L_{A2})}{1-M_7^2}\right], \quad TS3_{1,2} = j\sin\left[\frac{k(L_2+L_{A2})}{1-M_7^2}\right],$$
$$TS3_{2,1} = j\sin\left[\frac{k(L_2+L_{A2})}{1-M_7^2}\right], \quad TS3_{2,2} = \cos\left[\frac{k(L_2+L_{A2})}{1-M_7^2}\right], \tag{5b}$$

$$\begin{pmatrix} \bar{p}_8 \\ \rho_o c_o \bar{u}_8 \end{pmatrix} = \begin{bmatrix} TPOE2_{1,1} & TPOE2_{1,2} \\ TPOE2_{2,1} & TPOE2_{2,2} \end{bmatrix} \begin{pmatrix} \bar{p}_{12} \\ \rho_o c_o \bar{u}_{12} \end{pmatrix}, \tag{6}$$

$$\begin{pmatrix} \bar{p}_{12} \\ \rho_o c_o \bar{u}_{12} \end{pmatrix} = e^{-j\frac{M_{12} k L_{B2}}{1-M_{12}^2}} \begin{bmatrix} TS4_{1,1} & TS4_{1,2} \\ TS4_{2,1} & TS4_{2,2} \end{bmatrix} \begin{pmatrix} \bar{p}_{13} \\ \rho_o c_o \bar{u}_{13} \end{pmatrix}, \tag{7a}$$

where

$$TS4_{1,1} = \cos\left[\frac{kL_{B2}}{1-M_{12}^2}\right], \quad TS4_{1,2} = j\sin\left[\frac{kL_{B2}}{1-M_{12}^2}\right],$$
$$TS4_{2,1} = j\sin\left[\frac{kL_{B2}}{1-M_{12}^2}\right], \quad TS4_{2,2} = \cos\left[\frac{kL_{B2}}{1-M_{12}^2}\right], \tag{7b}$$

$$\begin{pmatrix} \bar{p}_{13} \\ \rho_o c_o \bar{u}_{13} \end{pmatrix} = \begin{bmatrix} 1 & 0 \\ 0 & S_{14}/S_{13} \end{bmatrix} \begin{pmatrix} \bar{p}_{14} \\ \rho_o c_o \bar{u}_{14} \end{pmatrix}, \tag{8}$$

$$\begin{pmatrix} \bar{p}_{14} \\ \rho_o c_o \bar{u}_{14} \end{pmatrix} = e^{-j\frac{M_{14}kL_2}{1-M_{14}^2}} \begin{bmatrix} TS5_{1,1} & TS5_{1,2} \\ TS5_{2,1} & TS5_{2,2} \end{bmatrix} \begin{pmatrix} \bar{p}_{15} \\ \rho_o c_o \bar{u}_{15} \end{pmatrix}, \tag{9a}$$

where

$$TS5_{1,1} = \cos\left[\frac{kL_2}{1-M_{14}^2}\right], \quad TS5_{1,2} = j\sin\left[\frac{kL_2}{1-M_{14}^2}\right],$$
$$TS5_{2,1} = j\sin\left[\frac{kL_2}{1-M_{14}^2}\right], \quad TS5_{2,2} = \cos\left[\frac{kL_2}{1-M_{14}^2}\right], \tag{9b}$$

$$\begin{pmatrix} \bar{p}_{15} \\ \rho_o c_o \bar{u}_{15} \end{pmatrix} = \begin{bmatrix} 1 & 0 \\ 0 & S_{16}/S_{15} \end{bmatrix} \begin{pmatrix} \bar{p}_{16} \\ \rho_o c_o \bar{u}_{16} \end{pmatrix}, \tag{10}$$

$$\begin{pmatrix} \bar{p}_{16} \\ \rho_o c_o \bar{u}_{16} \end{pmatrix} = e^{-j\frac{M_{16}kL_{B3}}{1-M_{16}^2}} \begin{bmatrix} TS6_{1,1} & TS6_{1,2} \\ TS6_{2,1} & TS6_{2,2} \end{bmatrix} \begin{pmatrix} \bar{p}_{17} \\ \rho_o c_o \bar{u}_{17} \end{pmatrix}, \tag{11a}$$

where

$$TS6_{1,1} = \cos\left[\frac{kL_{B3}}{1-M_{16}^2}\right], \quad TS6_{1,2} = j\sin\left[\frac{kL_{B3}}{1-M_{16}^2}\right],$$
$$TS6_{2,1} = j\sin\left[\frac{kL_{B3}}{1-M_{16}^2}\right], \quad TS6_{2,2} = \cos\left[\frac{kL_{B3}}{1-M_{16}^2}\right], \tag{11b}$$

$$\begin{pmatrix} \bar{p}_{17} \\ \rho_o c_o \bar{u}_{17} \end{pmatrix} = \begin{bmatrix} TPOC1_{1,1} & TPOC1_{1,2} \\ TPOC1_{2,1} & TPOC1_{2,2} \end{bmatrix} \begin{pmatrix} \bar{p}_{21} \\ \rho_o c_o \bar{u}_{21} \end{pmatrix}, \tag{12}$$

$$\begin{pmatrix} \bar{p}_{21} \\ \rho_o c_o \bar{u}_{21} \end{pmatrix} = e^{-j\frac{M_{21}k(L_{A3}+L_3)}{1-M_{21}^2}} \begin{bmatrix} TS7_{1,1} & TS7_{1,2} \\ TS7_{2,1} & TS7_{2,2} \end{bmatrix} \begin{pmatrix} \bar{p}_{22} \\ \rho_o c_o \bar{u}_{22} \end{pmatrix}, \tag{13a}$$

where

$$TS7_{1,1} = \cos\left[\frac{k(L_{A3}+L_3)}{1-M_{21}^2}\right], \quad TS7_{1,2} = j\sin\left[\frac{k(L_{A3}+L_3)}{1-M_{21}^2}\right],$$
$$TS7_{2,1} = j\sin\left[\frac{k(L_{A3}+L_3)}{1-M_{21}^2}\right], \quad TS7_{2,2} = \cos\left[\frac{k(L_{A3}+L_3)}{1-M_{21}^2}\right], \tag{13b}$$

The total transfer matrix assembled by multiplication is

$$\begin{pmatrix} \bar{p}_o \\ \rho_o c_o \bar{u}_o \end{pmatrix} = e^{-jk\left[\frac{M_o(L_1+L_{A1})}{1-M_o^2}+\frac{M_5 L_{B1}}{1-M_5^2}+\frac{M_7(L_2+L_{A2})}{1-M_7^2}+\frac{M_{12}L_{B2}}{1-M_{12}^2}+\frac{M_{14}L_2}{1-M_{14}^2}+\frac{M_{16}L_{B3}}{1-M_{16}^2}+\frac{M_{21}(L_{A3}+L_3)}{1-M_{21}^2}\right]}$$
$$\begin{bmatrix} TS1_{1,1} & TS1_{1,2} \\ TS1_{2,1} & TS1_{2,2} \end{bmatrix}\begin{bmatrix} TPOE1_{1,1} & TPOE1_{1,2} \\ TPOE1_{2,1} & TPOE1_{2,2} \end{bmatrix}\begin{bmatrix} TS2_{1,1} & TS2_{1,2} \\ TS2_{2,1} & TS2_{2,2} \end{bmatrix}$$
$$\begin{bmatrix} 1 & 0 \\ 0 & S_7/S_6 \end{bmatrix}\begin{bmatrix} TS3_{1,1} & TS3_{1,2} \\ TS3_{2,1} & TS3_{2,2} \end{bmatrix}\begin{bmatrix} TPOE2_{1,1} & TPOE2_{1,2} \\ TPOE2_{2,1} & TPOE2_{2,2} \end{bmatrix} \tag{14}$$
$$\begin{bmatrix} TS4_{1,1} & TS4_{1,2} \\ TS4_{2,1} & TS4_{2,2} \end{bmatrix}\begin{bmatrix} 1 & 0 \\ 0 & S_{14}/S_{13} \end{bmatrix}\begin{bmatrix} TS5_{1,1} & TS5_{1,2} \\ TS5_{2,1} & TS5_{2,2} \end{bmatrix}\begin{bmatrix} 1 & 0 \\ 0 & S_{16}/S_{15} \end{bmatrix}$$
$$\begin{bmatrix} TS6_{1,1} & TS6_{1,2} \\ TS6_{2,1} & TS6_{2,2} \end{bmatrix}\begin{bmatrix} TPOC1_{1,1} & TPOC1_{1,2} \\ TPOC1_{2,1} & TPOC1_{2,2} \end{bmatrix}\begin{bmatrix} TS7_{1,1} & TS7_{1,2} \\ TS7_{2,1} & TS7_{2,2} \end{bmatrix}\begin{pmatrix} \bar{p}_{22} \\ \rho_o c_o \bar{u}_{22} \end{pmatrix}$$

A simplified form in a matrix is expressed as

$$\begin{pmatrix} \bar{p}_o \\ \rho_o c_o \bar{u}_o \end{pmatrix} = \begin{bmatrix} T_{11}^* & T_{12}^* \\ T_{21}^* & T_{22}^* \end{bmatrix}\begin{pmatrix} \bar{p}_{22} \\ \rho_o c_o \bar{u}_{22} \end{pmatrix}. \tag{15}$$

3.2. Overall Sound Power Level

The sound transmission loss (*STL*) of a muffler is defined as [14]

$$STL_1(Q, f, RT_1, RT_2, RT_3, RT_4, RT_5, RT_6, RT_7, RT_8, RT_9, RT_{10}, RT_{11}, RT_{12}, RT_{13}, RT_{14}, RT_{15}, RT_{16}, RT_{17}, RT_{18}) \tag{16a}$$
$$= 20\log\left(\frac{\left|T_{11}^*+T_{12}^*+T_{21}^*+T_{22}^*\right|}{2}\right) + 10\log\left(\frac{S_o}{S_{22}}\right),$$

where

$RT_1 = (L_o - L_1 - L_3)/L_o$; $RT_2 = L_{Z2}/(L_o - L_1 - L_3 - 2L_2)$;

$$RT_3 = L_{B1}/L_{Z1};\ RT_4 = L_{C1}/(L_{Z1} - L_{B1});\ RT_5 = L_{B2}/L_{Z2};$$
$$RT_6 = L_{C2}/(L_{Z2} - L_{B2});\ RT_7 = L_{B3}/L_{Z3};\ RT_8 = L_{C3}/(L_{Z3} - L_{B3});\ RT_9 = D_1/D_o;$$
$$RT_{10} = D_2/D_o;\ RT_{11} = D_3/D_o;\ RT_{12} = D_4/D_o;\ RT_{13} = dh_1;$$
$$RT_{14} = \eta_1;\ RT_{15} = dh_2;\ RT_{16} = \eta_2;\ RT_{17} = dh_3;\ RT_{18} = \eta_3. \quad (16b)$$

The silenced octave sound power level emitted from a silencer's outlet is

$$SWL_i = SWLO_i - STL_i, \quad (17)$$

where

(1) $SWLO_i$ is the original *SWL* at the inlet of a muffler (or pipe outlet), and *i* is the index of the octave band frequency.

(2) STL_i is the muffler's *STL* with respect to the relative octave band frequency.

(3) SWL_i is the silenced *SWL* at the outlet of a muffler with respect to the relative octave band frequency.

Finally, the overall SWL_T silenced by a muffler at the outlet is

$$SWL_T = 10 * \log\left\{\sum_{i=1}^{n} 10^{SWL_i/10}\right\}$$
$$= 10 * \log\left\{10^{[SWLO(f_1) - STL(f_1)]/10} + 10^{[SWLO(f_2) - STL(f_2)]/10} + 10^{[SWLO(f_3) - STL(f_3)]/10} + \ldots + 10^{[SWLO(f_n) - STL(f_n)]/10}\right\}. \quad (18)$$

3.3. Objective Function

(A) STL maximization for a tone (f) noise

The objective function in maximizing the STL at a pure tone (*f*) is

$$OBJ_1 = STL(Q, f, RT_1, RT_2, RT_3, RT_4, RT_5, RT_6, RT_7, RT_8, RT_9, RT_{10}, RT_{11}, RT_{12}, RT_{13}, RT_{14}, RT_{15}, RT_{16}, RT_{17}, RT_{18}) \quad (19a)$$

The related ranges of the parameters are

$$Q = 0.02\ (m^3/s);\ L_o = 1.5(m);\ D_o = 0.5(m);\ L_2 = 0.05(m);$$
$$RT_1: [0.4,0.9];\ RT_2: [0.3,0.7];\ RT_3: [0.2,0.5];\ RT_4: [0.2,0.8];\ RT_5: [0.2,0.5];$$
$$RT_6: [0.2,0.8];\ RT_7: [0.2,0.5];\ RT_8: [0.2,0.8];\ RT_9: [0.2,0.8];\ RT_{10}: [0.2,0.8];$$
$$RT_{11}: [0.2,0.8];\ RT_{12}: [0.2,0.8];\ RT_{13}: [0.00175,0.007];$$
$$RT_{14}: [0.03,0.1];\ RT_{15}: [0.00175,0.007];\ RT_{16}: [0.03,0.1];$$
$$RT_{17}: [0.00175,0.007];\ RT_{18}: [0.03,0.1]. \quad (19b)$$

(B) SWL minimization for a broadband noise

To minimize the overall SWL_T, the objective function is

$$OBJ_2 = SWL_T(Q, RT_1, RT_2, RT_3, RT_4, RT_5, RT_6, RT_7, RT_8, RT_9, RT_{10}, RT_{11}, RT_{12}, RT_{13}, RT_{14}, RT_{15}, RT_{16}, RT_{17}, RT_{18}). \quad (20)$$

4. Model Check

Before performing the *GA* optimal simulation on mufflers, an accuracy check of the mathematical model on a one-chamber muffler with an expanded perforated intruding tube is performed by Wang *et al.* [15]. As indicated in Figure 4, the accuracy comparisons between theoretical data and analytical data are in agreement. Therefore, the model of a muffler with a perforated intruding tube is acceptable and adopted in the following optimization process.

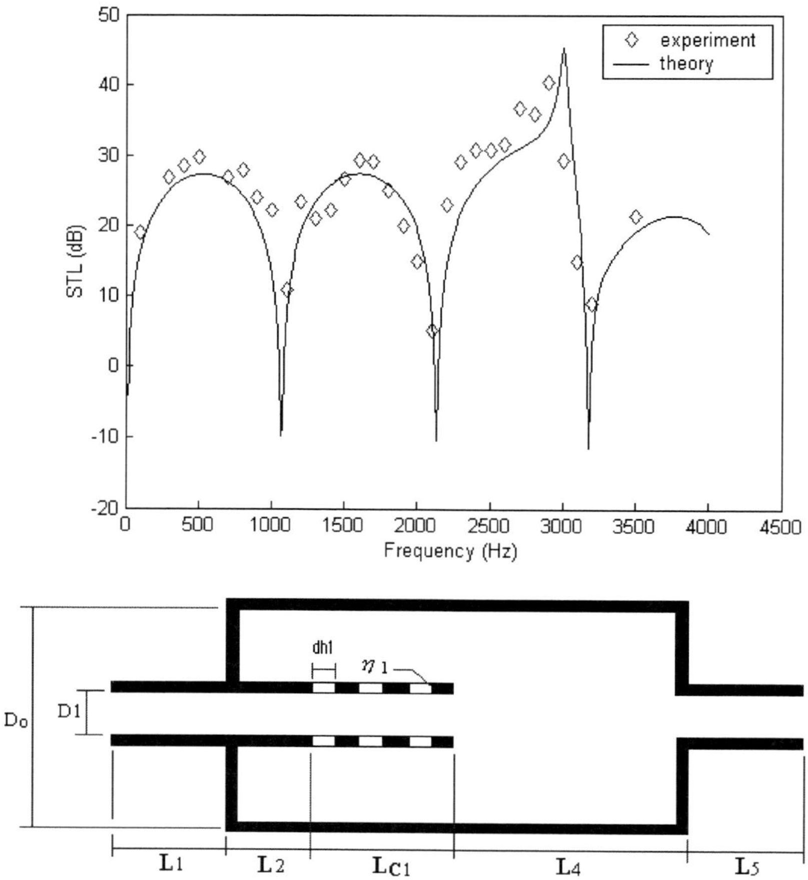

Figure 4. Performance of a one-chamber muffler equipped with an perforated intruding inlet tube [D_1 = 0.018(m), D_2 = 0.018(m), Do = 0.118 (m), L_4 = 0.08, L_2 = 0.0, L_{C1} = 0.08, L_1 = L_5 = 0.04, t_1 = 0.001(m), dh_1 = 0.003(m), η_1 = 0.03375, M_1 = 0.0]. [Experimental data is from Wang *et al.* [15]].

5. Case Studies

In this paper, the noise reduction of a fan system within a space-constrained room is exemplified and shown in Figure 1. The sound power level (*SWL*) inside the fan's outlet is shown in Table 1 where the overall *SWL* reaches 126.5 dB.

Table 1. Unsilenced *SWL* of a fan inside a duct outlet

Frequency - Hz	125	250	500	1000	2000	4000
SWLO - dB	125	120	114	102	92	90

It is obvious that four kinds of lower frequencies (125 Hz, 250 Hz, 500 Hz, 100 Hz) have higher noise levels (102~125 dB). To reduce the huge venting noise emitted from the fan's outlet, the noise elimination on four primary noises (125 Hz, 250 Hz, 500 Hz, 1000 Hz) using a three-chamber muffler hybridized with perforated intruding tubes (muffler A) is considered. The objective function on Eq.(18) becomes

$$SWL_T = 10 * \log\left\{\sum_{i=1}^{4} 10^{SWL_i/10}\right\}$$

$$= 10 * \log\left\{10^{[SWLO(f=125)-STL(f=125)]/10} + 10^{[SWLO(f=250)-STL(f=250)]/10} + 10^{[SWLO(f=500)-STL(f=500)]/10} + 10^{[SWLO(f=1000)-STL(f=1000)]/10}\right\}. \quad (21)$$

To obtain the best acoustical performance within a fixed space, numerical assessments linked to a *GA* optimizer are applied. Before the minimization of a broadband noise is performed, a reliability check of the *GA* method by maximization of the *STL* at a targeted one tone (600 Hz) has been performed.

As indicated in Figure 5, to evaluate the acoustic performance, four kinds of low back-pressure mufflers (muffler B: a one-chamber muffler hybridized with perforated intruding tubes; muffler C: a two-chamber muffler hybridized with perforated intruding tubes; muffler D: a three-chamber simple expansion muffler; and muffler E: a three-chamber muffler hybridized with non-perforated intruding tubes) are accessed and optimized. As shown in

Figures 1 and 2, the available space for a muffler is 0.5 m in width, 0.5 m in height, and 1.5 m in length. The flow rate (*Q*) and thickness of a perforated tube (*t*) are preset at 0.02 (m^3/s) and 0.001(m), respectively. The corresponding *OBJ* functions, space constraints, and the ranges of design parameters are summarized in Eqs. (19) ~ (20).

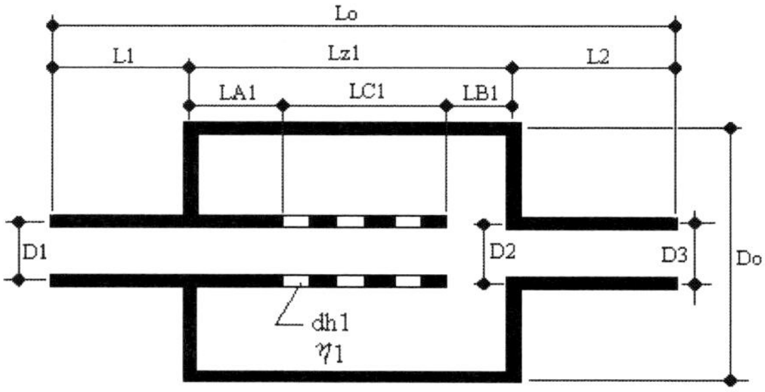

(1) Muffler B

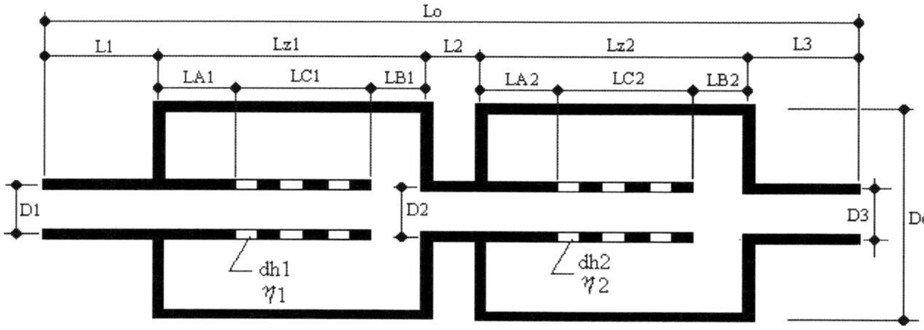

(2) Muffler C

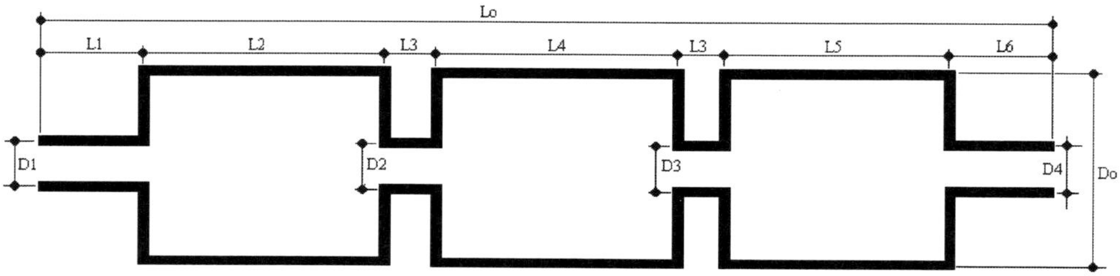

(3) Muffler D

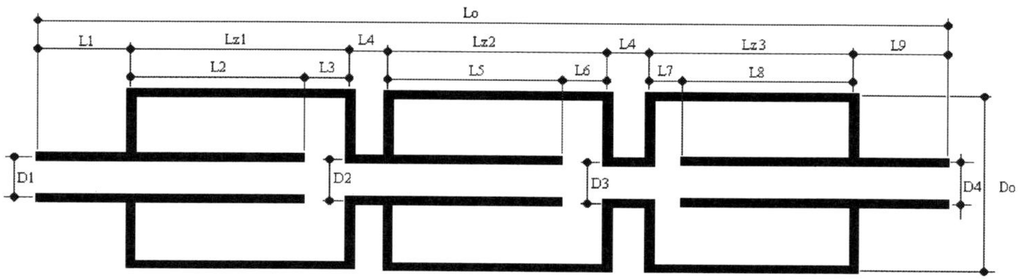

(4) Muffler E

Figure 5. Two kinds of low back-pressure mufflers [(1) muffler B: a one-chamber muffler hybridized with perforated intruding tubes; (2) muffler C: a two-chamber muffler hybridized with perforated intruding tubes; (3) muffler D: a three-chamber simple expansion muffler; and (4) muffler E: a three-chamber muffler hybridized with non-perforated intruding tubes].

6. Genetic Algorithm

The concept of Genetic Algorithms, first formalized by Holland [16] and then extended to functional optimization by D. Jong [17], involves the use of optimization search strategies patterned after the Darwinian notion of natural selection.

As the block diagram indicates in Figure 6, *GA* accomplishes the task of optimization by starting with a random "population" of values for the parameters of an optimization problem. Afterwards, a new "generation" with an improved value of the objection function is produced. In order to achieve the evolution of a new generation, the binary system, a representation of real numbers and integers, is used. In addition, by manipulating the strings, the operators of reproduction, crossover, mutation, and elitism are initiated sequentially. As indicated in Figure 7, to process the elitism of a gene, the tournament selection, a random comparison of the relative fitness from pairs of chromosomes, was applied.

During the *GA* optimization, one pair of offspring was generated from the selected parent by uniform crossover with a probability of *pc*. The applied mechanism of uniform crossover is depicted in Figure 8. By using the masked genes randomly generated, the gene information between parents will be internally exchanged if the mapping gene is 1.

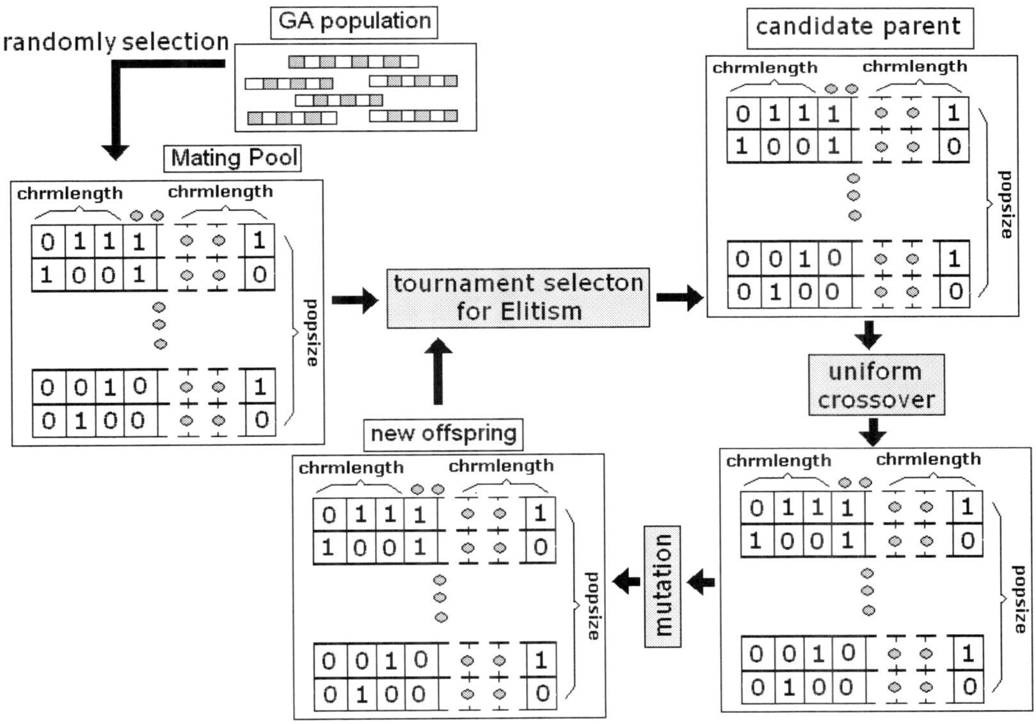

Figure 6. Operations in the *GA* method.

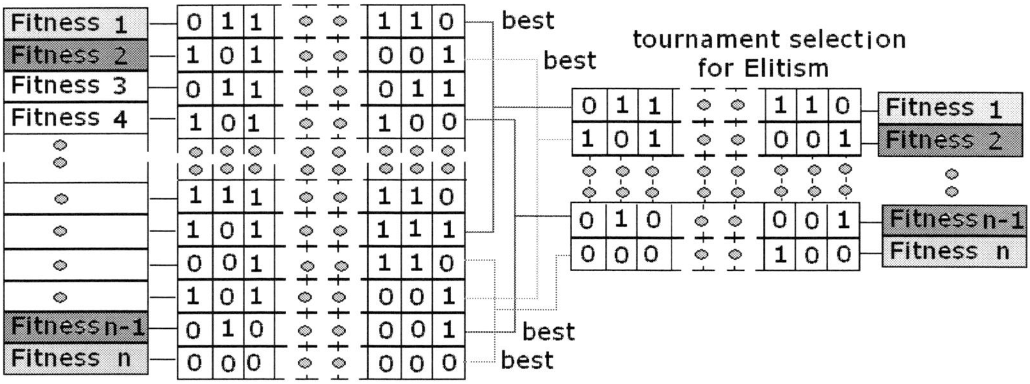

Figure 7. The scheme of elitism by tournament selection.

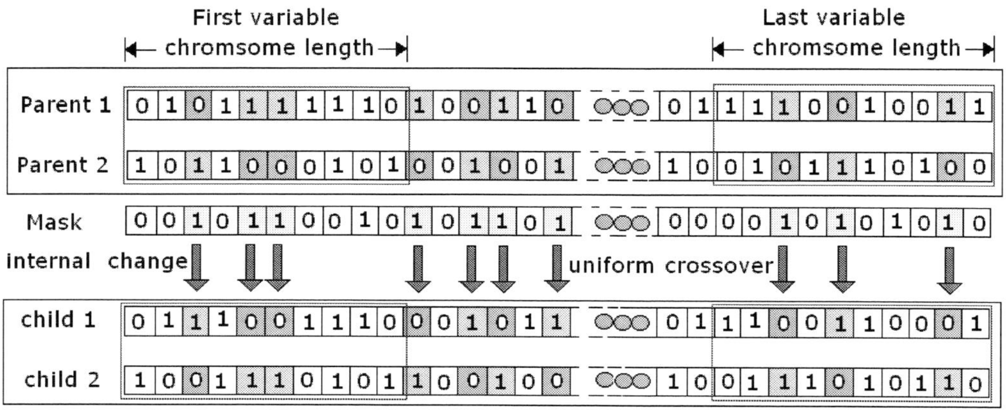

Figure 8. The scheme of uniform crossover.

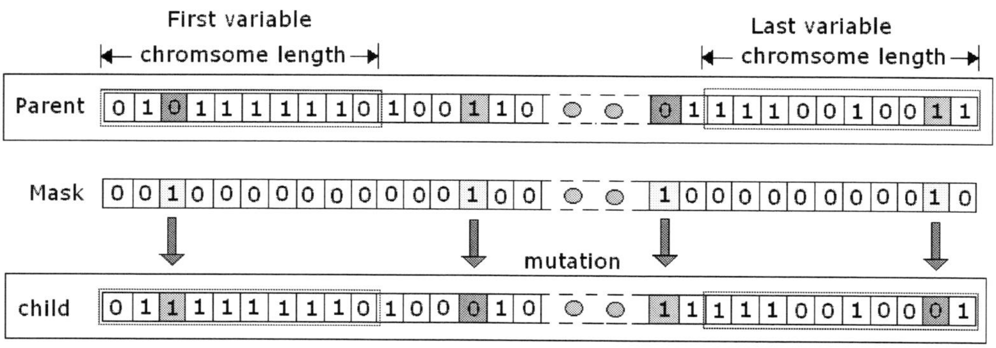

Figure 9. New random solution in a perturbed zone.

Genetically, mutation occurred with a probability of *pm* where the new and unexpected point was brought into the *GA* optimizer's search domain. A typical scheme of mutation is depicted in Figure 9. Likewise, by using masked genes randomly generated, the mapped gene will be converted from 1 to 0, or from 0 to 1, if the mapping gene is 1.

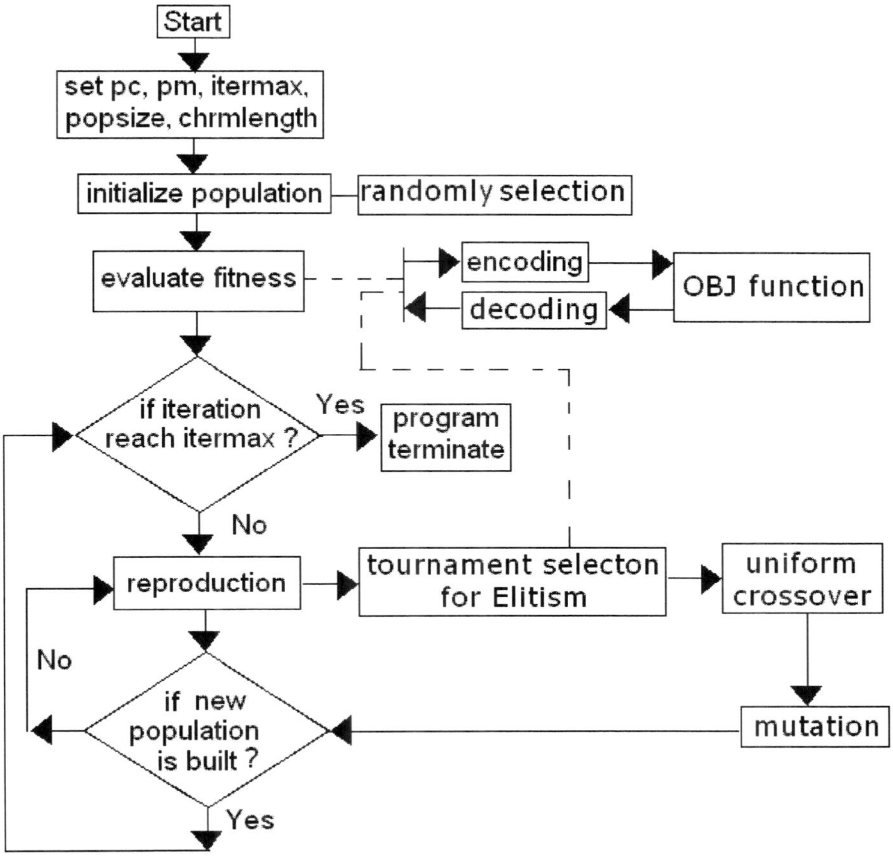

Figure 10. The block diagram of the *GA* optimization on mufflers.

To prevent the best gene from disappearing and to improve the accuracy of optimization during reproduction, the elitism scheme of keeping the best gene (one pair) in the parent generation with the tournament strategy was developed.

For the optimization of the objective function (*OBJ*), the design parameters of $(X_1, X_2, \ldots, X_k)$ were determined. When the *chrmlength* (the bit length of the chromosome) was chosen, the interval of the design parameter (X_k) with $[Lb, Ub]_k$ was mapped to the band of the binary value. The mapping system between the variable interval of $[Lb, Ub]_k$ and the kth binary chromosome of

$$[\underbrace{0\ 0\ 0\ 0 \bullet \bullet \bullet 0\ 0\ 0}_{chrmlength} \sim \underbrace{1\ 1\ 1\ 1 \bullet \bullet \bullet 1\ 1\ 1}_{chrmlength}]$$

was constructed. The encoding from x to *B2D* (binary to decimal) can be performed as

$$B2D_k = \text{integer}\left\{\frac{x_k - Lb_k}{Ub_k - Lb_k}(2^{chrmlength} - 1)\right\}. \tag{22}$$

The initial population was built up by randomization. The parameter set was encoded to form a string which represented the chromosome. By evaluating the objective function (*OBJ*), the whole set of chromosomes [$B2D_1$, $B2D_2$, …., $B2D_k$] that changed from binary form to decimal form was assigned a fitness by decoding the transformation system:

$$fitness = OBJ(X_1, X_2, ..., X_k); \tag{23a}$$

where

$$X_k = B2D_k*(Ub_k-Lb_k)/(2^{chrmlength}-1)+Lb_k. \tag{23b}$$

The process was terminated when a number of generations exceeded a pre-selected value of $iter_{max}$. The operations in the *GA* method are pictured in Figure 10.

7. Results and Discussion

7.1. Results

The accuracy of the *GA* optimization depends on six kinds of *GA* parameters including *elt* (*elitism*), *popsize* (number of population), *chrmlength* (bit length of the chromosome), $iter_{max}$ (maximum iteration), *pc* (crossover ratio), *pm* (mutation ratio). To achieve good optimization, the following parameters are varied step by step

elt (1 for yes); *popsize*(60, 90, 120); *chrmlength*(10, 15, 20); $iter_{max}$(50, 100);

pc(0.2, 0.5, 0.8, 0.9); *pm*(0.01, 0.05, 0.09).

Two results of optimization (one, pure tone noises used for *GA*'s accuracy check; and the other, a broadband noise occurring in a fan room) are described below.

7.1.1. Pure Tone Noise Optimization

By using Eqs. (19a)(19b), the maximization of the *STL* with respect to muffler B (a one-chamber muffler hybridized with perforated intruding tubes) at the specified pure tones (600Hz) was performed first. As indicated in Table 2, eleven sets of *GA* parameters are tried in the muffler's optimization. Obviously, the optimal design data can be obtained from the last set of *GA* parameters at (*popsize, chrmlength,* $iter_{max}$, *pc, pm*) = (120, 20, 100, 0.8, 0.05). Using the optimal design in a theoretical calculation, the optimal *STL* curves with respect to various *GA* parameters (*popsize, chrmlength,* $iter_{max}$, *pc, pm*) are plotted and depicted in Figures 11 and 12. As revealed in Figures 11 and 12, the *STL*s are precisely maximized at the desired frequencies.

To appreciate the acoustical performance with respect to various chambers, the shape optimization for a two-chamber muffler hybridized with perforated intruding tubes (muffler C) and a three-chamber muffler hybridized with perforated intruding tubes (muffler A) at the targeted tone (600 Hz) has been accessed using the *GA* parameters at (*popsize, chrmlength,* $iter_{max}$, *pc, pm*) = (120, 20, 100, 0.8, 0.05). The optimal *STL* curves with respect to various mufflers are plotted in Figure 13. Moreover, to investigate the acoustical influence for a three-chamber muffler with various internal tubes, muffler D (with simple expansion tubes) and muffler E (with non-perforated intruding tubes) have been optimized at the same target tone. The optimal *STL* curves of the three-chamber mufflers with respect to various internal tubes are plotted in Figure 14.

Table 2. Optimal *STL* for a one-chamber muffler with perforated intruding tubes at various *GA* parameters (muffler B: targeted tone of 600 Hz)

Item	GA parameters					Results							
	popsie	$Iter_{max}$	Chrm length	pc	pm	*RT*1	*RT*2	*RT*3	*RT*4	*RT*5	*RT*6	*RT*7	*STL* (dB)
1	60	50	10	0.2	0.01	0.6960	0.5727	0.7981	0.1310	0.2053	0.0666	0.0058	24.9
2	60	50	10	0.5	0.01	0.4481	0.3189	0.4632	0.3264	0.1224	0.0865	0.0069	25.0
3	60	50	10	0.8	0.01	0.6475	0.5923	0.4952	0.1064	0.2287	0.0931	0.0024	29.23
4	60	50	10	0.9	0.01	0.8564	0.5637	0.7759	0.2943	0.1400	0.0988	0.0061	28.8
5	60	50	10	0.8	0.05	0.5901	0.5817	0.4149	0.1001	0.4479	0.0596	0.0036	37.2
6	60	50	10	0.8	0.09	0.6842	0.1661	0.8778	0.1339	0.1201	0.0515	0.0059	33.3
7	60	50	15	0.8	0.05	0.6348	0.5456	0.3857	0.1001	0.1483	0.0842	0.0049	41.6
8	60	50	20	0.8	0.05	0.8841	0.2408	0.6964	0.1106	0.2348	0.0690	0.0067	44.5
9	90	50	20	0.8	0.05	0.6560	0.1935	0.8906	0.4060	0.1492	0.0848	0.0034	45.4
10	120	50	20	0.8	0.05	0.7572	0.5462	0.8418	0.1258	0.1130	0.0572	0.0055	48.5
11	**120**	**100**	**20**	**0.8**	**0.05**	0.2481	0.4775	0.7324	0.1167	0.1164	0.0604	0.0067	55.5

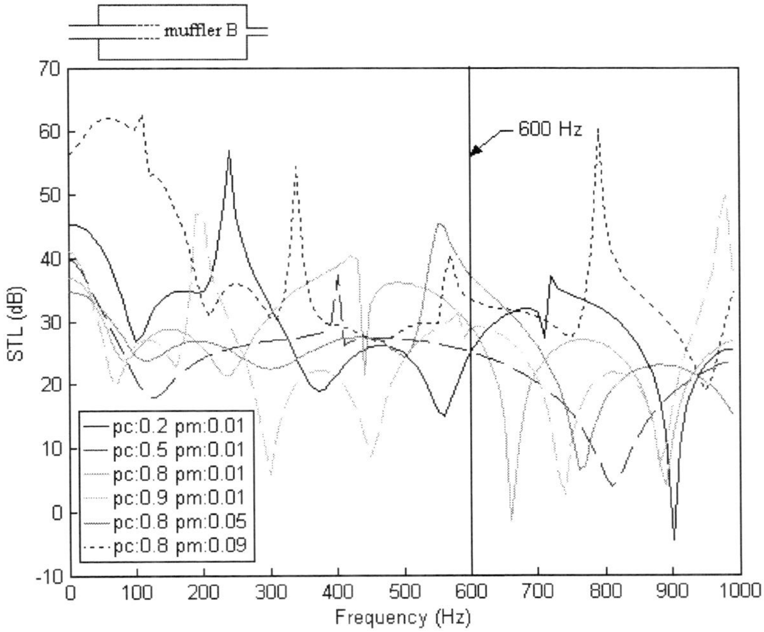

Figure 11. *STL* with respect to various *pc and pm* [muffler B: popsize = 60, chrmlength = 10, $iter_{max}$ = 50, target tone = 600 Hz].

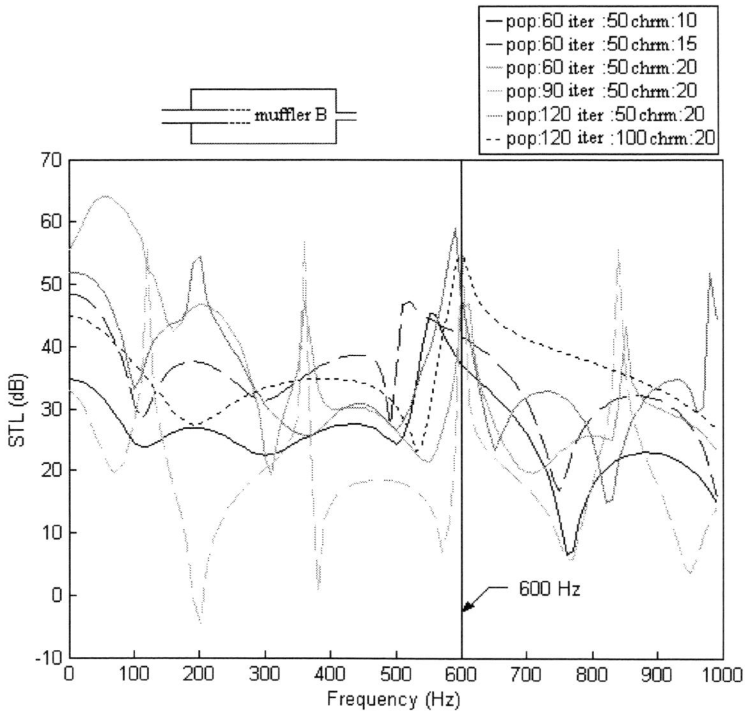

Figure 12. *STL* with respect to various *popsize, chrmlength, and iter*$_{max}$ [[muffler B: pc = 0.8, pm = 0.05, target tone = 600 Hz].

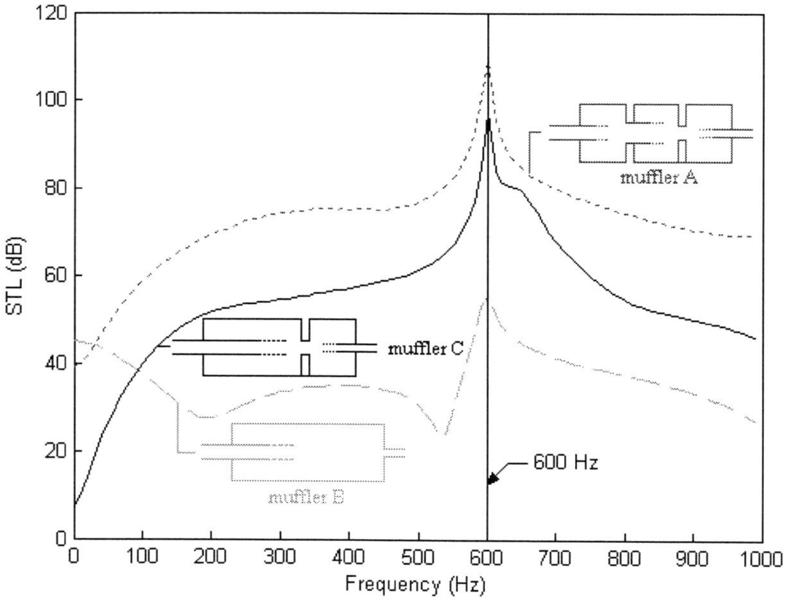

Figure 13. Comparison of the optimal *STL*s of three kinds of mufflers (mufflers A, B, and C) at target tone of 600 Hz.

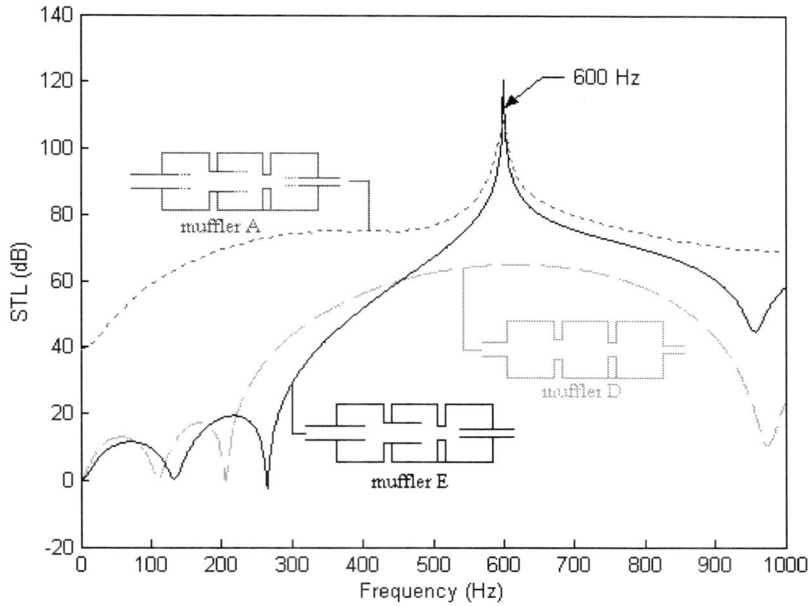

Figure 14. Comparison of the optimal *STLs* of three kinds of mufflers (mufflers A, D, and E at target tone of 600 Hz.

7.1.2. Broadband Noise Optimization

By using the above *GA* parameters, the muffler's optimal design data for five kinds of mufflers (muffler A: a three-chamber muffler hybridized with perforated intruding tubes; muffler B: a one-chamber muffler hybridized with perforated intruding tubes; muffler C: a two-chamber muffler hybridized with perforated intruding tubes; muffler D: a two-chamber simple expansion muffler [13]; and muffler E: a two-chamber muffler hybridized with non-perforated intruding tubes [12]) used to minimize the sound power level at the muffler's outlet is summarized in Table 3.

As illustrated in Table 3, the resultant sound power levels with respect to five kinds of mufflers have been reduced from 126.5 dB to 30.2 dB, 81.8 dB, 61.3 dB, 122.8 dB, and 113.1 dB. Using this optimal design in a theoretical calculation, the optimal *STL* curves with respect to various numbers of chambers and acoustical mechanisms of internal tubes are plotted and compared with the original *SWL* depicted in Figures 15 and 16.

7.2. Discussion

To achieve a sufficient optimization, the selection of the appropriate *GA* parameter set is essential. As indicated in Table 2, the best *SA* set at the targeted pure tone noise of 600 Hz has been shown. The related *STL* curves with respect to various *GA* parameters are plotted in Figures 11 and 12. Figures 11 and 12 reveal the predicted maximal value of the *STL* is located at the desired frequency. Therefore, using the *GA* optimization in finding a better design solution is reliable. To appreciate the acoustical performance with respect to various chambers, three kinds of mufflers (mufflers A~C) has been accessed using the same *GA* parameters. The comparison of optimal *STL* curves with respect to various mufflers is depicted in Figure 13.

Table 3. Comparison of acoustical performance with respect to five kinds of optimized mufflers within a same space-constrained situation (broadband noise)

Item	Muffler Type	Results						
1	three-chamber muffler equipped with perforated intruding tubes (muffler A)	*RT*1	*RT*2	*RT*3	RT4	RT5	*RT*6	SWL_T (dB)
		0.8960	0.6968	0.4976	0.7952	0.4976	0.7952	30.2
		RT7	RT8	RT9	RT10	RT11	RT12	
		0.4976	0.7952	0.7952	0.7952	0.7952	0.7952	
		RT13	RT14	RT15	RT16	RT17	RT18	
		0.006958	0.09944	0.006958	0.09944	0.006958	0.09944	
2	one-chamber muffler equipped with perforated intruding tubes (muffler B)	*RT**1	*RT**2	*RT**3		*RT**4		SWL_T (dB)
		0.6798	0.3029	0.3792		0.1042		81.8
		*RT*5*		*RT*6*		*RT*7*		
		0.1408		0.0520		0.0062		
3	two-chamber muffler equipped with perforated intruding tubes (muffler C)	*RT***1	*RT***2	*RT***3		*RT***4		SWL_T (dB)
		0.8994	0.4997	0.7993		0.7993		61.3
		*RT**5*		*RT**6*		*RT**7*		
		0.7993		0.006994		0.09992		
4	three-chamber simple expansion muffler (muffler D)	*RT****1		*RT****2		*RT****3		SWL_T (dB)
		0.4771		0.3617		0.2926		122.8
		*RT****4		*RT****5		*RT****6		
		0.2926		0.2926		0.2926		
5	three-chamber muffler equipped with non-perforated intruding tubes (muffler E)	*RT*****1	*RT*****2	*RT*****3	*RT****4*	*RT*****5		SWL_T (dB)
		0.6572	0.5058	0.3543	0.3543	0.3543		113.1
		*RT****6*	*RT*****7	*RT*****8	*RT*****9			
		0.5087	0.5087	0.5087	0.5087			

Note 1 (muffler B: for a one-chamber muffler equipped with perforated intruding tubes):
Lz = LA1 + LB1 + LC1; RT*1 = L z/Lo; RT*2 = L B1/ Lz; RT*3 = (Lz - L B1)/ LC1; RT*4 = D1/Do; RT*5 = D2/Do; RT*6 = dh1; RT*7 = η_1

Note 2 (muffler C: for a two-chamber muffler equipped with perforated intruding tubes):
RT**1 = 2Lz1/(Lo - L2); RT**2 = L B1/ Lz1; RT**3 = LC1/ (Lz1 - L B1); RT**4 = D1/Do; RT**5 = D2/Do; RT**6 = dh1; RT**7 = η_1 ; Lz2 = Lz1; LB2 = LB1; LC2 = LC1;LA2 = LA1; L3 = L1D3 = D1; dh2 = dh1; η_2 = η_1

Note 3 (muffler D: for a three-chamber simple expansion muffler):
Lz = L2 + L4 + L5 + 2L3; RT***1 = L z/Lo; RT***2 = L 4/(Lz - 2L3); RT***3 = D1/Do; RT***4 = D2/Do; RT***5 = D3/Do; RT***6 = D4/Do;

Note 4 (muffler E: for a three-chamber muffler equipped with non-perforated intruding tubes):
Lz = Lz1 + Lz2 + Lz3 + 2L4; RT****1 = L z/Lo; RT****2 = L z2/(Lz - 2L4); RT****3 = L2/Lz1; RT****4 = L5/Lz2; RT****5 = L7/Lz3; RT****6 = D1/Do; RT****7 = D2/Do; RT****8 = D3/Do; RT****9 = D4/Do;

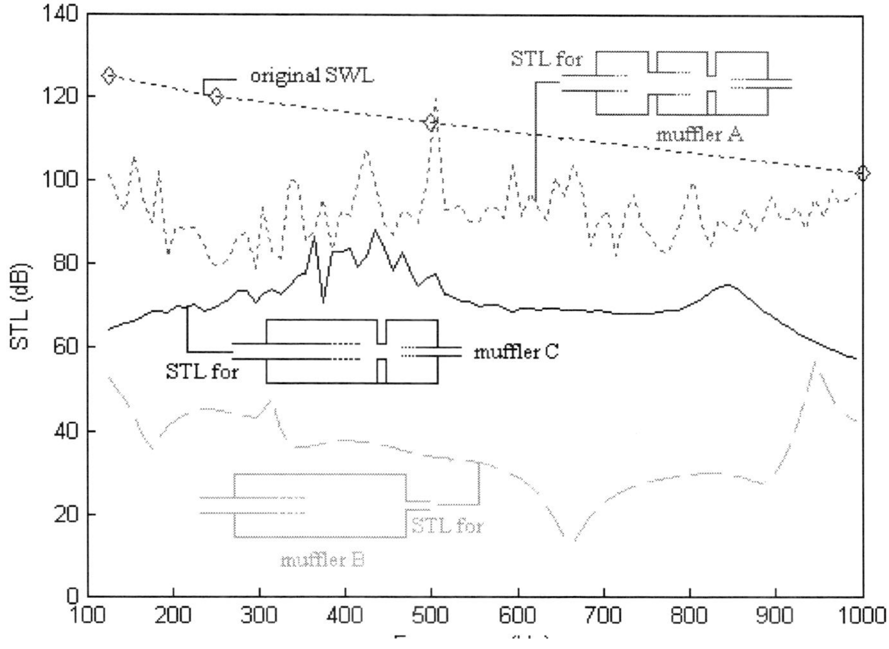

Figure 15. Comparison of the optimal *STL*s of three kinds of mufflers (mufflers A, B, and C) [broadband noise].

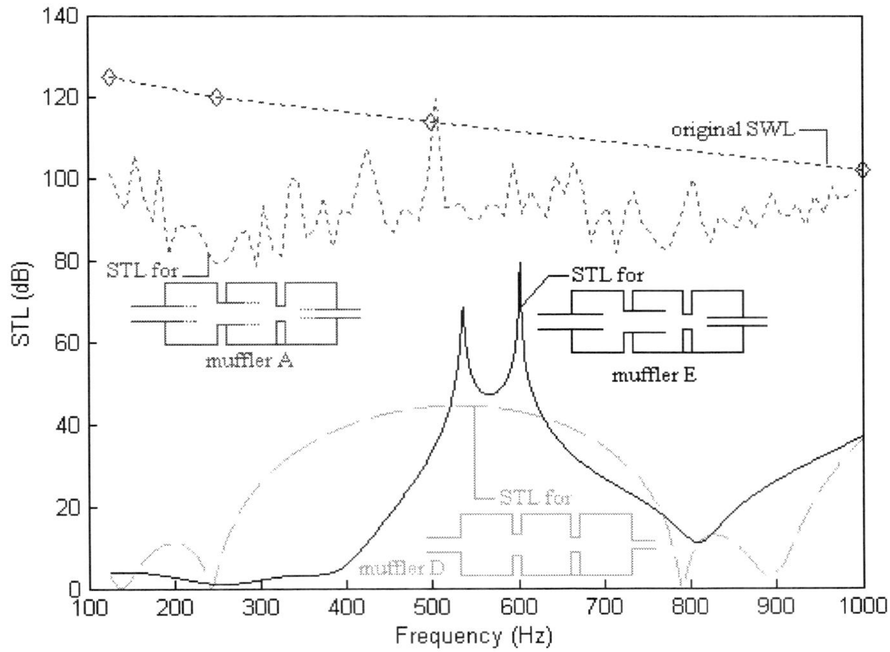

Figure 16. Comparison of the optimal *STL*s of three kinds of mufflers (mufflers A, D, and E) [broadband noise].

As indicated in Figure 13, the *STLs* of muffler A (three chambers), muffler B (one chamber), and muffler C (two chambers) reach 110 (dB), 55(dB), and 95 (dB). It is obvious that the muffler with more chambers will be superior to those with a small amount of chambers. Additionally, the comparison of optimal *STL* curves of three-chamber mufflers with various internal tubes is shown in Figure 14. As indicated in Figure 14, the *STLs* of muffler A (with perforated intruding tubes), muffler D (with simple expansion tubes), and muffler E (with non-perforated intruding tubes) at target tone (600 Hz) reach 120 (dB), 63(dB), and 120 (dB). It is found that the spectrum of *STL* for a muffler A equipped with the acoustical mechanism of perforated intruding tubes is superior to other mufflers.

In dealing with the broadband noise, the acoustical performance among five kinds of low back-pressure mufflers (mufflers A, B, C, D, and E) are shown in Table 3, Figure 15 and Figure 16. As can be observed in Table 3, the overall sound transmission loss of the three-chamber muffler equipped with perforated intruding tubes (muffler A) reaches 96.3 dB. However, the overall sound transmission loss of the one-chamber muffler with perforated intruding tubes (muffler B) and the two-chamber muffler with perforated intruding tubes (muffler C) are 44.7 dB and 65.2 dB. In addition, the overall sound transmission loss of the three-chamber simple expansion muffler (muffler D) and the three-chamber muffler with non-perforated intruding tubes (muffler E) are 3.7 dB and 13.6 dB.

The results shown in Figures 15 and 16 indicate that the three-chamber muffler hybridized with perforated intruding tubes is superior to the other mufflers. It can been seen that the muffler with more chambers will be superior to that equipped with a lesser amount of chambers; moreover, the acoustical performance for a muffler equipped with perforated intruding tubes will be better than those equipped with a simple expansion tube or non-perforated intruding tubes.

CONCLUSION

It has been shown that three-chamber mufflers hybridized with perforated intruding tubes can be easily and efficiently optimized within a limited space by using a decoupling technique, a plane wave theory, a four-pole transfer matrix, and a *GA* optimizer. As indicated in Table 2 and Figures 11~12, five kinds of *GA* parameters (*popsize, chrmlength,* $iter_{max}$, *pc*, *pm*) play essential roles in the solution's accuracy during *GA* optimization. Figures 13~14 indicate that the tuning ability established by adjusting design parameters of five kinds of mufflers (muffler A: a three-chamber muffler hybridized with perforated intruding tubes; muffler B: a one-chamber muffler hybridized with perforated intruding tubes; muffler C: a two-chamber muffler hybridized with perforated intruding tubes; muffler D: a three-chamber simple expansion muffler; and muffler E: a three-chamber muffler hybridized with non-perforated intruding tubes) is reliable.

In addition, the appropriate acoustical performance curve of five kinds of low back-pressure mufflers (mufflers A~E) has been assessed. As indicated in Table 3, the resultant SWL_T with respect to these mufflers is30.2 dB, 81.8 dB, 61.3 dB, 122.8 dB. Obviously, the acoustical mechanism using perforated intruding tubes inside the muffler's cavity has a best acoustical performance than those with no tubes and non-perforated intruding tubes. Moreover, the muffler with more chambers is also superior to those with fewer chambers.

Consequently, the approach used for the optimal design of the *STL* proposed in this study is quite efficient.

Acknowledgments

The author acknowledges the financial support of the National Science Council (NSC 97-2221-E-235-001, ROC).

APPENDIX A

Transfer Matrix of an Expanded Perforated Intruding Tube

As indicated in Figure 17, the perforated resonator is composed of an inner perforated tube and an outer resonating chamber. Based on Sullivan and Crocker's derivation [4], the continuity equations and momentum equations with respect to inner and outer tubes at nodes 1 and 2 are the following.

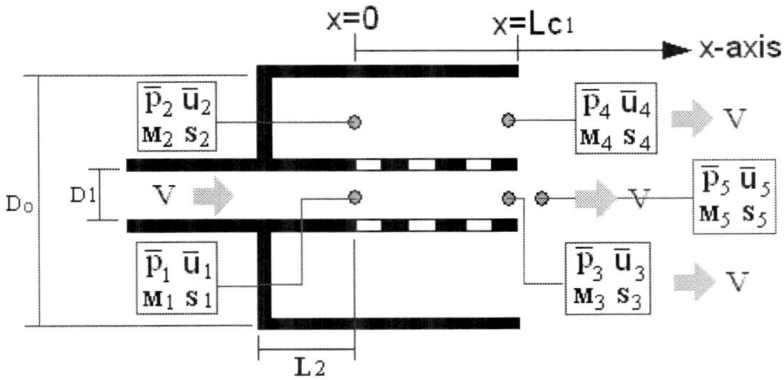

Figure 17. The acoustical mechanism of an expanded perforated intruding tube.

Inner tube: continuity equation

$$V_1 \frac{\partial \rho_1}{\partial x} + \rho_o \frac{\partial u_1}{\partial x} + \frac{4\rho_o}{D_1} u_{1,2} + \frac{\partial \rho_2}{\partial t} = 0 , \tag{A1}$$

momentum equation

$$\rho_o \left(\frac{\partial}{\partial t} + V_1 \frac{\partial}{\partial x} \right) u_1 + \frac{\partial p_1}{\partial x} = 0 . \tag{A2}$$

Outer tube: continuity equation

$$\rho_o \frac{\partial u_2}{\partial x} - \frac{4D_2\rho_o}{D_O^2 - D_2^2} u_{1,2} + \frac{\partial \rho_2}{\partial t} = 0 \quad , \tag{A3}$$

momentum equation

$$\rho_o \left(\frac{\partial}{\partial t} + V_2 \frac{\partial}{\partial x} \right) u_2 + \frac{\partial p_2}{\partial x} = 0 \quad . \tag{A4}$$

Assuming that the acoustic wave is a harmonic motion

$$p(x,t) = P(x) \cdot e^{j\omega t} . \tag{A5}$$

The isentropic process in the ducts yields

$$P(x) = \rho(x) \cdot c_o^2 . \tag{A6}$$

Assuming that the perforation along the inner tube is uniform ($d\varsigma / dx = 0$), the acoustic impedance of the perforation ($\rho_o c_o \ \varsigma$) is

$$\rho_o c_o \ \varsigma = \frac{p_1(x) - p_2(x)}{u_{1,2}(x)} , \tag{A7}$$

where ς is the specific acoustical impedance of the perforated tube. According to Sullivan [4] and Rao [7], the empirical formulations for the perforate with or without mean flow are adopted in this study.

Perforates with a stationary medium yield

$$\varsigma = [0.006 + jk(t + 0.75dh)] / \eta . \tag{A8a}$$

Perforates with a grazing flow yield

$$\varsigma = \left[7.337x10^{-3}(1 + 72.23M) + j2.2245x10^{-5}(1 + 51t)(1 + 204dh) f \right] / \eta , \tag{A8b}$$

where dh is the diameter of the perforated hole on the inner tube, t is the thickness of the inner perforated tube, and η is the porosity of the perforated tube.

The available ranges of above parameters are

$$\text{M: } 0.05 \leqq M \leqq 0.2, \tag{A8c}$$

$$\eta\text{: } 0.03 \leqq \eta \leqq 0.1, \tag{A8d}$$

$$\text{t: } 0.001 \leqq t \leqq 0.003, \tag{A8e}$$

dh: 0.00175≦ dh ≦0.007. (A8f)

Substituting Eqs. (A5) ~ (A7) into Eqs. (A1) ~ (A4) yields

$$\rho_o c_o \frac{du_1}{dx} = -\left[jkp_1 + \frac{V}{c_o} \cdot \frac{dp_1}{dx} + \frac{4 \cdot (p_1 - p_2)}{D_2 \varsigma} \right], \tag{A9}$$

$$\rho_o c_o \frac{du_2}{dx} = -\left[jkp_2 - \frac{4D_2 \cdot (p_1 - p_2)}{(D_o^2 - D_2^2 \varsigma)} \right], \tag{A10}$$

$$\rho_o c_o \left(jku_1 + \frac{V}{c_o} \cdot \frac{du_1}{dx} \right) = -\frac{dp_1}{dx}, \tag{A11}$$

$$j\rho_o c_o k u_2 = -\frac{dp_2}{dx}. \tag{A12}$$

Eliminating u_1 and u_2 by the differentiation of Eq.(A11) and the substitution of Eq. (A12) yields

$$\left[\left(1 - M_1^2\right) \frac{d^2}{dx^2} - 2jM_1 k \frac{d}{dx} + k^2 \right] p_1 - \frac{4}{D_2 \varsigma} \left[M_1 \frac{d}{dx} + jk \right] (p_1 - p_2) = 0, \tag{A13}$$

$$\left[\frac{d^2}{dx^2} + k^2 \right] p_2 + j \frac{4D_2}{(D_o^2 - D_2^2)\varsigma} (p_1 - p_2) = 0, \tag{A14}$$

where $M_1 = \frac{V_1}{c_o}$.

Alternatively, Eqs.(A13) and (A14) can also be expressed as

$$\begin{bmatrix} D^2 + \alpha_1 D + \alpha_2 & \alpha_3 D + \alpha_4 \\ \alpha_5 D + \alpha_6 & D^2 + \alpha_7 D + \alpha_8 \end{bmatrix} \begin{bmatrix} p_1 \\ p_2 \end{bmatrix} = \begin{bmatrix} 0 \\ 0 \end{bmatrix}, \tag{A15a}$$

where

$$D = \frac{d}{dx}; \qquad \alpha_1 = -\frac{jM_1}{1 - M_1^2} \left(2k - j \frac{4}{D_2 \varsigma} \right); \qquad \alpha_2 = \frac{1}{1 - M_1^2} \left(k^2 - j \frac{4k}{D_2 \varsigma} \right);$$

$$\alpha_3 = \frac{M_1}{1 - M_1^2} \cdot \frac{4}{D_2 \varsigma}; \; \alpha_4 = \frac{jM_1}{1 - M_1^2} \cdot \frac{4k}{D_2 \varsigma}; \; \alpha_5 = 0;$$

$$\alpha_6 = \frac{j4kD_2}{(D_o^2 - D_2^2)\varsigma}; \; \alpha_7 = 0; \; \alpha_8 = k^2 - \frac{j4kD_2}{(D_o^2 - D_2^2)\varsigma}. \tag{A15b}$$

Developing Eq. (A15a) yields

$$p_1'' + \alpha_1 p_1' + \alpha_2 p_1 + \alpha_3 p_2' + \alpha_4 p_2 = 0\,, \tag{A16a}$$

$$\alpha_5 p_1' + \alpha_6 p_1 + p_2'' + \alpha_7 p_2' + \alpha_8 p_2 = 0\,. \tag{A16b}$$

Let

$$p_1' = \frac{dp_1}{dx} = y_1\,,\; p_2' = \frac{dp_2}{dx} = y_2\,,\; p_1 = y_3\,,\; p_2 = y_4\,. \tag{A17}$$

According to Eqs. (A16) and (A17), the new matrix between {y'} and {y} is

$$\begin{bmatrix} y_1' \\ y_2' \\ y_3' \\ y_4' \end{bmatrix} = \begin{bmatrix} -\alpha_1 & -\alpha_3 & -\alpha_2 & -\alpha_4 \\ -\alpha_5 & -\alpha_7 & -\alpha_6 & -\alpha_8 \\ 1 & 0 & 0 & 0 \\ 0 & 1 & 0 & 0 \end{bmatrix} \begin{bmatrix} y_1 \\ y_2 \\ y_3 \\ y_4 \end{bmatrix} \tag{A18a}$$

which can be briefly expressed as

$$\{y'\} = [\mathrm{NN}]\{y\}. \tag{A18b}$$

Let

$$\{y\} = [\Omega\Omega]\{\Gamma\Gamma\} \tag{A19a}$$

which is

$$\begin{bmatrix} dp_1/dx \\ dp_2/dx \\ p_1 \\ p_2 \end{bmatrix} = \begin{bmatrix} \Omega\Omega_{1,1} & \Omega\Omega_{1,2} & \Omega\Omega_{1,3} & \Omega\Omega_{1,4} \\ \Omega\Omega_{2,1} & \Omega\Omega_{2,2} & \Omega\Omega_{2,3} & \Omega\Omega_{2,4} \\ \Omega\Omega_{3,1} & \Omega\Omega_{3,2} & \Omega\Omega_{3,3} & \Omega\Omega_{3,4} \\ \Omega\Omega_{4,1} & \Omega\Omega_{4,2} & \Omega\Omega_{4,3} & \Omega\Omega_{4,4} \end{bmatrix} \begin{bmatrix} \Gamma\Gamma_1 \\ \Gamma\Gamma_2 \\ \Gamma\Gamma_3 \\ \Gamma\Gamma_4 \end{bmatrix}, \tag{A19b}$$

$[\Omega\Omega]_{4x4}$ is the model matrix formed by four sets of eigen vectors $\Omega\Omega_{4x1}$ of $[\mathrm{NN}]_{4x4}$.

Substituting Eq. (A19) into (A18) and then multiplying $[\Omega\Omega]^{-1}$ by both sides yields

$$[\Omega\Omega]^{-1}[\Omega\Omega]\{\Gamma\Gamma'\} = [\Omega\Omega]^{-1}[\mathrm{NN}][\Omega\Omega]\{\Gamma\Gamma\}. \tag{A20}$$

Set

$$[\chi\chi] = [\Omega\Omega]^{-1}[\mathrm{NN}][\Omega\Omega] = \begin{bmatrix} \beta\beta_1 & 0 & 0 & 0 \\ 0 & \beta\beta_2 & 0 & 0 \\ 0 & 0 & \beta\beta_3 & 0 \\ 0 & 0 & 0 & \beta\beta_4 \end{bmatrix}, \tag{A21}$$

where $\beta\beta_i$ is the eigen value of $[\mathrm{NN}]$.

Eq.(A19) can be thus rewritten as

$$\{\Gamma\Gamma'\}=[\chi\chi]\{\Gamma\Gamma\}. \qquad \text{(A22)}$$

Obviously, Eq.(A21) is a decoupled equation. The related solution then becomes

$$\Gamma\Gamma_i = ff_i e^{\beta\beta_i x}. \qquad \text{(A23)}$$

Using Eqs.(A2),(A4),(A19) and (A23), the relationship of acoustic pressure and particle velocity becomes

$$\begin{bmatrix} p_1(x) \\ p_2(x) \\ \rho_o c_o u_1(x) \\ \rho_o c_o u_2(x) \end{bmatrix} = \begin{bmatrix} \mathrm{HH}_{1,1} & \mathrm{HH}_{1,2} & \mathrm{HH}_{1,3} & \mathrm{HH}_{1,4} \\ \mathrm{HH}_{2,1} & \mathrm{HH}_{2,2} & \mathrm{HH}_{2,3} & \mathrm{HH}_{2,4} \\ \mathrm{HH}_{3,1} & \mathrm{HH}_{3,2} & \mathrm{HH}_{3,3} & \mathrm{HH}_{3,4} \\ \mathrm{HH}_{4,1} & \mathrm{HH}_{4,2} & \mathrm{HH}_{4,3} & \mathrm{HH}_{4,4} \end{bmatrix} \begin{bmatrix} ff_1 \\ ff_2 \\ ff_3 \\ ff_4 \end{bmatrix}, \qquad \text{(A24a)}$$

where

$$\mathrm{HH}_{1,i} = \Omega\Omega_{3,i} e^{\beta\beta_i x}; \ \mathrm{HH}_{2,i} = \Omega\Omega_{4,i} e^{\beta\beta_i x}; \ \mathrm{HH}_{3,i} = -\frac{\Omega\Omega_{1,i} e^{\beta\beta_i x}}{jk + M_1 \beta\beta_i};$$

$$\mathrm{HH}_{4,i} = -\frac{\Omega\Omega_{2,i} e^{\beta\beta_i x}}{jk}. \qquad \text{(A24b)}$$

Substituting x = 0 and x = *L*c into Eq. (A24) yields

$$\begin{bmatrix} p_1(0) \\ p_2(0) \\ \rho_o c_o u_1(0) \\ \rho_o c_o u_2(0) \end{bmatrix} = [\mathrm{HH}(0)] \begin{bmatrix} ff_1 \\ ff_2 \\ ff_3 \\ ff_4 \end{bmatrix}, \qquad \text{(A25a)}$$

$$\begin{bmatrix} p_1(L_C) \\ p_2(L_C) \\ \rho_o c_o u_1(L_C) \\ \rho_o c_o u_2(L_C) \end{bmatrix} = [\mathrm{HH}(L_C)] \begin{bmatrix} ff_1 \\ ff_2 \\ ff_3 \\ ff_4 \end{bmatrix}. \qquad \text{(A25b)}$$

Combining with Eqs. (A25a) and (A25b), the resultant relationship of acoustic pressure and particle velocity between x = 0 and x = *L*c becomes

$$\begin{bmatrix} p_1(0) \\ p_2(0) \\ \rho_o c_o u_1(0) \\ \rho_o c_o u_2(0) \end{bmatrix} = [\mathrm{TT}] \begin{bmatrix} p_1(L_C) \\ p_2(L_C) \\ \rho_o c_o u_1(L_C) \\ \rho_o c_o u_2(L_C) \end{bmatrix}, \tag{A26a}$$

where

$$[\mathrm{TT}] = [\mathrm{HH}(0)]\,[\mathrm{HH}(L_C)]^{-1} = \begin{bmatrix} TT_{1,1} & TT_{1,2} & TT_{1,3} & TT_{1,4} \\ TT_{2,1} & TT_{2,2} & TT_{2,3} & TT_{2,4} \\ TT_{3,1} & TT_{3,2} & TT_{3,3} & TT_{3,4} \\ TT_{4,1} & TT_{4,2} & TT_{4,3} & TT_{4,4} \end{bmatrix}. \tag{A26b}$$

Let

$$p_1(0) = \overline{p}_1;\ p_1(L_c) = \overline{p}_3;\ u_1(0) = \overline{u}_1;\ u_1(L_c) = \overline{u}_3;$$
$$p_2(0) = \overline{p}_2;\ p_2(L_c) = \overline{p}_4;\ u_2(0) = \overline{u}_2;\ u_2(L_c) = \overline{u}_4.$$

Eq. (A26) can be expressed as $\overline{p}_1, \overline{p}_2, \overline{p}_3, \overline{p}_4, \overline{u}_1, \overline{u}_2, \overline{u}_3$ and $\overline{u}_4$

$$\begin{bmatrix} \overline{p}_1 \\ \overline{p}_2 \\ \rho_o c_o \overline{u}_1 \\ \rho_o c_o \overline{u}_2 \end{bmatrix} = [\mathrm{TT}] \begin{bmatrix} \overline{p}_3 \\ \overline{p}_4 \\ \rho_o c_o \overline{u}_3 \\ \rho_o c_o \overline{u}_4 \end{bmatrix}. \tag{A27}$$

The equation of mass continuity between point 3 and point 5 with a mean fl ow is expressed in equation (A28) as

$$\begin{aligned} &c_o \rho_o S_3 \overline{u}_3 + S_3 M_3 \overline{p}_3 \\ &= c_o \rho_o S_5 \overline{u}_5 + c_o \rho_o S_4 \overline{u}_4 + S_5 M_5 \left(\overline{p}_5 - \frac{p_o}{C_v} \frac{R K_e M_5 Y_5}{p_o} \frac{v_{c,5} - M_5 p_{c,5} / Y_5}{1 - M_5^{\,2}} \right) \end{aligned} \tag{A28a}$$

or

$$\begin{aligned} &c_o \rho_o S_3 \overline{u}_3 + S_3 M_3 \overline{p}_3 \\ &= c_o \rho_o S_5 \overline{u}_5 + c_o \rho_o S_4 \overline{u}_4 + S_5 M_5 \left(\overline{p}_5 - \frac{p_o (\gamma - 1) K_e M_5 Y_5}{p_o} \frac{v_{c,5} - M_5 p_{c,5} / Y_5}{1 - M_5^{\,2}} \right) \end{aligned} \tag{A28b}$$

where

$$K_e = \left[\frac{S_5}{S_3} - 1 \right]^2 ; \; Y_5 = \frac{c_o}{S_5} . \tag{A28c}$$

The concept of static enthalpy deduced by Munjal [28] is described as

$$\begin{bmatrix} p_{c,5} \\ v_{c,5} \end{bmatrix} = \begin{bmatrix} 1 & M_5 Y_5 \\ \frac{M_5}{Y_5} & 1 \end{bmatrix} \begin{bmatrix} \overline{p}_5 \\ \rho_o S_5 \overline{u}_5 \end{bmatrix} . \tag{A29}$$

Substituting Eq. (A29) into Eq. (A28) yields

$$\begin{aligned} & c_o \rho_o S_3 \overline{u}_3 + S_3 M_3 \overline{p}_3 \\ = & c_o \rho_o S_5 \overline{u}_5 \left[1 - \frac{(\gamma - 1)}{c_o} Y_5 K_e S_5 M_5^2 \right] + \left(M_5 S_5 \overline{p}_5 + c_o \rho_o \overline{u}_4 S_4 \right) \end{aligned} \tag{A30a}$$

or

$$c_o \rho_o S_3 \overline{u}_3 + S_3 M_3 \overline{p}_3 = c_o \rho_o S_5 \overline{u}_5 \left[1 - \mathrm{Y}_e \right] + \left(M_5 S_5 \overline{p}_5 + c_o \rho_o \overline{u}_4 S_4 \right) , \tag{A30b}$$

where

$$\mathrm{Y}_e = \frac{(\gamma - 1)}{c_o} Y_5 K_e S_5 M_5^2 . \tag{A30c}$$

The equation of momentum for a steady flow is

$$\begin{aligned} & S_3 \overline{p}_3 + 2 \rho_o S_3 V_3 \overline{u}_3 + S_3 M_3^2 \overline{p}_3 \\ = & -c_{11} \left(\begin{aligned} & S_5 \overline{p}_5 + 2 \rho_o S_5 \overline{u}_5 + \\ & S_5 M_5^2 \left[\overline{p}_5 - (\gamma - 1) k_e M_5 Y_5 \frac{v_{c,5} - M_5 p_{c,5} / Y_5}{1 - M_5^2} \right] \end{aligned} \right) - c_{12} S_4 \overline{p}_4 , \end{aligned} \tag{A31a}$$

where

$$c_{11} = -1 ; \; c_{12} = 1 . \tag{A31b}$$

Substituting Eq. (A30) into Eq. (A31) yields

$$S_3\left(1+M_3^2\right)\overline{p}_3+2\rho_o c_o S_3 M_3 \overline{u}_3+c_{11}\left(S_5+S_5 M_5^2\right)\overline{p}_5$$
$$=-c_{11}\left(\frac{2S_5}{c_o}-\frac{(\gamma-1)K_e M_5^3 Y_5 S_5^2}{c_o}\right)\rho_o c_o \overline{u}_5-c_{12}S_4\overline{p}_4. \tag{A32}$$

The equation of energy conservation for a steady flow is

$$\overline{p}_3+\rho_o V_3\overline{u}_3=\overline{p}_5+\rho_o V_5\overline{u}_5+K_e\rho_o V_5\overline{u}_5 \tag{A33a}$$

or

$$\overline{p}_3+\rho_o V_3\overline{u}_3=\overline{p}_5+(1+K_e)\rho_o V_5\overline{u}_5. \tag{A33b}$$

The rigid wall at boundary yields

$$\frac{\overline{p}_2}{\rho_o c_o \overline{u}_2}=-j\cot(kL_2) \tag{A34a}$$

or

$$\overline{p}_2 X=\rho_o c_o \overline{u}_2\,;\,X=-j\tan(kL_2). \tag{A34b}$$

Expanding Eq. (A27) yields

$$\overline{p}_3=TT_{1,1}\overline{p}_1+TT_{1,2}\overline{p}_2+TT_{1,3}\rho_o c_o\overline{u}_1+TT_{1,4}\rho_o c_o\overline{u}_2, \tag{A35a}$$

$$\overline{p}_1=TT_{2,1}\overline{p}_1+TT_{2,2}\overline{p}_2+TT_{2,3}\rho_o c_o\overline{u}_1+TT_{2,4}\rho_o c_o\overline{u}_2, \tag{A35b}$$

$$\rho_o c_o\overline{u}_3=TT_{3,1}\overline{p}_1+TT_{3,2}\overline{p}_2+TT_{3,3}\rho_o c_o\overline{u}_1+TT_{3,4}\rho_o c_o\overline{u}_2, \tag{A35c}$$

$$\rho_o c_o\overline{u}_4=TT_{4,1}\overline{p}_1+TT_{4,2}\overline{p}_2+TT_{4,3}\rho_o c_o\overline{u}_1+TT_{4,4}\rho_o c_o\overline{u}_2. \tag{A35d}$$

Plugging Eq. (A34b) into Eqs. (A35a), (A35b), (A35c) and (A35d) yields

$$\overline{p}_3=TT_{1,1}\overline{p}_1+(TT_{1,2}+TT_{1,4})\overline{p}_2+TT_{1,3}\rho_o c_o\overline{u}_1, \tag{A36a}$$

$$\overline{p}_4=TT_{2,1}\overline{p}_1+(TT_{2,2}+TT_{2,4}X)\overline{p}_2+TT_{2,3}\rho_o c_o\overline{u}_1, \tag{A36b}$$

$$\rho_o c_o\overline{u}_3=TT_{3,1}\overline{p}_1+(TT_{3,2}+TT_{3,4}X)\overline{p}_2+TT_{3,3}\rho_o c_o\overline{u}_1, \tag{A36c}$$

$$\rho_o c_o \bar{u}_4 = TT_{4,1}\bar{p}_1 + (TT_{4,2} + TT_{4,4})\bar{p}_2 + TT_{4,3}\rho_o c_o \bar{u}_1 . \quad \text{(A36d)}$$

By plugging Eqs.(A30b)(A32)(A33b) into Eqs.(A36a)(A36b)(A36c)(A36d) and eliminating the parameters $\bar{p}_2, \bar{u}_2$, $\bar{p}_3, \bar{u}_3, \bar{p}_4, \bar{u}_4$, the simplified equations yield

$$\bar{p}_1 = \frac{NN_3 NN_6 - NN_7 NN_2}{NN_1 NN_6 - NN_5 NN_2}\bar{p}_5 + \frac{NN_4 NN_6 - NN_8 NN_2}{NN_1 NN_6 - NN_5 NN_2}\rho_o c_o \bar{u}_5 , \quad \text{(A37a)}$$

$$\rho_o c_o \bar{u}_1 = \frac{NN_3 NN_5 - NN_7 NN_1}{NN_1 NN_6 - NN_5 NN_2}\bar{p}_5 + \frac{NN_4 NN_5 - NN_8 NN_1}{NN_1 NN_6 - NN_5 NN_2}\rho_o c_o \bar{u}_5 , \quad \text{(A37b)}$$

where

$NN_1 = W_2 W_5 - W_1 W_6$; $NN_2 = W_3 W_6 - W_2 W_7$; $NN_3 = W_2 W_8 - W_6$;
$NN_4 = W_2 W_9 - W_4 W_6$; $NN_5 = W_2 W_{10} - W_1 W_{11}$; $NN_6 = W_3 W_{11} - W_2 W_{12}$;
$NN_7 = -W_{11} - W_2 W_{13}$; $NN_8 = -W_4 W_{11} - W_2 W_{14}$; $W_1 = TT_{1,1} + TT_{3,1} M_3$;
$W_2 = TT_{1,2} + TT_{1,4} X + (TT_{3,2} + XTT_{3,4}) M_3$; $W_3 = TT_{1,3} + TT_{3,3} M_3$;
$W_4 = (1 + K_e) M_5$; $W_5 = TT_{3,1} S_3 + TT_{1,1} S_3 M_3 - TT_{4,1} S_4$;
$W_6 = (TT_{3,2} + XTT_{3,4}) S_3 + (TT_{1,2} + XTT_{1,4}) S_3 M_3 - (TT_{4,2} + XTT_{4,4}) S_4$;
$W_7 = TT_{3,3} S_3 + TT_{1,3} S_3 M_3 - TT_{4,3} S_4$;
$W_8 = M_5 S_5$; $W_9 = (1 - Y_e + M_5) S_5$;
$W_{10} = TT_{1,1}(1 + M_3^2) S_3 + 2TT_{3,1} S_3 M_3 + C_{12} TT_{2,1} S_4$;
$W_{11} = (TT_{1,2} + XTT_{1,4})(1 + M_3^2) S_3 + 2(TT_{3,2} + XTT_{3,4}) S_3 M_3 + C_{12}(TT_{2,2} + XTT_{2,4}) S_4$;
$W_{12} = TT_{1,3}(1 + M_3^2) S_3 + 2TT_{3,3} S_3 M_3 + C_{12} TT_{2,3} S_4$;
$W_{13} = C_{11}(S_5 + S_5 M_5^2)$;
$W_{14} = C_{11}\left[2 S_5 M_5 - S_5 M_5^3 (\gamma - 1) K_e\right]$. (A37c)

Combining Eqs(A37a,b) into a matrix form yields

$$\begin{bmatrix} \bar{p}_1 \\ \rho_o c_o \bar{u}_1 \end{bmatrix} = \begin{bmatrix} TPOE_{1,1} & TPOE_{1,2} \\ TPOE_{2,1} & TPOE_{2,2} \end{bmatrix} \begin{bmatrix} \bar{p}_5 \\ \rho_o c_o \bar{u}_5 \end{bmatrix}, \quad \text{(A38a)}$$

where

$$TPOE_{1,1} = \frac{NN_3 NN_6 - NN_7 NN_2}{NN_1 NN_6 - NN_5 NN_2}; \; TPOE_{1,2} = \frac{NN_4 NN_6 - NN_8 NN_2}{NN_1 NN_6 - NN_5 NN_2};$$

$$TPOE_{2,1} = \frac{NN_3 NN_5 - NN_7 NN_1}{NN_1 NN_6 - NN_5 NN_2}; \; TPOE_{2,2} = \frac{NN_4 NN_5 - NN_8 NN_1}{NN_1 NN_6 - NN_5 NN_2}. \tag{A38b}$$

APPENDIX B

Transfer Matrix of a Contracted Perforated Intruding Tube

Similarly, the perforated intruding tube shown in Figure 18 is symmetrical to the above perforated intruding outlet tube. As derived in Eqs. (A1)~(A36) in appendix A, the acoustical pressure and acoustical particle velocity at points 10 and 14 is

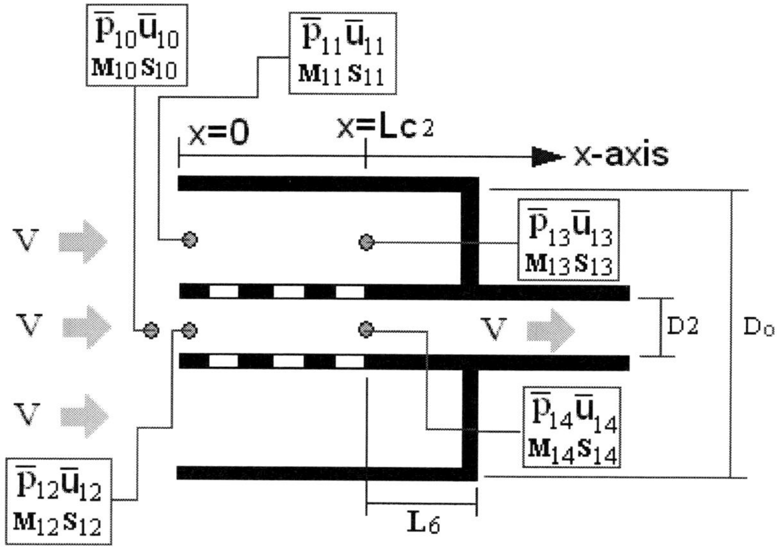

Figure 18. The acoustical mechanism of a contracted perforated intruding tube.

$$\bar{p}_{11} = (TTT_{1,1}\bar{p}_{13} + TTT_{1,3}XX)\bar{p}_{13} + TTT_{1,2}\bar{p}_{14} + TTT_{1,4}\rho_o c_o \bar{u}_{14}, \tag{B1a}$$

$$\bar{p}_{12} = (TTT_{2,1} + TTT_{2,3}XX)\bar{p}_{13} + TTT_{2,2}\bar{p}_{14} + TTT_{2,4}\rho_o c_o \bar{u}_{14}, \tag{B1b}$$

$$\rho_o c_o \bar{u}_{11} = (TTT_{3,1} + TTT_{3,3}XX)\bar{p}_{13} + TTT_{3,2}\bar{p}_{14} + TTT_{3,4}\rho_o c_o \bar{u}_{14}, \tag{B1c}$$

$$\rho_o c_o \bar{u}_{12} = (TTT_{4,1} + TTT_{4,3}XX)\bar{p}_{13} + TTT_{4,2}\bar{p}_{14} + TTT_{4,4}\rho_o c_o \bar{u}_{14}. \tag{B1d}$$

By substituting and eliminating parameters ($\bar{p}_{11}, \bar{u}_{11}, \bar{p}_{12}, \bar{u}_{12}, \bar{p}_{13}, \bar{u}_{13}$) in Eq.(B1), the simplified equations become

$$\bar{p}_{10} = \frac{NNN_7 NNN_2 - NNN_3 NNN_6}{NNN_5 NNN_2 - NNN_1 NNN_6} \bar{p}_{14} + \frac{NNN_8 NNN_2 - NNN_4 NNN_6}{NNN_5 NNN_2 - NNN_1 NNN_6} \rho_o c_o \bar{u}_{14}, \tag{B2a}$$

$$\rho_o c_o \bar{u}_{10} = \frac{NNN_7 NNN_1 - NNN_3 NNN_5}{NNN_6 NNN_1 - NNN_2 NNN_5} \bar{p}_{14} + \frac{NNN_8 NNN_1 - NNN_4 NNN_5}{NNN_6 NNN_1 - NNN_2 NNN_5} \rho_o c_o \bar{u}_{14}, \tag{B2b}$$

where

$$AA_1 = TTT_{1,1} + (1 + K_C) M_{12} TTT_{3,1};$$

$$AA_2 = (TTT_{1,2} + XX_2 TTT_{1,4}) + (1 + K_C) M_{12} (TTT_{3,2} + XX_2 TTT_{3,4});$$

$$AA_3 = TTT_{1,3} + (1 + K_C) M_{12} TTT_{3,3}; AA_4 = M_{10};$$

$$BB_1 = \left[\left(1 - \frac{(\gamma-1) K_C M_{12}^2}{1 - M_{12}^2}\right) + \frac{(\gamma-1) K_C M_{12}^4}{1 - M_{12}^2}\right] S_{12} TTT_{3,1} + M_{12} S_{12} TTT_{1,1} + S_{11} TTT_{4,1},;$$

$$BB_2 = \left[\left(1 - \frac{(\gamma-1) K_C M_{12}^2}{1 - M_{12}^2}\right) + \frac{(\gamma-1) K_C M_{12}^4}{1 - M_{12}^2}\right] \left[TTT_{3,2} + XX_2 TTT_{3,4}\right] S_{12} + M_{12} S_{12} \left(TTT_{1,2} + XX_2 TTT_{1,4}\right) + S_{12} \left(TTT_{4,2} + XX_2 TTT_{4,4}\right);$$

$$BB_3 = \left[\left(1 - \frac{(\gamma-1) K_C M_{12}^2}{1 - M_{12}^2}\right) + \frac{(\gamma-1) K_C M_{12}^4}{1 - M_{12}^2}\right] S_{12} TT_{3,3} + M_{12} S_{12} TT_{1,3} + S_{11} TT_{4,3},;$$

$$BB_4 = S_{10} M_{10}, BB_5 = S_{10},$$

$$BB_6 = C_{21} \left(S_{12} + S_{12} M_{12}^2\right) TTT_{1,1} + C_{21} \left[2 S_{12} S_{12} - S_{12} M_{12}^3 (\gamma - 1) K_C\right] TTT_{3,1} + C_{22} S_{11} TTT_{2,1},$$

$$BB_7 = C_{21}\left(S_{12} + S_{12}M_{12}^2\right)\left(TTT_{1,2} + XX_2 TTT_{1,4}\right)$$
$$+C_{21}\left[2S_{12}M_{12} - S_{12}M_{12}^3(\gamma-1)K_C\right]\left(TTT_{3,2} + XX_2 TTT_{3,4}\right)$$
$$+C_{22}S_{11}\left(TTT_{2,2} + XX_2 TTT_{2,4}\right),$$
$$BB_8 = C_{21}\left(S_{12} + S_{12}M_{12}^2\right)TTT_{1,3}$$
$$+C_{21}\left[2S_{12}M_{12} - S_{12}M_{12}^3(\gamma-1)K_C\right]TTT_{3,3} + C_{22}S_{11}TTT_{2,3},$$
$$BB_9 = S_{10}(1+M_{10}^2), \qquad BB_{10} = 2S_{10}M_{10} \qquad NNN_1 = BB_2 - AA_2 BB_4,$$
$$NNN_2 = AA_4 BB_2 - AA_2 BB_5, \qquad NNN_3 = AA_1 BB_2 - AA_2 BB_1,$$
$$NNN_4 = AA_3 BB_2 - AA_2 BB_3, \qquad NNN_5 = BB_7 + AA_2 BB_9,$$
$$NNN_6 = AA_4 BB_7 + AA_2 BB_{10}, \; NNN_7 = AA_1 BB_7 - AA_2 BB_6,$$
$$NNN_8 = AA_3 BB_7 - AA_2 BB_8. \tag{B2c}$$

Combining Eqs(B2a,b) into a matrix form yields

$$\begin{bmatrix} \bar{p}_{10} \\ \rho_o c_o \bar{u}_{10} \end{bmatrix} = \begin{bmatrix} TPOC_{1,1} & TPOC_{1,2} \\ TPOC_{2,1} & TPOC_{2,2} \end{bmatrix} \begin{bmatrix} \bar{p}_{14} \\ \rho_o c_o \bar{u}_{14} \end{bmatrix}, \tag{B3a}$$

where

$$TPOC_{1,1} = \frac{NNN_7 NNN_2 - NNN_3 NNN_6}{NNN_5 NNN_2 - NNN_1 NNN_6};$$
$$TPOC_{1,2} = \frac{NNN_8 NNN_2 - NNN_4 NNN_6}{NNN_5 NNN_2 - NNN_1 NNN_6}; TPOC_{2,1} = \frac{NNN_7 NNN_1 - NNN_3 NNN_5}{NNN_6 NNN_1 - NNN_2 NNN_5}$$
$$TPOC_{2,2} = \frac{NNN_8 NNN_1 - NNN_4 NNN_5}{NNN_6 NNN_1 - NNN_2 NNN_5}. \tag{B3b}$$

REFERENCE

[1] Davis, D. D., Stokes, J. M., Moorse, L., Theoretical and experimental investigation of mufflers with components on engine muffler design, *NACA Report*, pp. 1192 (1954).

[2] Prasad, M. G., A note on acoustic plane waves in a uniform pipe with mean flow, *Journal of Sound and Vibration*, 95(2), pp. 284-290 (1984).

[3] Prasad, M. G., Crocker, M. J., Studies of acoustical performance of a multi-cylinder engine exhaust muffler system, *Journal of Sound and Vibration*, 90(4), pp.491-508 (1983).

[4] Sullivan, J. W., Crocker, M. J., Analysis of concentric tube resonators having unpartitioned cavities, *Acous. Soc. Am.*, 64, pp. 207-215 (1978).

[5] Sullivan, J. W., A method of modeling perforated tube muffler components I: theory, *Acous. Soc. Am.*, 66, pp. 772-778 (1979).

[6] Sullivan, J. W., A method of modeling perforated tube muffler components II: theory, *Acous. Soc. Am.*, 66, pp. 779-788 (1979).

[7] Rao, K. N., Munjal, M. L., A generalized decoupling method for analyzing perforated element mufflers, Nelson Acoustics Conference, Madison (1984).

[8] Jayaraman, K., Yam, K., Decoupling approach to modeling perforated tube muffler component, *Acous. Soc. Am.,* 69(2), pp. 390-396 (1981).

[9] Munjal, M. L., Rao, K. N., Sahasrabudhe, A. D., Aeroacoustic analysis of perforated muffler components, *Journal of Sound and Vibration*, 114(2), pp. 173-88 (1987).

[10] Peat, K. S., A numerical decoupling analysis of perforated pipe silencer elements, *Journal of Sound and Vibration*, 123(2), pp. 199-212 (1988).

[11] Wang, C. N., A numerical scheme for the analysis of perforated intruding tube muffler components, *Applied Acoustics*, 44, pp. 275-286 (1995).

[12] Yeh, L. J., Chang, Y. C., Chiu, M. C., Application of genetic algorithm to the shape optimization of a constrained double-chamber muffler with extended tubes, *Journal of Marine Science and Technology*, 12(3), pp. 189-199 (2004).

[13] Chang, Y. C., Yeh, L. J., Chiu, M. C., Shape optimization on double-chamber mufflers using genetic algorithm, *Proc. ImechE Part C: Journal of Mechanical Engineering Science,* 10, pp. 31-42 (2005).

[14] Munjal, M. L., *Acoustics of ducts and mufflers with application to exhaust and ventilation system design*, John Wiley & Sons, New York (1987).

[15] Wang, C. N., *The application of boundary element method in the noise reduction analysis for the automotive mufflers*, Doctor thesis, Taiwan University (1992).

[16] Holland J., Adaptation in natural and artificial system. *Ann Arbor: University of Michigan Press* (1975).

[17] Jong D., An analysis of the behavior of a class of genetic adaptive systems, Doctoral Thesis, Department of Computer and Communication Sciences, *Ann Arbor,* University of Michigan (1975).

In: Sound Waves: Propagation, Frequencies and Effects ISBN 978-1-61470-001-2
Editors: V. Abagnali and G. Fabbri

Chapter 5

THE SOUND VELOCITY INTO TURBULENT FLOW

S. S. Rybanin
Institute of Problems of Chemical Physics RAS
Chernogolovka, Russia

Abstract

The sound velocity into the turbulent medium depends both on the sound velocity connecting with the usual molecular transport of the impulse and on the turbulent pulsation of the flow velocity too. Its value is higher than the usual sound velocity into the laminar medium as the turbulence transports the impulse additionally by its way. In this paper some generalization of the previous works touching with the problems of the sound waves propagation into turbulent medium is presented.

INTRODUCTION

The purpose of this paper is to summarize some investigations devoted to the problem of the definition and the calculation of the sound velocity into the turbulent flow [1]-[3]. This problem is important not only by itself but it is connected with the formulation of some principles of the choice of the stationary velocity of the detonation of explosives from the infinite numbers which are possible from point of view of the conservation lows. For the stationary detonation for which the flow of its products is laminar, this problem was been solved yet in previous centuries by Chapman [4] and Jouguet [5] in accordance with them the detonation velocity should have minimal possible value allowed by the conservation lows [4] and the detonation products velocity relative to its front is equal to the sound one [5]. Due to instability of the detonation complex, that is, the shock wave and following after it burning, the flow of the detonation products is turbulent, that is, extremely non-stationary and has a pulsating nature. Taking into account this circumstance White [6] has formulated the averaged conservation equations for the case of isotropic turbulence and found the minimal value of the velocity of "turbulent" detonation demonstrating the correctness of Chapman's rule [4] only. The correctness Jouguet's rule [5] for the "turbulent" detonation was demonstrated by Rybanin [1] which could calculate the sound velocity into the turbulent flow. Thus, the fundamental problem of the detonation velocity selection into the turbulent flow was solved by means of the introduction of the conception of the sound velocity in the turbulent medium. It is possible to give some examples of other problems for which the study of the spread of the slow disturbances such as the sound waves into the turbulent medium is important too (see for instance references [6] and [7]). In this paper some generalization of the previous author's works [1], [2] touching with the problems of the sound waves propagation into turbulent medium is presented.

Conservation Equations for Turbulent Flow

The conservation equations which will be used for the analysis of the sound waves spread into non-stationary adiabatic flow are expressed in the tensor form as [7], [8]:

Conservation of mass:

$$\frac{\partial \rho}{\partial t} = -\frac{\partial (\rho U_k)}{\partial x_k}. \tag{1}$$

Conservation of momentum:

$$\frac{\partial (\rho U_i)}{\partial t} = -\frac{\partial (\delta_{ik} p + \rho U_i U_k)}{\partial x_k}. \tag{2}$$

Conservation of energy:

$$\frac{\partial}{\partial t}\left(\frac{\rho U_i^2}{2} + \rho E\right) = -\frac{\partial}{\partial x_k}\left[\rho U_k\left(\frac{U_i^2}{2} + E + pV\right)\right]. \tag{3}$$

where ρ is the density, t is the time, x_k is the spatial coordinate (k = 1, 2, 3), U_k is the k-component of the flow velocity, δ_{ik} is the symbol equal to 1 at $i = k$ and equal to 0 at $i \neq k$, p is the pressure, E is the specific internal energy, V is the specific volume ($V = 1/\rho$).

To describe the turbulent flow we shall assume that all physical parameters can be presented in the following traditional view [7], [8]: $A = \overline{A} + A'$ where the bar and the prime denote the time-average and fluctuating components, respectively. It is proposed as usually [8] that

$$\overline{A} = \frac{1}{\tau}\int_0^{\tau} A(t+\eta)d\eta, \tag{4}$$

where the time of averaging τ is proposed to be greater in comparison with the characteristic time of the turbulence.

We shall further assume that mean flow velocity $\overline{U_1}$ is parallel to the x_1-axis and the transverse derivatives of any time-averaged quantity vanish. Let the fluctuating velocity be $U_1' = U_2' = U_3' = u$ $(\overline{U_i'} = 0)$, and density ρ, pressure p, and temperature T be $\rho = \overline{\rho} + \rho'$, $p = \overline{p} + p', T = \overline{T} + T'$, respectively. Following White [6], in averaged equations only terms of second order in the velocity fluctuations are retained, and isotropic turbulence is assumed $(\overline{u_i u_j} = \frac{1}{3}\overline{u^2}\delta_{ij})$. The values $\overline{\rho' u_y}$ and $\overline{\rho' u_z}$, and by symmetry $\overline{\rho' u_x}$, are of higher order are hence neglected. It is the further assumed that p', ρ', and T' are the same order and that terms such as $\overline{T' u_i}$ may be neglected in the present approximation.

Then for the mean motion the conservation of mass, momentum, and energy along x_1-axis are expressed in the first approximation by

$$\frac{\partial \overline{\rho}}{\partial t} = -\frac{\partial(\overline{\rho}\overline{U_1})}{\partial x_1}, \tag{5}$$

$$\frac{\partial(\overline{\rho}\overline{U_1})}{\partial t} = -\frac{\partial}{\partial x_1}\left[\left(\overline{p} + \overline{\rho}\overline{u^2}\right) + \overline{\rho}\overline{U_1}^2\right], \tag{6}$$

$$\frac{\partial}{\partial t}\left(\frac{\overline{\rho}\overline{U_1}^2}{2} + \frac{3}{2}\overline{\rho}\overline{u^2} + \overline{\rho}\overline{E}\right) = -\frac{\partial}{\partial x_1}\left[\overline{\rho}\overline{U_1}\left(\frac{\overline{U_1}^2}{2} + \frac{3}{2}\overline{u^2} + \overline{E} + \left(\overline{p} + \overline{\rho}\overline{u^2}\right)\overline{V}\right)\right]. \tag{7}$$

The equations (5)-(7) can be interpreted as one describing a medium having the effective` pressure $\overline{p_\Sigma}$ equal to

$$\overline{p_\Sigma} = \overline{p} + \overline{\rho}\overline{u^2} \tag{8}$$

and the effective specific internal energy $\overline{E_\Sigma}$ equal to

$$\overline{E_\Sigma} = \overline{E} + \frac{3}{2}\overline{u^2}. \tag{9}$$

Using the nomenclatures (8)-(9) the equations (6)-(7) can be rewritten as

$$\frac{\partial(\overline{\rho}\overline{U_1})}{\partial t} = -\frac{\partial}{\partial x_1}\left[\overline{p_\Sigma} + \overline{\rho}\overline{U_1}^2\right], \tag{10}$$

$$\frac{\partial}{\partial t}\left(\frac{\overline{\rho}\overline{U_1}^2}{2} + \overline{\rho}\overline{E_\Sigma}\right) = -\frac{\partial}{\partial x_1}\left[\overline{\rho}\overline{U_1}\left(\frac{\overline{U_1}^2}{2} + \overline{E_\Sigma} + \overline{p_\Sigma}\overline{V}\right)\right]. \tag{11}$$

Thus, after introducing the effective pressure (8) and the effective specific internal energy (9) the conservation equations put on the classic form.

It follows from Eqs. (5) – (7) that for the adiabatic flow when $\frac{dS}{dt} = 0$, where S is the specific entropy, the following correlation is valid

$$\frac{d\left(\overline{E} + \frac{3}{2}\overline{u^2}\right)}{dt} + \left(\overline{p} + \overline{\rho}\overline{u^2}\right)\frac{d\overline{V}}{dt} = 0. \tag{12}$$

But simultaneously the basic thermo dynamical equation for the averaging parameters is valid in according with that for the adiabatic conditions we can describe

$$\frac{d\overline{E}}{dt} + \overline{p}\frac{d\overline{V}}{dt} = 0. \tag{13}$$

It can be shown from Eqs. (12) and (13) that there is the following equation describing the dependence the velocity of the turbulent pulsation on time for the adiabatic conditions

$$\frac{d\left(\frac{3}{2}\overline{u^2}\right)}{dt} + \overline{\rho}\overline{u^2}\frac{d\overline{V}}{dt} = 0. \tag{14}$$

The Sound Velocity Into Turbulent Flow

Applying the standard methods [7] to calculate the sound velocity into the turbulent flow c_t from the conservation equations of mass and momentum (5), (10)-(11) we obtain

$$c_t = \left[\left.\frac{\partial \overline{p}_\Sigma}{\partial \overline{\rho}}\right|_S\right]^{\frac{1}{2}} = \left[\left.\frac{\partial\left(\overline{p} + \overline{\rho}\overline{u^2}\right)}{\partial \overline{\rho}}\right|_S\right]^{\frac{1}{2}}. \tag{15}$$

It follows from Eq. (14) that

$$\left.\frac{\partial(\overline{\rho}\overline{u^2})}{\partial \overline{\rho}}\right|_S = \frac{5}{3}\overline{u^2}. \tag{16}$$

Taking into account Eq. (16) we obtained from Eq. (15) the final expression for the calculation of the sound velocity c_t into the turbulent flow as

$$c_t = \left[\left. \frac{\partial \overline{p}}{\partial \overline{\rho}} \right|_S + \frac{5}{3}\overline{u^2} \right]^{\frac{1}{2}} = \left(c^2 + \frac{5}{3}\overline{u^2} \right)^{\frac{1}{2}}, \tag{17}$$

where $c = \left(\left. \frac{\partial \overline{p}}{\partial \overline{\rho}} \right|_S \right)^{\frac{1}{2}}$ is the sound velocity connecting with the usual molecular transport of the impulse.

It should be mentioned here that the similar relation for the calculation of the sound velocity into the turbulent flow was obtained in the work [3] though the authors of the sited paper used some different method for the averaging of the conservation equations.

CONCLUSION

Thus, the sound velocity into the turbulent medium c_t (17) depends on the sound velocity connecting with the usual molecular transport of the impulse c and on the turbulent pulsation of the flow velocity u too. It is higher than the usual sound velocity c into the laminar medium as the turbulence transports the impulse additionally by its way. Physically the result obtained is quite obvious and should be valid for more complex models of turbulence.

REFERENCES

[1] Rybanin, S. S. *Combust. Explo. Shock Waves*. 1966, vol. 2, 29-35.

[2] Rybanin, S. S. *On the Theory of Non-classical Detonation*. PhD thesis, Moscow Physical and Technical Institute, Dolgoprudnij, SSSR, 1966.

[3] Trofimov V. S.; A. N. Dremin *Fizika Goreniya I Vzriva*. 1966, vol. 3, 19-30.

[4] Chapman, D. L. *Philos. Mag.* 1899, vol. 47, 90-104.

[5] Jouguet, E. *Mathem. J.* 1904, vol. 6, 5.

[6] White, D. R. *Phys. Fluids*. 1961, vol. 4, 465-480.

[7] Landau, L. D.; Lifshits *Hydrodynamic*; Nauka, Moscow, 1986.

[8] Hinze, J. O. *Turbulence*; McGraw-Hill Book Company, Inc., New York Toronto London, 1959.

In: Sound Waves: Propagation, Frequencies and Effects ISBN 978-1-61470-001-2
Editor: V. Abagnali and G. Fabbri

Chapter 6

INFRASOUND GENERATION BY TURBULENT CONVECTION

Mariam Akhalkatsi and Grigol Gogoberidze*
Institute of Theoretical Physics, Ilia State University
3/5, Cholokashvili Ave., Tbilisi 0162, Georgia
Center for Fusion, Space and Astrophysics
Department of Physics, The University of Warwick
Coventry CV4,7AL, United Kingdom

Abstract

Variety of artificial and geophysical processes radiate infrasound, acoustic waves with frequencies below the normal limit of human hearing. Infrasonic signals are manifestations and/or precursors of extreme geophysical events or anthropogenic processes and, due to the property of infrasound to travel for great distances relatively undiminished, could be quite useful in advanced warning and monitoring purposes. Several universities and institutions around the world have research programs for studying acoustic generation and propagation mechanisms as well as for development of instrumentations and techniques for observation. Observations conducted over the last decades revealed that strong convective storms, such as supercells, that have cloud tops greater than 14 km or create a hook echo and are capable of producing strong tornadoes, generate significant infrasound in a passband from 0.5 to 2.5 Hz, with peak frequencies between 0.5 and 1 Hz, substantially before $(0.5-1$ hrs) tornado appearance. Broad and smooth spectra of observed infrasound radiation indicates that turbulence is the most promising sources of the radiation. In this chapter we review properties and characteristics of atmospheric infrasound waves. We study acoustic radiation from turbulent convection using Lighthill's acoustic analogy and taking into account the effects of stratification, temperature fluctuations and moisture in the air. It is shown that in saturated moist air turbulence in addition to the Lighthill's quadrupole and dipole sources of sound (related to stratification and temperature fluctuations), there exist monopole sources related to heat and mass production during the condensation of moisture. We determine the acoustic power of these monopole sources and show that radiation of a monopole source related to the nonstationary heat production

*E-mail address: gogober@geo.net.ge

during the condensation of moisture is dominant for typical parameters of strong convective storms. The results are in good qualitative agreement with the main observed infrasound characteristics e.g. total acoustic power and characteristic frequency. We perform spectral analysis of this source and give quantitative explanation of the high correlation between intensity of infrasound generated by supercell storms and later tornado formation. It is shown that low lifting condensation level (LCL) and high values of convective available potential energy (CAPE), which are known to favor significant tornadoes, also lead to a strong enhancement of supercells low frequency acoustic radiation. This qualitative analysis indicate the potential for infrasonic detection systems to determine potentially tornadic storms and improve tornado forecast.

PACS 43.28.Ra, 43.28.Dm, 92.60.Qx.
Keywords: Infrasound, Aeroacoustics, Turbulent convection, Moisture, Supercell, Tornado.

1. Introduction

Sound, the pressure wave which propagate through compressible media such as air or liquid and is also transmitted through solids, is composed of large range of frequencies, but humans typically can perceive only limited spectrum of acoustic wave energy. Sound waves having frequencies between approximately 20 Hz and 20 kHz are normally audible to humans, although these limits are not strictly defined. Younger humans can hear some high-pitched sounds. Under ideal conditions and at very high volume human listener will be able to perceive acoustic waves as low as 12 Hz.

The frequencies higher than the upper limit (20 kHz) of human hearing characterize the ultrasound regime and sound with frequencies below the minimum normal frequency of human audibility (20 Hz) is called infrasound. Many animals can detect and produce sounds beyond the human limits. Some animals (dogs, cats, dolphins, bats and mice) can hear ultrasound and bats use ultrasound's property of revealing details about the inner structure of the medium for orientation purposes. Animals use infrasonic waves going through the earth by natural disasters for detecting danger. Some of them (whales, elephants, hippopotamuses, rhinoceros, giraffes, okapi, and alligators) are known to use infrasound for long-distance communication and migrating birds might have capability of using naturally generated infrasound, from sources such as turbulent airflow over mountain ranges, as a navigational aid.

More precisely, frequency range just below 20 Hz (the threshold where human hearing and feeling cross over) is known as near-infrasound and frequencies between 0.01 Hz to 1 Hz define the range of infrasound. Below about 0.01 Hz (acoustic cut-off frequency) atmospheric waves are designated acoustic/gravity waves to indicate that gravity as well as pressure acts as a restoring force for fluid oscillations. These waves are no longer exactly longitudinal and at lower frequencies (around 0.002 Hz) they transform into internal atmospheric-gravity waves.

Infrasound is radiated from a variety of geophysical and man-made processes, each having different signature, frequency ranges, intensity and duration. Civilization sources include subsonic airplanes, supersonic aircrafts flying through the sound barrier (sonic booms), spacecraft launches, explosions (both chemical and nuclear), large industrial fires and gas exhausts, machinery such as diesel engines and wind turbines. These artificial infrasound signals usually vastly differ from natural continuous or pulsed signals generated by earthquakes (when the earth's tectonic plates are displaced), avalanches (when changes of state between ice and liquid occur), severe weather (turbulent convection, tornadoes, lightning, microbursts and sprites), volcanic eruptions (turbulence, explosions and other processes), standing ocean waves (microbaroms generated by non-linear ocean-wave interactions), meteors (from sonic booms, when entering the atmosphere at supersonic speeds and from explosions), airflow over rough topography (mountains), geomagnetic activity (correlated with solar flares and magnetic storms), polar aurora, solar eclipses (from displacement of large volumes of air), turbulence aloft, calving of icebergs and glaciers, tsunamis, forest fires, waterfalls and coastal surf. Most of geophysical infrasound signals are already well identified.

Infrasound waves having amplitudes exceeding the sensitivity threshold of the human organism to infrasound (about $0.2 - 0.6$ Pa) are known to be biologically effective and induce negative physiological and psychological effects in humans (Gavreau 1968; Delyukov and Didyk 1999). Experiments show that infrasound can cause nervous feelings of revulsion or fear, anxiety, extreme sorrow and chills, uneasiness or disorientation and feelings of pressure on the chest. Low frequency pressure oscillations can influence human mental activity by causing significant changes in attention and short-term memory functions, performance rate, mental processing flexibility and even could be partly responsible for meteorosensitivity in humans (Delyukov and Didyk 1999). Tandy and Lawrence (1998) suggested that infrasonic vibrations of 19 Hz, very close to the resonant frequency of the eyeball, can cause odd sensations and produce an optical illusion that people attribute to supernatural events such as ghosts.

2. General Characteristics of Atmospheric Infrasound

Infrasonic signals may provide valuable information on some important anthropogenic and a range of geophysical processes on a global scale and could be quite useful in advanced warning and monitoring of extreme geophysical events due to the property of infrasound to travel for great distances relatively undiminished. Audible sound is strongly attenuated by atmospheric viscosity and thermal conduction. 90 per cent of the energy of a 1 kHz signal is absorbed after traveling 7 km at sea level (Cook 1962), but 1 Hz infrasonic signals are detectable on the distance of about 3000 km from the source and can have amplitudes in the range of 10^{-2} Pa to 10^{2} Pa at the receiver, due to the reason that the classical acoustic absorption coefficient is proportional to the square of the wave frequency.

Earth's atmosphere consists of several layers each with specific characteristics such as temperature or composition. Air pressure and density decrease with altitude in the atmosphere, however temperature has a complicated profile, with the sequences of negative and positive temperature gradients, separated by narrow boundaries of constant temperature. The vertical temperature profile determines four principal layers of Earth's atmosphere, troposphere, stratosphere, mesosphere and thermosphere (from bottom to top) and the fifth main layer is the outermost layer called exosphere. Several other layers are distinguished by other properties.

Infrasound waves propagate with the speed of sound, which depends on the composition, physical properties and conditions of the medium the waves pass through and is a fundamental property of the material. The speed of sound in gases also depends on temperature and in 20°C air at the sea level is approximately 343 m/s. Different composition and temperature of atmospheric layers as well as upper-atmospheric winds affect the effective sound speed in the atmosphere and this dependence is described by the following expression (Gossard and Hooke 1975)

$$c_{seff} = \sqrt{\gamma_g r T} + \mathbf{n} \cdot \mathbf{u}, \tag{1}$$

where γ_g is the ratio of specific heats, r is the gas constant and $\mathbf{n} \cdot \mathbf{u}$ is the projection of the wind velocity $\mathbf{u}$ on the direction of wave propagation.

As a result, the propagation of sound waves strongly depends on the structure of the atmosphere. Temperature changes with altitude and upper atmospheric winds act as a waveguide by refracting the acoustic ray paths and trap much of the acoustic energy, which allows acoustic waves to travel great distances along the earth's surface (Georges and Beasley 1977). Positive temperature gradients in the upper atmosphere bend waves back to earth. For example, the waves propagating at elevation angles (from horizontal) less than about 60° are bent back to the earth's surface by positive temperature gradients near 30 and 100 km altitudes (Georges and Beasley 1977). Upper tropospheric and stratospheric wind component along the sound ray path alters their geometry by changing the wave's phase velocity and transverse components change the azimuthal direction the waves travel (Georges and Beasley 1977). Influenced by these two effects, acoustic waves bounce back and forth many times between the earth's surface and atmospheric layers, propagate almost horizontally and are detectable hundreds to thousands kilometers away from the source region.

Dependence of infrasound propagation speed on upper atmospheric winds and temperature profile results also in the formation of the acoustic audibility and silence zones on the ray path (Kulichkov 1992; Jones *et al.* 2004). In addition, seasonal changes, tides, and both planetary and internal gravity waves as well as diffraction and scattering from the irregularities in the middle atmosphere, significantly affect the conditions and characteristics of infrasound propagation (Kulichkov 2003).

Decrease of temperature with atmospheric height in the lower and middle atmosphere causes decrease of refractive index. Due to this reason, infrasound waves of natural or artificial origin are focused to propagate upward from the source and the main part of acoustic energy is transferred to the upper atmosphere (Lastovicka 2006). As a result, infrasound waves are greatly affected by high-altitude winds, which refract their ray paths, as it is mentioned above, but some waves escape from middle atmosphere and a significant amount of infrasonic energy is transmitted up to great heights (above 300 km) in the ionosphere. Due to this feature infrasound is considered to be more efficient in coupling between lower and upper atmosphere and in transferring energy to ionospheric heights than other type of waves - planetary, tidal and gravity waves (Lastovicka 2006). Ionosphere is a magnetized plasma environment, partially ionized by solar radiation, which envelopes the earth and forms the interface between the atmosphere and space. Upward propagating waves affect the ionosphere, mainly the most persistently ionized F2-layer at $250 - 400$ km and might be responsible for observed variations and anomalies, which cannot be explained by solar and geomagnetic contributions (Rishbeth 2006). Hickey (2001) calculated that infrasonic waves can quite significantly heat the thermosphere and ionosphere. Except of large disturbances during geomagnetic storms, ionospheric effects of infrasonic waves are detectable in the upper atmosphere by ground-based Doppler sounders that observe the motions of the ionized layers. They are distinguished from the ionospheric "background" by their monochromatic appearance and short periods, in contrast to the longer periods and irregular appearance of most background fluctuations (Georges 1973; Sindelarova *et al.* 2009). Spectral features of infrasound observed in the ionosphere differ from surface observations due to atmospheric filtering and resonant interactions between infrasound waves and the atmospheric temperature structure. The later can cause, for example, fine-structure peaks and narrow-band characteristics in ionospheric observations of severe thunderstorm broad-band infrasound (Jones and Georges 1976).

3. History of Infrasound Studies

The first observations of naturally occurring infrasound that was ever recorded using instruments were in the aftermath of the 1883 eruption of the Krakatoa volcano in Indonesia, the most severe violent volcanic explosion on Earth in modern times. The explosion was heard throughout the area and beyond, over 1/3rd of the earth's surface, with reports of roar or cannon-like sounds being heard nearly 3.000 miles (4.828 km) from its point of origin (Pierce and Posey 1971). Infrasonic waves from the tremendous explosions of Krakatoa circled the globe at least seven times, shattered windows hundreds of miles away and were recorded by barographs throughout the world. Barometric records demonstrated for the first time the ability of low frequency sound to propagate for thousands of kilometers. The eruption of Krakatoa triggered a series of interdisciplinary geophysical studies of the pressure signals produced by volcanos.

A huge meteor exploded presumably 8 km above the earth's surface in Siberia near the Tunguska River in 1908 and destroyed the taiga forest at a distance of twenty five kilometers around the epicenter. The Tunguska impact, which was technically an air burst explosion rather than a ground collision, had the force of a multi-megaton nuclear blast. Seismic waves from Great Siberian Meteor were registered in Russia and Europe on seismic stations and produced fluctuations in atmospheric pressure were strong enough to be detected by then recently invented barographs (recording aneroid barometers) (Shaw and Dines 1904) in Britain. Whipple (1930) analyzed recordings made by microbarographs (high resolution barographs) and published the first microbarograms (recordings of the barometric pressure over time).

The study of long-range sound characteristics at long distances from different sources started in 19th century and in 20th century this phenomenon was used for exploring the temperature and wind profiles of the middle and upper atmosphere (Fujiwara 1914; Lindemann and Dobson 1922; Whipple 1939). Data obtained from acoustic methods revealed strong high-altitude winds and the regions of increased temperature in the upper atmosphere, which explained the observed refraction of sound in the upper atmosphere and zones of audibility and inaudibility caused by the characteristics of atmospheric temperature and wind profiles (Whipple 1935).

Since the early 1950s, after World War II, studies of infrasound waves of both natural and man-made origin have become important in developing the methods for detecting, monitoring and classifying atmospheric nuclear explosions at great distances and global infrasonic monitoring network was set in many countries. Infrasound research included development of observational techniques, such as the methods of infrasound recordings and source energy estimation, data analysis for identifying valid signals from nuclear tests and wind noise reduction methods. Theoretical models were developed to determine the intensity and spectrum of sound waves of different origin and to predict infrasound propagation in the atmosphere. Infrasonic stations with improved microbarographs and measurement systems recorded acoustic, acoustic-gravity and Lamb waves from nuclear tests and also revealed wide variety of geophysical and artificial sources of infrasound (Donn et al. 1963; Donn and Posmentier 1967; Posmentier 1967).

In 1963 the Partial Test Ban Treaty (PTBT) was signed by the Soviet Union, the United States and the United Kingdom, prohibiting nuclear testing in the atmosphere, oceans and space, which resulted in diminished interest in infrasound science and technology. In early 1970s, after the deployment of satellite-based nuclear detection systems, the science of atmospheric infrasound was mainly focused on understanding the origin and structure of natural infrasound. Although, since 1996, construction of infrasound monitoring stations has become a part of the Comprehensive Nuclear Test Ban Treaty (CTBT) global alarm system and infrasound monitoring is now one of the four technologies used by the International Monitoring System (IMS) along with seismological measurements, hydroacoustics, radionuclide measurements. Revival of scientific interest in infrasonic technology is mainly related to the ability of infrasound studies to identify the location of the atmospheric nuclear explosion. Although, since underground nuclear explosions also generate infrasound waves, combined use of both the infrasound and the seismological technologies allows for better information gathering and analysis of possible underground tests as well. IMS infrasonic network stations are distributed uniformly over the surface of the globe and located in a wide variety of environments, where possible well away from sources of infrasonic background noise in order to improve the signals reception (Christie and Campus 2010). Specialized software detects infrasound signals, categorizes and identifies the most significant infrasound signals at each individual station of the global IMS infrasound network. Vast amount of monitoring data from this network are received by the Preparatory Commission for the Comprehensive Nuclear-Test-Ban Treaty Organization (CTBTO), than continuously transmitted via satellite or virtual private network (VPN) to the International Data Center (IDC) in Vienna, Austria, stored and referenced in the IDC database, automatically processed and analyzed in near real time by specialized software and automatic event bulletin is produced together with seismic, hydroacoustic and radionuclide technologies. The final and fully operational configuration of global IMS infrasound network will consist of 60 array stations situated strategically in 35 countries around the world and will be much more sensitive and far larger system than any previously operated network. The IDC infrasound data are also very useful for scientific applications and could expand knowledge and help in identifying of various types of infrasound sources for use in geophysical hazard-warning systems.

4. Instrumentation and Processing Techniques

Over the years improved instruments, optimized systems and processing software for infrasonic detection and identification have been developed and theoretical studies of low frequency sound generation mechanisms and dynamics have been conducted. Parallel application of different remote observing sensors (e.g. Doppler radar, satellite, seismic) with infrasonic systems provides increased potential for collecting important information concerning various anthropogenic and geophysical processes, detection, monitoring and warning of hazardous geophysical phenomena and testing theoretical models of infrasonic generation mechanisms. Several universities and a number of institutions around the world have infrasonic research programs for studying acoustic propagation and generation mechanisms as well as for development of instrumentations and array processing techniques.

The fundamental principles for development and deployment of infrasonic detection and monitoring instruments and technology have been studied for many years (Haubrich 1968). The design and construction of an proper infrasound monitoring system depends on a large number of factors, including the number and configuration of the array of detectors, sensitivity of the infrasound detectors at all frequencies of interest, coherence of signals between sensors, techniques for accurate estimate of signal azimuth for use in source location algorithms, automatic signal-detection and incoming data processing software as well as efficient background noise reducing systems.

Modern infrasound pressure sensors (microbarographs) at professional infrasound observatories have the ability to detect frequencies less than 10^{-3} Hz and pressure variations down to 10^{-5} Pascals (Whitaker and Mutschechner 1997). Pressure sensitivity of these detectors at infrasound frequencies is comparable to the human ear sensitivity at audio frequencies, which has the threshold of about $2 \cdot 10^{-5}$ Pa. Detection threshold limits of infrasound pressure sensors are determined by background noise, such as wind noise, turbulence or weather-related changes (Bedard and Georges 2000). Advances in infrasound monitoring technology and signal array processing resulted in lower detection thresholds, more accurate location estimates, increased confidence in signal parameters retrieved and improved ability to distinguish between various signal types.

Typical infrasound waves are characterized by space and time coherence and maintain a similar wave-form as they propagate over the array of detectors, show little or no change in either amplitude or frequency and after traveling distances of tens or hundreds of wavelengths have nearly plane wave fronts (Bedard *et al.* 2004a). On the other hand, nonacoustic pressure fluctuations are highly uncorrelated in space and increase in intensity at lower frequency. Array processing using these features of infrasound improves detectability of infrasound and allows reliable detection of even small-amplitude infrasonic signals by filtering wind and turbulent noise with efficient spatial filters and wind-noise reducing systems.

Infrasound observatory typically consists of array of several elements about 100 m on a side. In each element infrasound waves arrive at the ports on the twelve porous hoses, which extend from the center capsule and are designed to reduce and filter wind-generated noise, while long-wave infrasonic signals are not affected by this filter (Bedard *et al.* 2004a). Hoses connect the sensitive barometer, microbarometer, in capsule to the outside air. Microbarometer measures with high precision changes in the air pressure that are produced by coherent response of infrasonic waves over all the ports, while incoherent pressure fluctuations are averaged over space. A greater number of array elements results in a higher signal-to-noise ratio by reducing noise through signal summation, therefore arrays with larger number of microbarometers are usually built in areas that are exposed to strong winds. Apart from background noise reduction, an array of detectors allows to determine the direction of infrasound wave by time delays of arriving signals between each pair of elements over the array. An additional development of a wind barrier ("eddy fence") around the array raises the atmospheric boundary layer and breaks up the wind shear layer into small, random eddies (Bedard *et al.* 2004a). This method provides all-weather detection capability of infrasonic observatory.

5. Infrasound from Strong Convective Storms

In the 1970s Geoacoustics Group at National Atmospheric Administration (NOAA) in the USA found a connection between low frequency pressure waves and distant severe weather systems. Two kind of waves, wavelike fluctuations in the ground-based radio soundings of the ionospheric F region and ground level infrasonic pressure fluctuations recorded with arrays of sensitive microbarographs, displayed similar phenomenologies and were associated with severe convective storms (Jones and Georges 1976; Georges 1973). It was found that spectral features of observed ionospheric infrasound, such as fine-structure peaks and narrow-band characteristics, differ from broad-band infrasound observations on the ground. This differences were explained by atmospheric filtering and resonant interactions between infrasound waves and the atmospheric temperature structures (Jones and Georges 1976). Georges (1973) assumed that ionospheric and surface infrasound waves are different manifestations of the same broad band emission. Observations revealed that strong convective storms with cloud tops greater than 14 km produce significant infrasound having frequencies between 0.1 and 0.5 Hz and amplitudes up to about 0.5 Pa with mean duration of about two hours. Georges (1973) estimated the acoustic power to be as high as 10^7 watts. Although, emitted acoustic power is only small portion of convective storm's energy budget.

The most severe kind of convective storm is a supercell, strong thunderstorm with a deep, continuously rotating updraft (a mesocyclone), continuous downdraft and persistent updraft-downdraft couplet. Of the four classifications of convective storms (supercell, single-cell, multi-cell, and squall line), supercells are potentially the most dangerous, the largest and quasi-steady-state storms as they can last for many hours and generate extreme weather by producing copious amounts of large and damaging hail, torrential, flood-producing rainfall, strong straight-line winds, substantial downbursts, frequent lightnings and tornadoes. They are most frequent in the Great Plains of the United States, northeastern India and eastern Australia, which are the continental mid-latitude areas that lie to the east of a mountain range and poleward of a warm ocean. Supercells occasionally occur also in Central Europe (Sindelarova *et al.* 2009) and many other mid-latitude regions.

Supercells develop from the tilting of the horizontal vortices associated with the high vertical shear of the strong environmental winds. Strong updrafts lift the air turning about a horizontal axis and cause this air to turn about a vertical axis. This forms the rotating updraft, mesocyclone, usually $2 - 10$ km across. Cloud top of this severe storm can break through the troposphere and reach into the lower levels of the stratosphere. Strong environmental winds at the top of updraft blow out the cooled air and precipitation is not falling through the updraft. Thus, downdrafts and updraft become separated. This is the reason of the supercells being quasi-steady and severe.

Supercells usually produce tornadoes, the most violent of the atmospheric storms. Tornado is an intensively rotating column of air, about 100 m in diameter, with wind speeds of approximately 130 m/s. Only few types of clouds typically spawn tornadoes and, although, only 30 per cent of supercells produce tornadoes, they are responsible for almost all violent, large and long-lived tornadoes, which cause serious damage to inhabitants. Advance warning of tornadoes is extremely important issue for scientists. Satellite data and Doppler weather radar observations are used in order to determine the structure of storms and their potential to cause severe weather. Pulse-Doppler radar, a modern weather radar, which uses the Doppler effect to determine the relative velocity of objects, detects storm location and measures the radial velocity of the winds in a storm. Thus, this radar can detect the rotation in storms from more than several kilometers. The "hook echo" appendage on weather radar, produced by precipitation wrapped around the supercell because of rotating updraft, indicates the presence of a mesocyclone and also the great possibility of supercell being tornadic. Whereas the mesocyclone signatures are well detected by Doppler radar, for distant storms only high areas are observable and radar data resolution also decreases with distance. Consequently, tornado may develop more quickly than radar can detect meteorological situations leading to tornadogenesis. Apart from that, whereas the development and meteorological conditions conducive to formation of thunderstorms and mesocyclones are fairly well understood, the exact mechanisms by which tornadoes form in supercell storms (tornadogenesis) is still not well known and difficulties remain with distinguishing between tornadic and non-tornadic mesocyclones. Detection of infrasound from supercells could be useful for severe weather and tornado advanced warning and monitoring, especially for long ranges, where radar scan capability is limited, and between radar scan intervals.

Several sources of infrasound are associated with strong convective storms and accompanying meteorological events. In addition to infrasound generated by mesocyclone, low frequency waves are detected from microbursts, lightning, sprites and tornadoes.

Microburst is an intense very localized column of cold air, less than 4 km in diameter, with outward winds of up to 75 m/s. It develops from thunderstorm-generated downdrafts, descends and accelerates within minutes and when microburst reaches the surface, a ring vortex forms around it, which radially spreads away from the point of impact with the ground. This divergence of the wind, unlike convergent winds in a tornado, is the signature of the microburst. They are usually short-lived (couple of seconds), but scale suddenness of a microburst makes it a serious hazard for aviation, especially for aircrafts at low altitude during takeoff and landing, due to the low-level wind shear caused by its gust front. Hardin and Pope (1989) estimated the spectrum of microburst infrasonic radiation to be in the range 220 Hz and it can be detected at relatively small distances from the thunderstorm.

Thunderstorm lightning, an atmospheric electrical discharge, generates pulse-like infrasound waves in a passband from 0.5 to 20 Hz (Dessler 1973; Beasley *et al.* 1976; Few 1985) with the maximum range for the detection of about 50 km. Lightning stroke initiates a sonic shock and produces the acoustic remnant, thunder. Infrasound radiated by lightning is a low-frequency portion of the acoustic spectrum of thunder, which has the acoustic spectral peaks in audible range (Bass 1980). Greatest electrical activity, and, consequently, short-lived infrasound emission, is usually associated with the mature stages of storm formation (Beasley *et al.* 1976).

Sprites are high-altitude transient, luminescent electromagnetic events occurring between the tops of some thunderstorms and the upper atmosphere. They generate high-frequency short and impulsive infrasound signatures (Liszka 2008).

Proto-tornadic structures, funnel clouds and tornadoes radiate infrasound in the frequency range from about 1 to 20 Hz. Based on coincident radar measurements of tornados, which show a strong relationship between funnel diameter and infrasound frequency, it is usually supposed that these infrasound waves are generated by tornadic vortex. The possible generation mechanisms include radial vibrations of the vortex core (Abdullah 1966), the corotation of suction vortices (Powell 1960; Georges 1976; Mitchell *et al.* 1992), turbulence in the vortex boundary layer (Tatom *et al.* 1995) and Rossby waves of a tornado-like vortex (Bedard and Georges 2000; Schecter *et al.* 2008). Bedard (2005) eliminated boundary layer turbulence as possible mechanism, because the typical frequency of the tornado/boundary interaction is higher than observed infrasound frequency and, in addition, tornado infrasound is observed from concentrated regions of rotation aloft. Acoustic waves produced by two corotating vortices tend to be significantly lower (less than 0.1 Hz), than observed $0.1 - 10$ Hz frequency range, but sufficient number suction vortices can produce higher frequency infrasound. The radial vibration model proposed by Abdullah (1966) predicts the fundamental frequency of radially symmetric radial vibration to be inversely proportional to core radius:

$$f = \frac{207}{a}, \quad (2)$$

where a is the radius of maximum winds measured in meters (Bedard and Georges 2000). A vortex with radius of about 200 m will produce a frequency of 1 Hz. Nicholls *et al.* (2004) conducted simulation with an a strong vortex of 60 m/s and obtained 1 Hz infrasound by perturbing vortex from cyclostrophic balance.

Schecter *et al.* (2008) simulated tornado-like vortex with Regional Atmospheric Modeling System (RAMS) and showed that Rossby waves (caused by radial gradient of basic-state axial vorticity) of a sufficiently intense tornados, with maximum wind speeds exceeding a modest threshold, can generate 0.110 Hz infrasound, but their production can be inhibited because of strong suppression of Rossby waves by eddy viscosity or critical layers in a monotonic vortex. Thus, the prevailing source of tornado infrasound, best consistent with infrasonic data, is not yet resolved (Schecter *et al.* 2008).

It has to be noted that in addition to acoustic radiation, severe weather systems generate much lower frequency gravity waves (Bowman and Bedard 1971) as well as Lamb waves, thermal acoustic waves from energy release of growing thunderstorm, and these waves could be valuable indicators of storm processes (Pielke *et al.* 1993; Nicholls and Pielke 1994a; Nicholls and Pielke 1994b; Nicholls and Pielke 2000; Bedard 2005).

To be effective, the warning system must give early enough warning of extreme storm phenomena. The capability of advance warning of tornadoes by Doppler radar and infrasonic observatory is limited to 10 minutes lead time if the goal is the detection of a funnel or tornado vortex, because of a number of factors. The tornado touchdown time after funnel forms is only several minutes. In addition, the radar measurements depend on a distance to the tornado, the type and orientation of the funnel and obstacles blocking radar detection.

Inability in discriminating between tornadic and non-tornadic supercells is also a gaps in the current Doppler radar capabilities. Warning value of tornado vortex infrasound detection has fundamental limitation due to the relatively slow speed of acoustic wave propagation. At an average transit speed of about 300 m/s, the infrasonic waves take about 5 min to travel 100 km (Georges and Greene 1975).

However, infrasound from storm systems often precedes an observed tornado by up to an hour (Georges and Greene 1975; Bedard *et al.* 2004a; Bedard *et al.* 2004b). The evidence that infrasound is emitted well before the tornado formation could help to avoid the effects of propagation delay and increase warning lead time and detection accuracy of severe weather events. The potential value of infrasound detection as an aid in predicting tornadogenesis depends on determining what signals mean in terms of storm structure, tornado evolution and tornadogenesis and identifying the correlation between low frequency infrasound signals from supercell storms and later tornado formation. Combination of this study with the infrasonic measurements could increase the potential in receiving useful information from an infrasound detecting system for discriminating between supercells that produce tornadoes and those that do not. Together with climatological studies of thunderstorms this information can help in identifying environments supportive of significant tornadoes and reduce false alarms from non-tornadic supercells.

Optimized operational infrasonic system consists of more than one observatory in order to determine not only a direction of infrasonic signal arrival, but also to triangulate and locate the apparent source of the emissions. The continuous Infrasound Network (ISNet) developed by the National Oceanic Atmospheric Administration's (NOAA) Environmental Technology Laboratory (ETL) in Boulder, Colorado, USA, includes three acoustic arrays, at Boulder Atmospheric Observatory in Erie, CO and National Weather Service offices at Goodland, KS and Pueblo, CO. This infrasonic system operates continuously, requires no storm chasing and covers the region of the central United States known as Tornado Alley. Tornado Alley is located on the high plains between the Appalachian and Rocky Mountains, where severe convective thunderstorms and extreme meteorological events, including tornadoes are common through any time between March and November, but are concentrated in the spring, and peak in activity in the late afternoon and evening. This is mostly due to the unique geography of the continent that allows for frequent collisions of warm and cold air to form the right pre-existing conditions for severe weather. Advanced warning of severe weather events is extremely important issue and responsibility of the National Weather Service.

Infrasound measurements conducted by ISNeT operation and collocated with a Doppler radar provide a unique dataset for comparing infrasonic measurements with well-observed storm kinematics. Infrasonic systems developed for ISNeT to measure infrasound originating from regions of severe weather operate at frequencies in the range 0.5 to 10 Hz and regional range scales of hundreds of kilometers or less (Bedard *et al.* 2004a). In contrast, past global monitoring networks observations were focused on acoustic frequency passbands below 0.5 Hz because of the low-frequency content of large nuclear explosions (typically below about 0.2 Hz) (Bedard *et al.* 2004b) and the arrays were designed for detections at continental scales of thousands of kilometers.

Bedard *et al.* (2004b) summarized and analyzed data of all significant infrasonic signals collected in 2003 from severe thunderstorms during the ISNeT operation, in parallel with data from the past measurements. They made comparisons between infrasonic and Doppler radar measurements for over one hundred cases with a variety of signal types. These studies revealed that almost all strong convective storms that have cloud tops greater than 14 km or create a hook echo and which are capable of producing hail and strong tornadoes generate significant infrasound in a passband from 0.5 to 2.5 Hz, with peak frequencies between 0.5 and 1 Hz (e.g. Bedard 2005; Bedard *et al.* 2004a). Bedard (2004a) concluded that radiation of infrasound in this passband is not a natural consequence of all thunderstorms. Infrasound radiation of a tornadic thunderstorm is much stronger than infrasound of a non-severe weather system and this radiation is strongly correlated with formation of tornadoes by supercell storms (Georges 1973). Although, infrasound emission in this frequency range is not related with tornado itself and is caused by convective processes that precede tornado formation. Therefore, mesocyclones may be a primary sources for this radiation (Szoke *et al.* 2004). Summarizing differences between predicted acoustic signal arrival times from reported tornados and start times of infrasonic detection Bedard *et al.* (2004b) also confirmed that infrasound is usually produced substantially before ($0.5-1$ hrs) reports of tornadoes and that other sound generation processes, not related to tornadic vortices, could be responsible for infrasound emission in a passband from 0.5 to 2.5 Hz. They also provided statistics comparing the infrasonic bearing sectors and the tornado vortex location and concluded that, although the storm environment wind and temperature gradient could be responsible for bearing deviations, there remains the possibility that another storm feature, not related to tornadic vortices, could radiate infrasound from another location within the storm (Bedard 2005).

Over the years, several potential sound generation mechanisms were compared with measured characteristics of infrasound (Georges and Greene 1975; Georges 1976; Beasley *et al.* 1976; Bedard and Georges 2000; Akhalkatsi and Gogoberidze 2009). Such mechanisms include release of latent heat, dipole radiators, boundary layer turbulence, lightning, electrostatic sources, vortex sound (radial vibrations and the co-rotation of suction vortices) and turbulence (Lighthill's quadrupole source, dipole radiation related to temperature inhomogeneities and monopole sources related to the moisture of the air). Georges (1976) eliminated many sources as likely candidates and concluded that vortex sound is the most likely model. Bedard (2005) also found that the radial vibration model (Abdullah 1966) is most consistent with infrasonic data. But infrasound from proto-tornadic structures, funnel clouds and tornadoes in the frequency range from about 1 to 20 Hz is radiated after the vortex is already formed.

As it was mentioned above, other storm related generation mechanisms also can not explain infrasound emission in a passband from 0.5 to 2.5 Hz substantially before ($0.5-1$ hrs) reports of tornadoes. The typical frequency of the vortex boundary layer turbulence is higher than observed infrasound frequency. Infrasonic signals from lightning or electrostatic sources have short, impulsive nature and show no correlation with 0.5 to 2.5 Hz frequency range of tornadic storm.

Nicholls *et al.* (2004) performed a numerical simulation of a non-supercell tornado storm, using Regional Atmospheric Modeling System (RAMS) and suggested that the occurrence of high frequency infrasound coincided with the development of considerable small-scale turbulence that may had caused small-scale latent heating fluctuations, although they did not find any significant contribution of latent heat release.

Broad and smooth spectra of observed infrasound emission from strong convective storms indicate that turbulence is the most promising sources of this radiation. Turbulent convection in a stratified, moist atmosphere initiates several acoustic sources: Lighthill's quadrupole source, dipole radiation related to temperature inhomogeneities, monopole sources related to nonstationary heat and mass production during the condensation of moisture, dipole source related to stratification, the dipole and quadrupole sources related to inhomogeneity of background velocity profile and monopole source related to variability of adiabatic index. All these sources will be discussed in detail later, here the main characteristics are briefly discussed.

Lighthill's quadrupole source of sound is provided by the interaction of turbulent vortices. The acoustic power of Lighthill's source was estimated by Proudman (1952) and for the infrasound radiation from convective storms this study usually leads to the underestimation of the acoustic power and requires extraordinary high characteristic velocity of the turbulence (Georges and Greene 1975; Gossard and Hooke 1975). Schecter *et al.* (2008) also concluded that the amplitude of the infrasound (at 5 km) can exceed the estimated 0.25 Pa threshold of nonsevere weather value only if the Mach number of the turbulence is sufficiently high for the amplitude of the radiation pressure field, i.e. the characteristic velocity of the turbulence is greater than about 40 m/s, when the characteristic length scale is less than a few hundred meters, which would be extraordinary in any terrestrial storm system.

The dipole radiation related to temperature inhomogeneities (thermo-acoustic source) is of the same order as radiation of Lighthill's quadrupole source (Akhalkatsi and Gogoberidze 2009). The dipole source related to stratification as well as the dipole and quadrupole sources related to inhomogeneity of background velocity are very inefficient sources of sound (Akhalkatsi *et al.* 2004). The dipole source related to stratification is weaker then acoustic power of Lighthills quadrupole source (Akhalkatsi *et al.* 2004) and the monopole source related to variability of adiabatic index has negligible acoustic power (Howe 2001).

Akhalkatsi and Gogoberidze (2009) showed that for low Mach number turbulent convection and typical parameters of strong convective storms, infrasound radiation should be dominated by a monopole acoustic source related to the nonstationary heat production during the condensation of moisture. The total power of this source is of order 10^7 watts, two orders higher than thermo-acoustic and Lighthill's quadrupole radiation power, and is in qualitative agreement with observations of strong convective storms (Bowman and Bedard 1971; Georges 1973; Georges and Greene 1975; Bedard and Georges 2000;). The acoustic power of the source related to nonstationary gas mass production is negligible compared to the radiation initiated by heat production.

Infrasound waves emitted by monopole source related to heat production during the condensation of moisture interact with the atmosphere and could be attenuated by various processes, but also could be farther amplified while traveling though the saturated moist region of supercell storm. Naugolnykh and Rybak (2008) showed that the presence of a finite amplitude infrasound waves causes modulation of the dynamic stability of a supersaturated vapor containing water droplets and, consequently, condensation in the region of reduced stability. The linear resonance interaction of sound waves with modulated heat release can lead to effective amplification of sound waves at low frequencies of interest.

Theoretical results require future verification by realistic numerical simulations of infrasound radiation from tornadic thunderstorm as well as field measurements of acoustic radiation from severe thunderstorms. Schecter *et al.* (2008) simulated the infrasound radiation from a single-cell non-tornadic thunderstorm in a shear-free moist environment. The dominant infrasound in the $0.1-10$ Hz frequency band appeared to radiate from the vicinity of the melting level, where diabatic processes involving hail were active and this passband accounted for most of the structure of the simulated thunderstorm signal during the period of measurement. Although, the barely resolved $0.52.5$ Hz infrasound from the simulated thunderstorm was an order of magnitude lower than the corresponding infrasound that is observed from severe weather. They suggested that severe weather signals would be an order of magnitude higher than the simulated thunderstorm signal and speculated that a larger and more violent storm, perhaps one that includes a tornado, would be necessary to reproduce the observed $0.52.5$ Hz infrasound.

Observed high correlation between intensity of low frequency infrasound signals from supercell storms and the probability of later tornado formation will be qualitatively explained below. It will be showed that acoustic power of a monopole source related to the moisture of the air strongly depends on the same parameters that are the most promising in discriminating between nontornadic and tornadic supercells according to the recent study of tornadogenesis (Markowsky and Richardson 2009). Particularly, low lifting condensation level (LCL) and high values of convective available potential energy (CAPE), which are known to favor significant tornadoes, also lead to a strong enhancement of supercells low frequency acoustic radiation. This qualitative analysis indicate the potential for infrasonic detection systems to determine potentially tornadic storms. ISNeT data combined with the information from Doppler Radar may help to improve tornado forecast and reduce false alarms from non-tornadic supercells.

6. Lighthill's Acoustic Analogy

Lighthill's acoustic analogy (Lighthill 1952) represents the basis for understanding of the sound generation by turbulent flows. In this section formulation of the acoustic analogy is introduced, Lighthill's quadrupole source is discussed and mathematical methods used for farther analysis are considered.

Aeroacoustics is concerned with sound generated by aerodynamic forces or motions originating in a flow rather than by the externally applied forces or motions of classical acoustics. Thus, the sounds generated by vibrating violin strings and loudspeakers, i.e. produced by the vibration of solids, fall into the category of classical acoustics, whereas sound generated by the unsteady aerodynamic forces on propellers or by turbulent flows fall into the domain of aeroacoustics. The airflow may contain fluctuations as a result of instability. At low Reynolds numbers, when viscous forces are larger than inertial forces (laminar flow) these fluctuations give a regular eddy pattern which is responsible for the sound produced by musical wind instruments. At high Reynolds numbers inertial forces are dominant and initial fluctuations result in an irregular turbulent motion (turbulent flow) which is responsible for the roar of the wind and of jet aeroplanes. The theory of aerodynamic sound is concerned with pressure fluctuations that occur far from the source where the amplitude of the motion is small and the effects of compressibility and finite propagation speed of the disturbances are important. This region is called acoustic field. The pressure (and density) fluctuations are weak in this region and satisfy the acoustic wave equation.

The study of flow-generated acoustic waves began with Gutin's theory of propeller noise, which was developed in 1936. He obtained a theoretical expression for the sound produced by a propeller in static operation as a function of tip speed, number of blades, thrust and torque, and the dimensions of the propeller, which was valid at distances large compared with the propeller diameter.

Lighthill (1952, 1954) developed a theory for the sound radiated into free space and thus he neglected neighboring resonators and all effects of reflection, diffraction, absorbtion or scattering by solid boundaries. Ignoring the influence of boundaries on the production of sound as opposed to the production of vorticity reduces the aerodynamic sound problem to the study of mechanisms that convert kinetic energy of rotational motions into acoustic waves involving longitudinal vibrations of fluid particles.

Due to the non-linearity of the governing equations it is very difficult to predict the sound production by fluid flows. This occurs typically for flows with high Reynolds numbers, for which non-linear inertial terms in the equation of motion are much larger than the viscous terms. Lighthill introduced the idea of calculating the far-field sound generated by unsteady flow with an acoustic analogy to deal with the problem of jet noise. The Lighthill's idea provides an approximation by assuming that the source term is in some sense known or that it can at least be modeled in an approximate fashion. In the lighthill's analogy, the fully nonlinear problem is taken to be analogous to the problem of sound propagating in a linear acoustic medium at rest subject to an external forcing that represents the turbulent source. Lighthill reformulated the Navier–Stokes equation into an exact, inhomogeneous wave equation whose source terms are important only within the turbulent (vortical) regions. Sound is expected to be such a very small component of the whole motion that, once generated, its back-reaction on the source region may then be determined by neglecting the production and propagation of the sound.

There are two principal source types in free nonsaturated vortical flows: a quadrupole, whose strength is determined by the unsteady Reynolds stress and a dipole, which is important when mean mass density variations occur within the source region. Proudman (1952) derived an equation for the radiated acoustic power per unit mass of the quadrupole source:

$$N \sim M^5, \tag{3}$$

where M is the turbulent Mach number. In Proudman's (1952) analysis, the equation for acoustic power was derived assuming Gaussian statistics with normal joint probability distributions for the turbulent velocities and their first two time derivatives. As was pointed out by Lighthill, his equations imply that there is an exact analogy between the density fluctuations in any real flow and those produced by a quadrupole source in an ideal (non-moving) acoustic medium.

Lighthill's acoustic analogy would be inappropriate if the Mach number is large enough for compressibility to be important in the source flow, when the source flow is coupled to a resonator, such as an organ pipe, when solid boundaries or when bubbles are present in the case of liquids. His ideas were subsequently extended by Curle (1955), Powell (1960), and Ffowcs Williams and Hall (1970) to include the effects of solid boundaries. These extensions include Gutin's (1936) analysis for propeller noise and, in fact, provide a complete theory of aerodynamically generated sound that can be used to predict blading noise as well as jet noise.

While Lighthill's strategy turned out to be remarkably successful in predicting the gross features of the sound radiation from turbulent air jets, engine manufacturers needed a much more sensitive tool with the capability of predicting how even relatively small changes in the flow would affect the radiated sound. This resulted in a number of attempts to improve the Lighthill approach. Early efforts were focused on accounting for mean flow interaction effects. Phillips (1960), Lilley (1974), and many others rearranged the Navier–Stokes equations into the form of an inhomogeneous convective or moving-medium wave equation rather than the inhomogeneous stationary-medium wave equation originally proposed by Lighthill. But these methods appeared to be incapable of predicting the changes in the sound field that occur when noise suppression devices are deployed and therefore couldn't be used to evaluate the acoustic performance of these devices.

Goldstein (2003) rewrote the Navier–Stokes equations into the general set of linearized inhomogeneous Euler equations (in convective form) but with modified dependent variables. The source terms are exactly the same as those that would result from externally imposed shear-stress and energy-flux perturbations and the equations are therefore exactly the same as the Navier–Stokes equations, but with the viscous stress perturbation replaced by an appropriate Reynolds stress and the heat flux perturbation replaced by an appropriate enthalpy flux. The "basic flow" about which the equations are linearized can be any solution to a very general class of inhomogeneous Navier–Stokes equations with arbitrarily specified source strengths. His method put the classical approaches to the jet noise problem on a more rational basis and also extended in new directions. The rewritten Navier–Stokes equations remained nonlinear, but the nonlinearity was effectively contained in the generalized Reynolds stresses and enthalpy flux - which also contained contributions from the base-flow sources.

The acoustic-analogy-type approaches and their extensions roughly correspond to treating the generalized stresses and enthalpy flux as known source terms that can be estimated or modeled. This doesn't imply that acoustic analogy equations can provide an unambiguous identification of sources, except for the generalized incompressible flows. These equations are only useful when the "base flow" is reasonably close to the actual fluid motion and, in most cases, can only serve as a guide for identifying and ultimately modeling the apparent sources of sound. Since the sound is just a by-product of all the processes occurring in the flow, it is highly unlikely that "true" sound sources can be identified in any realistic turbulent motion.

Validation of Lighthill's acoustic analogy for studding the sound generation by turbulent flows has been shown by various experiments and numerical simulations (Whitmire and Sarkar 2000; Seror *et al.* 2001; Freund 2003; Panickar *et al.* 2005 and references therein), that are based on combined analytical-modeling method.

The most obvious approach to obtain meaningful predictions of the far field sound radiated by turbulence would be to use large-scale numerical simulation, i.e. DNS (direct numerical simulation). The number of mesh points needed to fully resolve any turbulent flow is proportional to the Reynolds number Re based on the characteristic length scale of the turbulent eddies raised to the nine-fourths power. But since typically Re is of the order of 10^5 to 10^7, this means that 10^{12} to 10^{15} grid points would be needed to resolve all of the relevant length scales. This implies that computing the far filed sound by DNS on a very large computational domain which includes both the turbulent source and the acoustic far field is unfortunately very expensive and problematic for even relatively simple flows.

An alternative strategy is to calculate the sound using a hybrid approach in which the turbulence is computed using a method such as DNS or LES (large eddy simulation), and the far field sound is calculated using an acoustic analogy. In various studies this method was used to calculate the sound from turbulence and compare acoustic-analogy predictions with theoretical and experimental results. Sarkar and Hussaini (1993) computed the sound from decaying isotropic turbulence using a hybrid DNS/Lighthill acoustic-analogy approach. Witkowska *et al.* (1995) also computed the sound from isotropic turbulence for forced and unforced cases using both DNS and LES to evaluate the turbulent source in the Lighthill acoustic analogy. Lilley (1994) derived an alternative analytical method of determining the radiated acoustic power per unit mass of the Lighthill's quadrupole source and evaluated his analytical results using statistics of the Lighthill source obtained from the DNS databases of Sarkar and Hussaini (1993) and Lilley (1994). These studies show that the hybrid acoustic-analogy method can be used to compute the acoustic source and obtain sound radiated by isotropic turbulence.

Validation of various forms of the acoustic analogy for different flow configurations have been performed by comparing the sound calculated from direct computations or exact analytical solutions with acoustic-analogy predictions. Mitchell *et al.* (1992) and Colonius *et al.* (1994) studied the sound radiated by compressible co-rotating vortex pair and the scattering of sound waves from a compressible viscous vortex, respectively. Colonius *et al.* (1994) validate the Lilley acoustic analogy for a forced, two-dimensional, compressible shear layer by comparing DNS results with acoustic-analogy predictions. Mitchell *et al.* (1997) validate the Lighthill acoustic analogy by comparison with DNS results for axisymmetric, nonturbulent subsonic and supersonic jets. In these studies the emphasis was to investigate the sound from large coherent structures rather than the effects of smaller turbulence scales on the radiated sound. Bastim *et al.* (1995) calculated the sound from a subsonic turbulent plain jet using the hybrid approach. Freund (1999) performed a DNS of a jet with Mach number equal to 0.9 and Reynolds number equal to 3600 and analyzed the acoustic sources in the jet. Whitmire and Sarkar (2000) computed sound from a turbulent flow using DNS and compared their results with acoustic-analogy predictions. They considered a three-dimensional region of forced turbulent flow with a small turbulent Mach number so that the source is spatially compact (i.e. the turbulence integral scale is much smaller than the acoustic wavelength). Seror *et al.* (2001) studied the problem of the estimation of the noise by forced isotropic turbulence using hybrid LES/Lighthill analogy approach. Freund (2003) computed turbulent statistics that are relevant to jet noise modeling using a previously validated simulation database of a cold jet with Mach number $M = 0.9$. Panickar *et al.* (2005) examined instability mode switching in various supersonic jet configurations that involve resonant acoustics (situations where flow instabilities are enhanced by feedback).

All these studies verified the ability of the Lighthill acoustic analogy to predict sound generated by a three-dimensional turbulent source containing many length and time scales.

Lighthill's method of calculating the aerodynamic emission of sound waves in a homogeneous medium was extended by Stein (1967) to calculate the acoustic and gravity-wave emission by turbulent motions in a stratified atmosphere. In the solar convective region the characteristic size of turbulent eddies is considered to be comparable to the scale height of the stratification produced by gravity. In a stratified environment gravity as well as pressure acts as a restoring force for fluid oscillations, and two types of waves - acoustic and gravity waves occur, depending on the dominant restoring force. Stein's analysis showed, that the stratification cuts off the acoustic radiation at low Mach numbers and for typical parameters of the solar convective region gravity-wave emission is much more efficient than acoustic.

Goldreich and Kumar (1990) studied acoustic and gravity wave generation by turbulent convection in a plane parallel, stratified atmosphere that consists of two semi-infinite layers, the lower being adiabatic and polytropic and the upper being isothermal. They estimated efficiencies for the conversion of the convective energy flux into both trapped and propagating waves and calculated the total emissivities for the different wave types. Their theoretical results obtained for the amplitudes and linewidths of the solar p-Modes match the observational ones in the upper part of the solar convection zone. This agreement supports the hypothesis that the solar p-Modes are stochastically excited by turbulent convection.

6.1. Mathematical Formulation

Lighthill (1952) introduced acoustic analogy approach to calculate acoustic radiation from relatively small regions of turbulent flow embedded in an infinite homogeneous fluid in which the speed of sound and the density are constant. The dynamics of the flow is governed by the continuity and momentum equations:

$$\frac{D\rho}{Dt} + \rho \nabla \cdot \mathbf{v} = \rho q, \tag{4}$$

$$\rho \frac{D\mathbf{v}}{Dt} = -\nabla p + \nabla \sigma + f, \tag{5}$$

where $\mathbf{v}$, ρ and p are velocity, density and pressure respectively; σ is a viscous stress tensor. $D/Dt \equiv \partial/\partial t + \mathbf{v} \cdot \nabla$ is Lagrangian time derivative; An external volume flow source q within the fluid and an externally applied volume force f are added to continuity and Navier–Stokes equations respectively in order to clearly show the nature of different sources. It is assumed that these source terms cause no entropy production.

Using the continuity equation (4) we can obtain the momentum equation in conservation form:

$$\frac{\partial \rho v_i}{\partial t} + \frac{\partial (\rho v_i v_j)}{\partial x_j} = -\frac{\partial p}{\partial x_j} + \frac{\partial \sigma}{\partial x_j} + \rho q v_i + f_i. \tag{6}$$

Taking the time derivative of the mass equation (4) and subtracting from this the divergence of the momentum equation (6) we obtain:

$$\frac{\partial^2 \rho}{\partial t^2} = \frac{\partial^2 (\rho v_i v_j - \sigma_{ij})}{\partial x_i \partial x_j} + \frac{\partial^2 p}{\partial x_i^2} + \frac{\partial (\rho q)}{\partial t} - \frac{\partial (\rho q v_i + f)}{\partial x_i}. \tag{7}$$

Subtracting on both sides of this equation a term $c_s^2(\partial^2 \rho / \partial x_i^2)$ provides this equation the form of the inhomogeneous wave equation:

$$\frac{\partial^2 \rho}{\partial t^2} - {c_s}^2 \frac{\partial^2 \rho}{\partial x_i^2} = \frac{\partial^2 (\rho v_i v_j - \sigma_{ij})}{\partial x_i \partial x_j} + \frac{\partial^2 (p - c_s^2 \rho)}{\partial x_i^2} + \frac{\partial (\rho q)}{\partial t} - \frac{\partial (\rho q v_i + f)}{\partial x_i}, \tag{8}$$

where c_s is the speed of sound:

$$c_s \equiv \left(\frac{\partial p}{\partial \rho}\right)_s^{1/2}. \tag{9}$$

In general equation (8) is useless, as it is an equation with twelve unknowns (for a simple fluid σ_{ij} is symmetrical).

The key idea is to compare this equation with the equation for the perturbation of a uniform and stagnant fluid in the state (ρ_0, p_0). We consider an unsteady disturbance with characteristic length λ traveling at a propagation speed whose typical value is c_s through a fluid in which the velocity, pressure, and density are otherwise determined by the equations of a steady flow. Such a disturbance will introduce changes in velocity, pressure, density, entropy, and c_s^2 ($\mathbf{v}' = \mathbf{v} - \mathbf{v}_0, p' = p - p_0, \rho' = \rho - \rho_0, s' = s - s_0$) as it passes by a fixed observer. These changes will all occur on the time scale $T_p = 1/f$, where $f = c_s/\lambda$ is the characteristic frequency of the disturbance. We identify c_s as the speed of sound in the stagnant uniform fluid surrounding the listener. We can further define the perturbations ρ' and p' as the differences between the local values of ρ and p and the values of these quantities in the reference fluid surrounding the listener. In this generalization the amplitude of the disturbance measured be the magnitude of perturbations do not need to be small. As ρ_0 and p_0 are like c_s constant we can write Lighthill's equation (8) as:

$$\frac{\partial^2 \rho'}{\partial t^2} - {c_s}^2 \frac{\partial^2 \rho'}{\partial x_i^2} = \frac{\partial(\rho q)}{\partial t} - \frac{\partial(\rho q v_i + f)}{\partial x_i} + \frac{\partial^2 T_{ij}}{\partial x_i \partial x_j}, \tag{10}$$

where it is assumed that external volume flow source q does not involve any momentum injection and that the injected fluid has the same properties as the reference fluid (ρ_0, p_0). Hereafter the specific case with $\mathbf{v}_0 = 0$ is considered.

In Eq. (10)

$$T_{ij} = \rho v_i v_j + \delta_{ij}(p' - c_s^2 \rho') - \sigma_{ij} \tag{11}$$

is *Lighthill's turbulence stress tensor* and $\rho v_i v_j$ is the Reynolds stress. The second term is the excess of momentum transfer by the pressure over that in an ideal fluid of density ρ_0 and sound speed c_s. This is caused by wave amplitude nonlinearity and by mean density variations in the source flow. The viscous stress tensor σ_{ij} is linear in the perturbation quantities and properly accounts for the attenuation of the sound.

Higher order terms, having higher order derivatives aren't included in this equation, because they have negligible contribution, when far field approximation is used.

The left hand side of Eq. (10) is the wave operator of the homogeneous acoustic wave equation, which, in the absence of externally applied forces or moving boundaries, has only the trivial solution $\rho' = \rho - \rho_0 = 0$, because the radiation condition ensures that sound waves cannot enter from infinity. Lighthill's analogy implies an identification of the right hand side of this equation as a known source of sound. The sound generated in the real fluid may now be considered equivalent to that produced in an ideal, stationary acoustic medium that is forced by the source terms on the right hand side of Eq. (10). The problem of calculating the aerodynamic sound is therefore formally equivalent to solving this equation for the radiation into a stationary, ideal fluid produced by a distribution of source terms that vanish at large distances from the flow, i.e., the source region is very small relative to the wavelength of emanated sound. A source distribution satisfying this condition is said to be compact.

The first term on the right hand side of the wave equation is a monopole source of sound, which is produced by compact flow source and acts as if its entire strength were concentrated at a single point. The second term is a dipole source produced by an external volume force. This term can be treated as being composed of two equal-strength monopoles with opposite signs that have been brought together.

The third term, Lighthill's source, produces the acoustic field exactly equivalent to the emission of the quadrupole source, whose strength per unit volume is the Lighthill turbulence stress tensor T_{ij}. Quadrupole source can be thought as composed of two dipoles that are of equal strength but have opposite sign. Therefore, there is an exact analogy between the density fluctuations that occur in any real flow and the small amplitude density fluctuations that would result from a quadrupole source distribution (of strength T_{ij}) in a fictitious acoustic medium with sound speed c_s.

Lighthill's equation (10) is an exact consequence of the laws of conservation of mass and momentum and it must be satisfied by all real flows. Even for those flows that are sound-like T_{ij} accounts not only for the generation of sound, but also for all effects which occur whenever acoustic waves interact with moving flows (self-modulation due to acoustic nonlinearity, convection by the flows, refraction due to sound speed variations, and attenuation due to thermal and viscous actions) and which, therefore, can not be predicted without some knowledge of the sound field. Nonlinear effects on propagation and dissipation are usually sufficiently weak to be neglected within the source region, although they may affect propagation to a distant observer. Convection and refraction of sound within and near the source flow can be important, for example when the sources are contained in a turbulent shear layer, or are adjacent to a large, quiescent region of fluid whose mean thermodynamic properties differ from those in the radiation zone. Effects of this kind are accounted for by contribution to T_{ij} that are linear in the perturbation quantities relative to a mean background flow. Thus, a knowledge of T_{ij} is, in effect, equivalent to solving the complete nonlinear equations governing the flow motion, which is virtually impossible for most flows of interest.

Nevertheless, we are usually content with approximate indications of the acoustic field magnitude and suggestions about its dependance on parameters and have no need for its highly accurate predictions. Moreover, there are certain types of flows where it is often possible to obtain fairly good estimates of T_{ij} and, consequently, good estimates of the sound field. In addition, acoustic analogy approach allows us to utilize the powerful tools of classical acoustics. Lighthill's analysis regards the source terms as a quantities about which we have at least some prior knowledge. Since aerodynamic sound sources of practical interest are very often acoustically compact, the far field solutions of the Lighthill's equation will automatically account for the extreme inefficiency of these sources and will provide reasonable estimates of the acoustic field even when they are not precisely known.

Thus, practical utility of Lighthill's equation rests on the regarding the right side of the equation as known source terms that vanish at large distances from the flow and on the hypothesis that all the effects, which occur whenever acoustic waves interact with moving flows, can be ignored. Below the reasonableness of these assumptions is shown.

Within the subsonic turbulent flow of relatively small spatial extent embedded in a uniform stationary atmosphere viscous stress σ_{ij} makes a much smaller contribution to T_{ij} than the Reynolds stress term $\rho v_i v_j$, because the ratio of these terms is of the order of magnitude of the Reynolds number ul/ν, which in virtually all applications of aerodynamic noise theory is quite large.

At sufficiently large distances from the flow acoustic approximation implies that velocity v_i is small and Reynolds stress term $\rho v_i v_j$ is negligible. Moreover, the effects of viscosity and heat conduction only cause a slow damping due to the conversion of acoustic energy into heat and have a significant effect only over very large distances. Thus, σ_{ij} can be entirely neglected for distances of propagation comparable to the wavelength.

Functional relationship between the pressure p, density ρ and the specific entropy s (entropy per unit mass) is given by assuming that the fluid maintains itself in a state of local thermodynamic equilibrium (i.e., that relaxation effects can be neglected). Then, since any thermodynamic property can be expressed as a function of any two others, i.e., by an equation of state:

$$p = p(\rho, s). \tag{12}$$

In the differential form it yields:

$$\frac{Dp}{Dt} = c_s^2 \frac{D\rho}{Dt} + \left(\frac{\partial p}{\partial s}\right)_\rho \frac{Ds}{Dt}. \tag{13}$$

Whenever the flow emanates from a region of uniform temperature and the heat transfer, which is of the same order of magnitude as the viscous effects, is relatively unimportant, the reference fluid remains uniform and stagnant so that the entropy is relatively constant and the equation of state (13) can be written as

$$p' = c_s^2 \rho'. \tag{14}$$

These assumptions show that the Reynolds stress T_{ij} is approximately equal to $\rho v_i v_j$ inside the flow and approximately equal to zero outside this region. When the mean density and sound speed are uniform, the density fluctuations in ρ produced by low Mach number, high Reynolds number fluctuations are of order $\rho_0 M^2$, and, hence, they are negligible. Thus,

$$T_{ij} \simeq \rho_0 v_i v_j. \tag{15}$$

Since only a very small fraction of the energy in the flow gets radiated as sound, it is reasonable to suppose that the source terms can be determined from measurements or estimates of the turbulence, without any prior knowledge of the sound field. Then the right side of Lighthill's Eq. (10) can be treated as known source terms.

6.2. Lighthill's Quadrupole Source

In the previous subsection Lighthill's acoustic analogy was introduced and it was shown that the problem of predicting the sound emission from a region of unsteady flow embedded in a uniform atmosphere can be reduced to the classical problem of predicting the sound field from known sources of limited spatial extent. In the absence of external volume flow source q (monopole terms) and an external volume force f (dipole terms) the leading order term in the sound field produced by a free turbulent flow is the Lighthill's quadrupole term. In this section solution of Lighthill's Eq. (10) for Lighthill's source is introduced.

Using Eqs. (10) and (14) we can rewrite the Lighthill's equation for quadrupole source to obtain

$$\frac{1}{{c_s}^2}\frac{\partial^2 p'}{\partial t^2} - \frac{\partial^2 p'}{\partial x_i^2} = \frac{\partial^2 T_{ij}}{\partial x_i \partial x_j}, \tag{16}$$

where we use the pressure fluctuations p' as aero-acoustic variable.

The pulse emitted from an acoustic source located at the point $\mathbf{y}$ travels the distance $\mathbf{R}$ in the time $\mathbf{R}/c_s$. Thus, the time at which the signal arriving at the point $\mathbf{x}$ at the time t was emitted from the point $\mathbf{y}$ is represented by $t - (\mathbf{R}/c_s)$. It is called the *retarded time*.

The most convenient way to obtain a better approximation for the effect of retarded time differences across the source region is to use the free space Green function

$$G_0(t, \mathbf{x}|\tau, \mathbf{y}) = \frac{\delta(t - \tau - |\mathbf{x} - \mathbf{y}|/c_s)}{4\pi {c_s}^2 |\mathbf{x} - \mathbf{y}|}, \tag{17}$$

which is solution of the equation:

$$\left(\nabla^2 - \frac{1}{{c_s}^2}\frac{\partial^2}{\partial t^2}\right) G_0 = -\delta(t - \tau)\delta(\mathbf{x} - \mathbf{y}) \tag{18}$$

and has the symmetry property

$$\frac{\partial G_0}{\partial x_i} = -\frac{\partial G_0}{\partial y_i} \tag{19}$$

for differentiation with respect to the source coordinate $\mathbf{y}$ and the observer coordinate $\mathbf{x}$.

The solution of Lighthill's equation (16) expressed in terms of the free space Green's function is given by

$$p'(\mathbf{x}, t) \approx c_s^2 \int_{-\infty}^{t} \int G_0(t, \mathbf{x}|\tau, \mathbf{y}) \frac{\partial^2(\rho_0 v_i v_j)}{\partial y_i \partial y_j} d^3\mathbf{y} d\tau. \tag{20}$$

By partial integration we can move the space derivative from the source term $\partial^2(\rho v_i v_j)/\partial y_i \partial y_j$ (which we do not know accurately) toward the well known Green's function. Using the symmetry property (19) we can replace the derivatives of G_0 with respect to the source coordinate $\mathbf{y}$ by derivatives with respect to the observer coordinate $\mathbf{x}$. As the integration is performed on the source coordinate, we can move this spatial derivatives out of the integral, and obtain the integral formulation

$$p'(\mathbf{x}, t) \approx \frac{\partial^2}{\partial x_i \partial x_j} \int \frac{\rho_0 v_i v_j (\mathbf{y}, t - |\mathbf{x} - \mathbf{y}|/c_s)}{4\pi |\mathbf{x} - \mathbf{y}|} d^3\mathbf{y}, \tag{21}$$

where the integration with respect to τ is carried out and $t - |\mathbf{x} - \mathbf{y}|/c_s$ is the retarded time.

Assuming a compact source and the far field ($|\mathbf{x}| \gg |\mathbf{y}|$) approximation we can use the following expansions:

$$|\mathbf{x} - \mathbf{y}| \approx |\mathbf{x}| - \frac{\mathbf{x} \cdot \mathbf{y}}{|\mathbf{x}|}, \tag{22}$$

$$\rho_0 v_i v_j(\mathbf{y}, t - |\mathbf{x} - \mathbf{y}|/c_s) \approx \rho_0 v_i v_j \left(t - \frac{|\mathbf{x}|}{c_s}\right) + \frac{\mathbf{x} \cdot \mathbf{y}}{c_s |\mathbf{x}|} \frac{\partial}{\partial t} \rho_0 v_i v_j \left(t - \frac{|\mathbf{x}|}{c_s}\right), \tag{23}$$

and using plane wave approximation for far field derivatives

$$\frac{\partial}{\partial x_i} \approx -\frac{x_i}{c_s |\mathbf{x}|} \frac{\partial}{\partial t}, \tag{24}$$

for the Lighthill source we obtain

$$p'(x, t) = -\frac{\rho_0 x_i x_j}{4\pi c_s^2 |\mathbf{x}|^3} \frac{\partial^2}{\partial t^2} \int v_i v_j \, d^3\mathbf{y}. \tag{25}$$

The order of magnitude of p' can be estimated in terms of the characteristic velocity v and length scale l of energy containing turbulent eddies in the source region. Fluctuations in $v_i v_j$ in different regions of the turbulent flow separated by distances greater then l tend to be statistically independent, and therefore generation of sound can be considered as a collection of F/l^3 independent eddies, where F is the volume occupied by the turbulence. The dominant frequency of the motion is $\sim v/l$, so the wavelength of the radiated sound is $\lambda \sim l/M_t$, where $M_t \equiv v/c_s \ll 1$ is turbulent Mach number. Therefore, each eddy is acoustically compact. Acoustic pressure generated by single eddy is $p'_1 \sim (l/|\mathbf{x}|)\rho_0 v^2 M_t^2$, and the acoustic power it radiates is $N_1 \sim 4\pi |\mathbf{x}|^2 p_1'^2/\rho_0 c_s \approx \rho_0 v^3 l^2 M_t^5$, which corresponds to Lighthill's eighth power law. For total acoustic power this yields Proudman's estimate (Proudman 1952),

$$N \sim \frac{\rho_0 v^8}{l c_s^5} F. \tag{26}$$

This equation shows that acoustic output of the turbulent system strongly depends on the rms turbulent velocity (proportional to its eights power) and inversely proportional to the characteristic lengthscale of the energy containing eddies.

7. Infrasound Generation by Tornadic Storms

In this section we study acoustic radiation from turbulent convection using Lighthill's acoustic analogy and taking into account the effects of stratification, temperature fluctuations and moisture in the air. Formulation of the generalized acoustic analogy (Goldstein 2003) implies: (i) dividing the flow variables into their mean and fluctuating parts; (ii) subtracting out the equation for the mean flow; (iii) collecting all the linear terms on one side of equations and the nonlinear terms on the other side; (iv) treating the latter terms as the known terms of sound.

7.1. General formalism

The dynamics of the convective motion of moist air is governed by the continuity, Euler, heat, humidity and ideal gas state equations:

$$\frac{D\rho}{Dt} + \rho \nabla \cdot \mathbf{v} = 0, \tag{27}$$

$$\rho\frac{D\mathbf{v}}{Dt} + 2\rho\mathbf{\Omega}\times\mathbf{v} = -\nabla p - \rho\nabla\Phi, \tag{28}$$

$$T\frac{Ds}{Dt} = -L_\nu\frac{Dq}{Dt}, \tag{29}$$

$$\rho = \frac{p}{RT}\frac{1}{1-q+q/\epsilon} = \frac{p}{RT}\frac{1}{1+aq}, \tag{30}$$

where $\mathbf{v}$, $2\mathbf{\Omega}\times\mathbf{v}$, ρ and p are velocity, Coriolis acceleration, density and pressure respectively; $D/Dt \equiv \partial/\partial t + \mathbf{v}\cdot\nabla$ is Lagrangian time derivative; L_ν is the latent heat of condensation and q is the mass mixing ratio of water vapor (humidity mixing ratio)

$$q \equiv \frac{\rho_\nu}{\rho}, \tag{31}$$

where ρ_ν is the mass of water vapor in unit volume; $\epsilon \equiv m_\nu/m_d \approx 0.622$ is the ratio of molecular masses of water and air; $a = 0.608$ and R is the universal gas constant.

In the set of Eqs. (27)-(30) diffusion and viscosity effects are neglected due to the fact that they have a minor influence on low frequency acoustic wave generation and propagation.

In this analysis we also assume $\Omega = 0$, as it is well known (Bluestein 1992) that Coriolis effects are negligible for mesoscale convective system dynamics. On the other hand, when the frequency of acoustic waves (Ω_a) satisfy the condition $\Omega_a \gg \Omega$, Coriolis effects also have negligible influence on acoustic wave dynamics.

The main idea of Lighthill's acoustic analogy is reformulation of the governing equations in a form suitable for the study of acoustic wave radiation process. To proceed in this direction one has to choose an appropriate "acoustic variable", that describes acoustic waves in the irrotational regions of the fluid. The generalized Bernoulli's theorem (Batchelor 1967) suggests that the total enthalpy

$$B \equiv E + \frac{p}{\rho} + \frac{v^2}{2} + \Phi, \tag{32}$$

where Φ is gravitational potential energy per unit mass, E is internal energy and $\nabla\Phi \equiv -\mathbf{g}$, is one of the possible appropriate choices (Howe 2001). B is constant in a steady irrotational flow and at large distances from acoustic sources perturbations of B represent acoustic waves.

To derive the acoustic analogy equation in terms of the total enthalpy it is useful to rewrite Euler's equation in the Crocco's form as follows

$$\rho\frac{D\mathbf{v}}{Dt} + \nabla B = -\omega\times\mathbf{v} + T\nabla s, \tag{33}$$

where ω is vorticity, T is temperature, s is specific entropy and

$$Tds = dE + pd\left(\frac{1}{\rho}\right) = dB - \frac{dp}{\rho} - d\Phi - d\left(\frac{v^2}{2}\right). \tag{34}$$

From the thermodynamic identity

$$d\rho = \left(\frac{\partial\rho}{\partial p}\right)_{s,q} dp + \left(\frac{\partial\rho}{\partial s}\right)_{p,q} ds + \left(\frac{\partial\rho}{\partial q}\right)_{s,p} dq, \tag{35}$$

where the subscripts serve as the reminders of the variables held constant, using Eqs. (30) we obtain

$$d\rho = \frac{1}{c_s^2}\,dp - \frac{\rho}{c_p}\,ds + \frac{a\rho}{1+aq}\,dq, \tag{36}$$

where

$$c_s \equiv \left(\frac{\partial p}{\partial \rho}\right)_{s,q}^{1/2}, \tag{37}$$

is the sound velocity and

$$c_p \equiv T\left(\frac{\partial s}{\partial T}\right)_{p,q} \tag{38}$$

is the specific heat of the air.

Eliminating the convective derivative of the density from Eq. (27) using Eq. (36) we have that

$$\frac{1}{\rho c_s^2}\frac{Dp}{Dt} + \nabla\cdot\mathbf{v} = \frac{1}{c_p}\frac{Ds}{Dt} + \frac{1}{1+aq}\frac{Dq}{Dt}. \tag{39}$$

Subtracting the divergence of Eq. (33) from the time derivative of Eq. (39) and using Eq (29) after long but straightforward calculations we obtain

$$\left(\frac{D}{Dt}\left(\frac{1}{c_s^2}\frac{D}{Dt}\right) - \frac{\nabla p\cdot\nabla}{\rho c_s^2} - \nabla^2\right) B = S_L + S_T + S_q + S_m + S_\gamma, \tag{40}$$

where $\gamma \equiv c_p/c_v$ is the ratio of specific heats and

$$S_L \equiv \left(\nabla + \frac{\nabla p}{\rho c_s^2}\right)\cdot(\omega\times\mathbf{v}), \tag{41}$$

$$S_T 7 \equiv -\left(\nabla + \frac{\nabla p}{\rho c_s^2}\right)\cdot(T\nabla s), \tag{42}$$

$$S_q \equiv \frac{\partial}{\partial t}\left(\frac{\gamma T}{c_s^2}\frac{Ds}{Dt}\right) + (\mathbf{v}\cdot\nabla)\left(\frac{T}{c_s^2}\frac{Ds}{Dt}\right), \tag{43}$$

$$S_m \equiv \frac{\partial}{\partial t}\left(\frac{a}{1+aq}\frac{Dq}{Dt}\right), \tag{44}$$

$$S_\gamma \equiv p\frac{\partial\gamma}{\partial q}\left(\frac{\partial q}{\partial t}(\mathbf{v}\nabla)p - \frac{\partial p}{\partial t}(\mathbf{v}\nabla)q\right). \tag{45}$$

Eq. (40) is suitable for the identification of different acoustic sources and the study of their acoustic output.

The nonlinear wave operator on the left of Eq. (40) is identical with that governing the propagation of sound in an irrotational, homentropic flow. Therefore the terms on the right may be identified as acoustic sources. Propagation of infrasound in the atmosphere was intensively studied by different authors (Ostashev *et al.* 2001) and references therein) and will not be considered in this chapter.

To further simplify the analysis of the acoustic output of different sources we make several standard assumptions:

(a) For acoustic wave generation process at low Mach number flow, all the convective derivatives in Eq. (40) can be replaced by time derivatives $\partial/\partial t$ (Goldstein 1976);

(b) For acoustic waves with the wavelength λ not exceeding the stratification length scale

$$\lambda \leq H \equiv \frac{c_s^2}{g} \approx 10^4 \mathrm{m}, \tag{46}$$

one can also neglect the influence of stratification on the acoustic wave generation process and consider background thermodynamic parameters in Eq. (40) as constants (Stein 1967).

(c) Neglecting nonlinear effects of acoustic wave propagation and scattering of sound by vorticity and taking into account that $M \ll 1$, for the acoustic pressure in the far field we have

$$p'(\mathbf{x}, t) \approx \rho_0 B(\mathbf{x}, t). \tag{47}$$

(d) Eq. (40) is equivalent to initial set of Eqs. (27)-(30) and therefore it describes not only acoustic waves, but also the instability wave solutions that are usually associated with large scale turbulent structures and continuous spectrum solutions related to "fine-grained" turbulent motions (Goldstein 2003; Goldstein 1984). In the presence of any kind of inhomogeneity, such as stratification or velocity shear, linear coupling between these perturbations is possible, and in principle acoustic waves can be generated by both instability waves and continuous spectrum perturbations. But in the case of low Mach number ($M \ll 1$) flows both kinds of perturbations are very inefficient sources of sound. The acoustic power is proportional to $e^{-1/2M^2}$ and $e^{-\pi\delta/2M}$ for instability waves and continuous spectrum perturbations respectively (Crighton and Huerre 1990). In the last expression δ is the ratio of length scales of energy containing vortices and background velocity inhomogeneity ($V/\partial_z V$). In the case of supercell thunderstorms $M \sim 0.1-0.15$ and $\delta \sim 10^{-2}$, therefore both linear mechanisms have negligible acoustic output and attention should be payed to sources of sound related to nonlinear terms and entropy fluctuations that will be studied next.

With these assumptions Eq. (40) simplifies and reduces to

$$\frac{1}{\rho_0}\left(\frac{1}{c_s^2}\frac{\partial^2}{\partial t^2} - \nabla^2\right) p' = S_L + S_T + S_\gamma + S_q + S_m, \tag{48}$$

with

$$S_L \approx \nabla \cdot (\omega \times \mathbf{v}), \tag{49}$$

$$S_T \approx -\nabla \cdot (T \nabla s), \tag{50}$$

$$S_\gamma = p\frac{\partial \gamma}{\partial q}\left(\frac{\partial q}{\partial t}(\mathbf{v}\nabla) p - \frac{\partial p}{\partial t}(\mathbf{v}\nabla) q\right), \tag{51}$$

$$S_m \approx \frac{a}{1 + aq}\frac{\partial^2 q}{\partial t^2}, \tag{52}$$

$$S_q \approx -\frac{\gamma L_\nu}{c_s^2}\frac{\partial^2 q}{\partial t^2}, \tag{53}$$

The first three terms on the right hand side of Eq. (48) represent well known sources of sound: the first term represents Lighthill's quadrupole source (Lighthill 1952); the second term is a dipole source related to temperature fluctuations (Goldstein 1976); S_γ is a monopole source related to variability of adiabatic index, that usually have negligible acoustic output (Howe 2001) and will not be considered in the present chapter; Eq. (48) shows that in the case of saturated moist air turbulence there exist two additional sources of sound. S_q and S_m are monopole sources related to nonstationary heat and mass production during the condensation of moisture, respectively.

7.2. Analysis of Different Sources

For estimation of different acoustic sources we follow the standard procedure (Goldstein 1976; Howe 2001). Namely, using a free space Green function of the wave equation

$$G(t,t',\mathbf{x},\mathbf{x}') = \frac{\delta(t-t'-|\mathbf{x}-\mathbf{x}'|/c_s)}{4\pi c_s^2|\mathbf{x}-\mathbf{x}'|}, \tag{54}$$

acoustic pressure fluctuations corresponding to a source S_i can be written as

$$p_i'(x,t) = \frac{1}{4\pi c_s^2}\int \frac{[S_i]_{t=t_*}}{|\mathbf{x}-\mathbf{x}'|} d^3\mathbf{x}', \tag{55}$$

where $t_* = t - |\mathbf{x}-\mathbf{x}'|/c_s$.

It has to be noted that using a free space Green function we neglect the effect of the acoustic wave reflection from the ground. In general this is not correct approximation for the study of low frequency acoustic wave dynamics in the atmosphere, but because we are interested only in the total acoustic power of the atmospheric turbulence the neglect of this effect is an adequate approximation.

Calculating acoustic radiation in the far field ($|\mathbf{x}| \gg |\mathbf{x}'|$), we can use following expansions

$$|\mathbf{x}-\mathbf{x}'| \approx |\mathbf{x}| - \frac{\mathbf{x}\cdot\mathbf{x}'}{|\mathbf{x}|}, \tag{56}$$

$$S_\alpha(t_*) \approx S_\alpha\left(t - \frac{|\mathbf{x}|}{c_s}\right) + \frac{\mathbf{x}\cdot\mathbf{x}'}{c_s|\mathbf{x}|}\frac{\partial}{\partial t}S_\alpha\left(t-\frac{|\mathbf{x}|}{c_s}\right), \tag{57}$$

and using plane wave approximation for far field derivatives

$$\frac{\partial}{\partial x_i} \approx -\frac{x_i}{c_s|\mathbf{x}|}\frac{\partial}{\partial t}, \tag{58}$$

for the Lighthill source we obtain

$$p_L'(x,t) = -\frac{\rho_0 x_i x_j}{4\pi c_s^2|\mathbf{x}|^3}\frac{\partial^2}{\partial t^2}\int v_i v_j d^3\mathbf{x}', \tag{59}$$

which corresponds to quadrupolar radiation field.

p_L' can be estimated in terms of the characteristic velocity v and length scale l of energy containing turbulent eddies. Fluctuations in $v_i v_j$ in different regions of the turbulent flow separated by distances greater then l tend to be statistically independent, and therefore generation of sound can be considered as a collection of F/l^3 independent eddies, where F is the volume occupied by the turbulence. The dominant frequency of the motion is $\sim v/l$, so the wavelength of the radiated sound is $\lambda \sim l/M_t$, where $M_t \equiv v/c_s \ll 1$ is turbulent Mach number. Therefore, each eddy is acoustically compact. Acoustic pressure generated by single eddy is $p'_{L1} \sim (l/|\mathbf{x}|)\rho_0 v^2 M_t^2$, and the acoustic power it radiates is $N_{L1} \sim 4\pi|\mathbf{x}|^2 p'^2_{L1}/\rho_0 c_s \approx \rho_0 v^3 l^2 M_t^5$, which corresponds to Lighthill's eighth power law. For total acoustic power this yields Proudman's estimate (Proudman 1952),

$$N_L \sim \frac{\rho_0 v^8}{l c_s^5} F. \tag{60}$$

Similar arguments can be used for the estimation of the acoustic power of a thermo-acoustical source S_T related to density (and therefore temperature) fluctuations, producing a dipole source (Howe 2001). The physics of this kind of acoustic radiation is the following: "hot spots" or "entropy inhomogeneities" behave as scattering centers at which dynamic pressure fluctuations are converted directly into sound. The acoustic power is (Akhalkatsi *et al.* 2004):

$$N_T \sim \frac{\rho_0 \Delta T^2 v^6}{l T^2 c_s^3} F = \frac{\Delta T^2}{M_t^2 T^2} N_L, \tag{61}$$

where ΔT denotes the rms of temperature fluctuations.

Acoustic sources S_q and S_m are related to the moisture of the air. They produce monopole radiation and physically have the following nature: suppose there exist two saturated air parcels of unit mass with different temperatures T_1 and T_2 and water masses $m_\nu(T_1)$ and $m_\nu(T_2)$. Mixing of these parcels leads to the condensation of water due to the fact that

$$2m_\nu(T_1/2 + T_2/2) < m_\nu(T_1) + m_\nu(T_2). \tag{62}$$

The water condensation leads to two effects important for sound generation: production of heat and decrease in the mass of the gass. Both of these effects are known to produce monopole radiation (Goldstein 1976; Howe 2001). Consequently, turbulent mixing of saturated air with different temperatures will not only lead to dipole thermo-acoustical radiation (61), but also to monopole radiation.

According to Eqs. (31) and (62) for the humidity mixing ratio fluctuation q'_s we have

$$q'_s = q_s(T + T') + q_s(T - T') - 2q_s(T). \tag{63}$$

In the limit $T'/T \ll 1$ this yields

$$q'_s \approx \frac{\partial^2 q_s}{\partial T^2} T'^2. \tag{64}$$

Substituting (43) into (55) and using (64), (56) and (57) we obtain

$$p'_q(\mathbf{x}, t) = -\frac{\rho_0 \gamma L_\nu}{4\pi c_s^2 |\mathbf{x}|} \frac{\partial^2 q_s}{\partial T^2} \frac{\partial^2}{\partial t^2} \int T'(\mathbf{x}', t) T'(\mathbf{x}', t) d^3\mathbf{x}', \tag{65}$$

which corresponds to a monopole radiation field.

For total acoustic power radiated by a monopole source related to moisture we have

$$N_q = \frac{4\pi|\mathbf{x}|^2}{\rho_0 c_s}\langle p'(\mathbf{x},t)p'(\mathbf{x},t)\rangle \sim \frac{\rho_0\gamma^2 L_\nu^2}{c_s^5}\left(\frac{\partial^2 q_s}{\partial T^2}\right)^2$$
$$\times \frac{\partial^4}{\partial t^4}\int d^3\mathbf{x}' d^3\mathbf{x}'' \langle T'(\mathbf{x}',t)T'(\mathbf{x}',t)T'(\mathbf{x}'',t)T'(\mathbf{x}'',t)\rangle. \tag{66}$$

Fluctuations of temperature in different regions of the turbulent flow separated by distances greater then length scale l of energy containing eddies are not correlated and therefore the integral in Eq.(66) can be estimated as $F_1 l^3 \Delta T^4$, where F_1 is the volume occupied by saturated moist air turbulence. Also taking into account that the characteristic timescale of the process is the turn over time of energy containing turbulent eddies l/v we finally obtain

$$N_q \sim \frac{\rho_0\gamma^2 L_\nu^2 \Delta q^2 M_t^4}{l c_s} F_1 = \frac{\gamma^2 L_\nu^2 \Delta q^2}{M_t^4 c_s^4}\frac{F_1}{F} N_L, \tag{67}$$

where Δq is the rms of humidity mixing ratio perturbations. For the acoustic power of the source related to gas mass production we obtain

$$N_m \sim \frac{\rho_0 a^2 c_s^3 \Delta q^2 M_t^4}{l} F_1 = \frac{a^2\Delta q^2}{M_t^4}\frac{F_1}{F} N_L. \tag{68}$$

7.3. Application to infrasound generation by tornadic convective storms

In this subsection we apply our findings to study infrasound generation by tornadic convective storms. Taking typical parameters of supercell storms to be $v \sim 5\ \mathrm{ms}^{-1}$, $\Delta T \sim 3^\circ$ K (Gossard and Hooke 1975; Bluestein 1992), $T = 270^\circ$ K and $c_s = 330$ m/s and using Eqs. (60) and (61) we see that the dipole radiation related to temperature inhomogeneities is of the same order as the radiation of Lighthill's quadrupole source (Akhalkatsi *et al.* 2004).

Combining Eqs.(67)-(68) and using $L_\nu \approx 2.5 \times 10^6\ \mathrm{m^2 s^{-2}}$ and $\gamma \approx 1.4$ we obtain

$$\frac{N_q}{N_m} \approx \left(\frac{\gamma L_\nu}{c_s^2}\right)^2 \approx 10^3, \tag{69}$$

Therefore the acoustic power of the source related to gas mass production is negligible compared to the radiation related to heat production.

Estimation of Δq is a bit more difficult. For saturation specific humidity we use Bolton's formula (Bolton 1980)

$$q_s \approx \frac{3.8}{p_0}\exp\left(\frac{17.67T_c}{T_c + 243.5}\right), \tag{70}$$

where $T_c = T - 273.15$ is the temperature in degree Celsius and p_0 is atmospheric pressure in mb. Taking Eq. (64) into account and using $p_0 \approx 800$ mb, we obtain

$$\Delta q \approx \frac{6.8\cdot 10^4}{(243.5 + T_c)^4}\exp\left(\frac{17.67T_c}{T_c + 243.5}\right)\Delta T^2 \equiv f(T_c)\frac{\Delta T^2}{T^2}. \tag{71}$$

Note that due to the numerator in the exponent, $f(T_c)$ strongly depends on temperature, e.g., $f(T_C = 10^\circ)/f(T_C = 0^\circ) \approx 2$.

Using Eqs. (60) and (67) we obtain

$$\frac{N_q}{N_L} \approx \left[\frac{\gamma L_\nu}{c_s^2}\right]^2 \left[\frac{\Delta T}{M_t T}\right]^4 \left[\frac{F_1}{F}\right]^2 f^2(T_c). \tag{72}$$

For our analysis we have assumed that $T_c = 0°\mathrm{C}$, $(f(0) \approx 1.66)$ and $F \approx 125\ \mathrm{km}^3$ (Georges and Greene 1975) and for the estimation of F_1 we note that the saturation level of atmospheric convection is usually at the height $\approx 1 - 1.5\ \mathrm{km}$. Also taking into account that $f(T_c)$ rapidly drops with the decrease of T_c, one can expect that the main acoustic radiation will be produced at the heights $(1.5 - 4)\ \mathrm{km}$ and consequently we assume $F_1 \approx 0.5F$, then Eq. (72) yields

$$\frac{N_q}{N_L} \approx 2 \times 10^2. \tag{73}$$

Therefore, we conclude that infrasound radiation of a supercell storm should be dominated by monopole sources related to heat production during water condensation. Also assuming that the constant of proportionality in Eq. (60) is equal to 100 (Goldstein 1976; Georges and Greene 1975) and $l \approx 10\ \mathrm{m}$, then the total radiation power is

$$N_q \approx 2.4 \times 10^7\ \mathrm{watts}, \tag{74}$$

in qualitative agreement with observations (Bowman and Bedard 1971; Bedard and Georges 2000; Georges and Greene 1975; Georges 1973).

As was mentioned previously, the characteristic frequency of the emitted acoustic waves is $\Omega \sim v/l$, and using characteristic values of the velocity and length scale we obtain $\tau \sim 10$ s for the period.

7.4. Spectrum of infrasound

In this subsection we perform detailed spectral analysis of a monopole source S_q related to heat production during the condensation of moisture. Assuming homogeneous and stationary turbulence we calculate the spectrum of acoustic radiation.

Dropping all other source terms (48), apart from S_q, reduces to the inhomogeneous wave equation

$$\frac{1}{\rho_0}\left(\frac{1}{c_s^2}\frac{\partial^2}{\partial t^2} - \nabla^2\right) p' = S_q. \tag{75}$$

Using standard methods for spectral analysis of the inhomogeneous wave equation (Goldstein 1976; Howe 2001; Gogoberidze *et al.* 2007), after a long but straightforward calculation, for the spectrum of temperature fluctuations $I(\mathbf{x}, \omega)$ we obtain

$$I(\mathbf{x}, \omega) = \frac{\omega^4 \pi \rho_0 \gamma^2 L_\nu^2}{2 c_s^5 |\mathbf{x}|^2} \left(\frac{\partial^2 q_s}{\partial T^2}\right)^2 \int \mathrm{d}^3\mathbf{x}' H\left(\mathbf{x}', \frac{\mathbf{x}}{|\mathbf{x}|}\omega, \omega\right), \tag{76}$$

where $H(\mathbf{x}', \mathbf{k}, \omega)$ is a spectral tensor of temperature fluctuations and represents a Fourier transform of a two point time delayed forth order correlation function of temperature fluctuation.

Equation (76) allows to calculate the spectrum of sound radiated by a monopole source related to the moisture, if statistical properties of the turbulent source can be determined. For our calculations we consider the Von Karman model of stationary and homogeneous turbulence, which is given by (Hinze 1975)

$$E_k = C_K \varepsilon^{2/3} k_0^{-5/3} \frac{(k/k_0)^4}{[1+(k/k_0)^2]^{17/6}} \tag{77}$$

for $k < k_d$ and $E_k = 0$ for $k > k_d$, where $2\pi/k_d$ is the dissipation length scale.

The Von Karman spectrum reduces to the Kolmogorov spectrum in the inertial interval ($k \geq k_0$), but also satisfactorily describes the energy spectrum in the energy containing interval, which is a dominant contributor to acoustic radiation.

As is known (Monin and Yaglom 1975), temperature fluctuations of homogeneous isotropic and stationary turbulence behaves like a passive scalar and therefore has the same spectrum as velocity fluctuation. Therefore for the spectral function of temperature fluctuation $F(\mathbf{k},\tau)$, which is spatial Fourier transform of temperature correlation function $\Theta(\mathbf{r},\mathbf{t}) = \langle \mathbf{T}'(\mathbf{x}+\mathbf{r},\mathbf{t})\mathbf{T}'(\mathbf{x},\mathbf{t})\rangle$, we assume

$$F(\mathbf{k},\tau) = \frac{Q_k}{4\pi k^2} f(\eta_k,\tau), \tag{78}$$

where

$$Q_k = \Delta T^2 k_0^{-1} \frac{(k/k_0)^4}{[1+(k/k_0)^2]^{17/6}}. \tag{79}$$

In equation Eq. (78) $\triangle T$ is rms of the temperature fluctuation, η_k is the autocorrelation function defined as

$$\eta_k = \sqrt{\frac{k^3 E_k}{2\pi}} \tag{80}$$

and $f(\eta_k,\tau)$ characterizes the temporal decorrelation of turbulent fluctuations, such that it becomes negligibly small for $\tau \gg 1/\eta_k$.

For $f(\eta_k,\tau)$ we use a square exponential time dependence (Kraichnan 1964)

$$f(\eta_k,\tau) = \exp\left(-\frac{\pi}{4}\eta_k^2\tau^2\right). \tag{81}$$

For a homogeneous turbulence two point time delayed forth order correlation function of temperature fluctuations $R(\mathbf{x}',\mathbf{x}'+\mathbf{r},\tau)$ is related to the temperature correlation function by means of the following relation (Monin and Yaglom 1975)

$$R(\mathbf{x}',\mathbf{x}'+\mathbf{r},\tau) = 2\left\langle T'(\mathbf{r},\tau)T'(\mathbf{r},\tau)\right\rangle\left\langle T'(\mathbf{r},\tau)T'(\mathbf{r},\tau)\right\rangle. \tag{82}$$

Using Eqs. (78),(80),(81) and the convolution theorem we find

$$H(\mathbf{k},\omega) = 2\int \mathrm{d}\mathbf{k}_1^3 \mathrm{d}\omega_1 g(\mathbf{k}_1,\omega_1) g(\mathbf{k}-\mathbf{k}_1,\omega-\omega_1), \tag{83}$$

where

$$g(\mathbf{k},\omega) = \frac{Q_k}{4\pi^2\eta_k}\exp\left(-\frac{\omega^2}{\pi\eta_k^2 k^2}\right). \tag{84}$$

For low Much number turbulence one can use the so called aeroacoustic approximation (Goldstein 1976), which for the Fourier spectrum implies that in Eq. (76) instead of $H\left(\mathbf{x}', \omega\mathbf{x}/|\mathbf{x}|, \omega\right)$ one can set $H\left(\mathbf{x}', 0, \omega\right)$.

Performing the integration with respect to frequency in Eq. (83) as well as angular variables in wave number space we obtain

$$H(0,\omega) = \frac{\sqrt{2}}{4\pi^2}\int\limits_{k_0}^{k_d} \mathrm{d}k \frac{Q_k^2}{k^2\eta_k}\exp\left(-\frac{\omega^2}{2\pi\eta_k^2}\right). \tag{85}$$

The aeroacoustic approximation simplifies finding asymptotic limits for the spectrum. In the low-frequency regime, taking the limit $\omega \to 0$ we obtain

$$H(0,\omega) \sim \frac{1}{10\pi^{3/2}}\frac{\Delta T^4}{k_0^4 M c_s}. \tag{86}$$

For the spectrum this means $I(\mathbf{x},\omega) \sim \omega^4$. Physically, these frequencies are lower than the lowest characteristic frequency in the problem, corresponding to the eddy turnover time on the energy containing scale.

At high frequencies $\omega \gg k_0 M R^{1/2}$, the integral in Eq. (85) is dominated by the contribution from its upper limit and we get

$$H(0,\omega) \sim \frac{3}{8\pi^{3/2}}\frac{\Delta T^4}{k_0^2}\frac{M c_s}{\omega^2}\exp\left(-\frac{\omega^2}{k_0^2 M^2 c_s^2 R}\right) \tag{87}$$

and

$$I(\mathbf{x},\omega) \sim \omega^2 \exp\left(-\frac{\omega^2}{k_0^2 M^2 c_s^2 R}\right). \tag{88}$$

The functional form of the high-frequency suppression is determined by the specific form of the time autocorrelation function of turbulence Eq. (81) (Kraichnan 1964), but for any autocorrelation the amplitude of emitted waves should be very small in this band. Physically, this limit corresponds to radiation frequencies which are larger than any frequencies of turbulent motions; consequently, no scale of turbulent fluctuations generates these radiation frequencies directly, and the resulting small radiation amplitude is due to the sum of small contributions from many lower-frequency source modes. Since the integral is dominated by the upper integration limit, the highest-frequency source fluctuations (which contain very little of the total turbulent energy) contribute most to this high-frequency radiation tail.

In the intermediate frequency regime, $k_0 M c_s < \omega < k_0 M c_s R^{1/2}$, the integral in Eq. (85) is dominated by the contribution around k_1, where $\eta_{k_1} \approx \omega$. The width of the dominant interval is $\Delta k_1 \sim k_1$. Physically, this implies that radiation emission at some frequency in this range is dominated by turbulent vortices of the same frequency. Consequently, for the inertial interval we obtain following estimate

$$H(0,\omega) \simeq \frac{1}{3\pi^{3/2}}\frac{\Delta T^4}{k_0^4 M c_s}\left(\frac{k_0 M c_s}{\omega}\right)^{15/2} \tag{89}$$

and

$$I(\mathbf{x}, \omega) \sim \omega^{-7/2}. \tag{90}$$

We performed numerical integration of Eq. (85) for the Von Karman model of turbulence and determined normalized spectrum of acoustic radiation to be

$$I_N(\nu) = \frac{4\sqrt{2}\pi c_s^2 |\mathbf{x}|^2}{\rho_0 \gamma^2 L_\nu^2 k_0 v_0^4 \Delta T^4 F_1} I(\mathbf{x}, \omega). \tag{91}$$

The normalized spectrum for characteristic length scale of energy containing eddies $l = 2\pi/k_0 = 15$ m and characteristic rms velocity $v_0 = 5$ m s^{-1} is presented in Fig. 1. As can be seen for these typical parameters the peak frequency of infrasound radiation is $\nu_{peak} \approx 0.8$ Hz.

As shown in the subsection 2.3. the peak frequency of acoustic radiation is inversely proportional to the turnover time of energy containing turbulent eddies $\nu_{peak} \sim v_0/l$, whereas total acoustic power is proportional to v_0^4, ΔT^4 and inversely proportional to l.

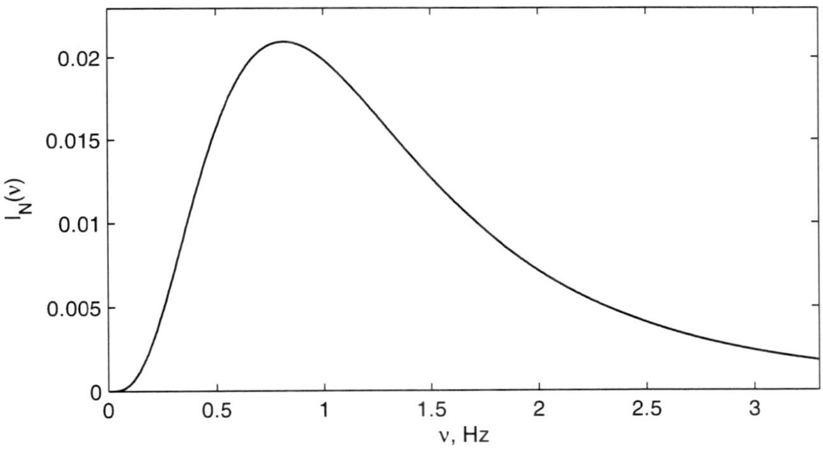

Figure 1. Normalized spectrum of acoustic radiation for Von Karman turbulence.

7.5. Infrasound correlation with tornadoes

Severe storm forecasting operations are based on several large scale environmental and storm (meso) scale kinematic and thermodynamic parameters (Lemon and Doswell 1979; Markowski *et al.* 1998; Markowski *et al.* 2002). These parameters are used to study the potential of severe weather, thunderstorm structure and organization and to discriminate between tornadic and nontornadic supercell environments. Recent climatological studies of thunderstorms using real-time radar data combined with observations of near-storm environment have been focused on the utility of various supercell and tornado forecast parameters (CAPE, Storm Relative Helicity – SRH, Bulk Richardson number – BRN and other parameters). Two parameters have been established to be the most promising in discriminating between nontornadic and tornadic supercells: boundary layer water vapor concentration (LCL hight) and low level vertical wind shear (Thompson and Edwards 2000; Rasmussen 2003; Markowski and Richardson 2009).

Examining a baseline climatology of parameters commonly used in supercell thunderstorm forecasting and research, Rasmussen and Blanchard (1998) established that the parameter that shows the most utility for discriminating between soundings of supercells with significant tornadoes and supercells without significant tornadoes is LCL height. The height of the LCL appeared to be generally lower for supercells with significant tornadoes than those without. Rasmussen and Blanchard (1998) also found that for storms producing large (at least 5 cm wide) hail only, without at least F2 strength tornadoes, the LCL heights were significantly higher than for ordinary thunderstorms. In their study half of the tornadic supercells soundings had LCLs below 800 m, while LCL heights above about 1200 m were associated with decreasing likelihood of significant tornadoes. They concluded that stronger evaporational cooling of moist downdraft leads to greater outflow dominance of storms in high LCL settings and low heights increase the likelihood of supercells being tornadic.

The work of Thompson and Edwards (2000) on assessing utility of various supercell and tornado forecasting parameters supports the finding that supercells above deeply mixed convective boundary layers, with relatively large dew point depressions and high LCLs, often do not produce tornadoes even in environments of large CAPE and/or vertical shear. They found the LCL to be markedly lower for supercells producing significant tornadoes than for those producing weak tornadoes, which were in turn lower than for nontornadic supercells. Particularly, no strong and violent tornadoes occurred for supercells with LCL > 1500 m.

Studying the relationship between Rear flank downdraft (RFD) thermodynamic characteristics and tornado likelihood Markowski *et al.* (2002) found that low LCL favors formation of significant tornadoes because the boundary layer relative humidity somehow alters the RFD and outflow character of supercells. Relatively warm and buoyant RFDs, which are supposed to be necessary for the genesis of significant tornadoes, were more likely in moist low-level environments than in dry low-level environments (Markowski *et al.* 2002). It appeared that relatively dry boundary layers, characterized by higher LCLs, support more low-level cooling through the evaporation of rain, leading to stronger outflow, which could decrease the likelihood of significant tornadoes in supercells. These are possible explanations for finding that the LCL height is generally lower in soundings associated with tornadic suppercells versus nontornadic (Rasmussen and Blanchard 1998; Thompson and Edwards 2000).

Thompson *et al.* (2003) reinforced the findings of previous studies related to LCL height as an important discriminator between significantly tornadic and nontornadic supercells and concluded that the differences in LCL heights across all storm groups studied were statistically significant, though the differences appeared to be operationally useful only when comparing significantly tornadic and nontornadic supercells. The lower LCL heights of the significantly tornadic storms supported the hypothesis of Markowski *et al.* (2002) that increased low-level relative humidity (RH) may contribute to increased buoyancy in the rear flank downdraft and an increased probability of tornadoes.

In idealized numerical simulations Markowski *et al.* (2003) investigated the effects of ambient LCL and the precipitation character of a rain curtain on the thermodynamic properties of downdraft, and ultimately on tornado intensity and longevity. The simulations were consistent with the observation that high boundary layer relative humidity values (i.e., low LCL height and small surface dewpoint depression) are associated with relatively warmer RFDs and more significant tornadogenesis than environments of relatively low boundary layer relative humidity. These findings of low LCL favoring significant tornadoes could explain observed high correlation between low frequency infrasound signals from supercell storm and later tornado formation.

Acoustic power of a monopole source related to the heat production during condensation of moisture can been estimated as (Akhalkatsi and Gogoberidze 2009):

$$N_q \sim \frac{\rho_0}{l c_s} [\gamma L_\nu]^2 \left[\frac{M_t \Delta T}{T}\right]^4 f^2(T_c) F_1, \tag{92}$$

where

$$f(T_c) \approx 6.8 \cdot 10^4 \frac{273.15 + T_c}{(243.5 + T_c)^4} \exp\left(\frac{17.67 T_c}{T_c + 243.5}\right). \tag{93}$$

and $T_c = T - 273.15$ is the temperature in degree Celsius.

Due to the numerator in the exponent, $f(T_c)$ strongly depends on temperature, e.g., $f(T_C = 10^\circ)/f(T_C = 0^\circ) \approx 2$ and according to multiplier $f^2(T_c)$ in Eq. (92), increase of saturated air temperature causes rapid enhancement of total acoustic power radiated by a monopole source.

On the other hand, low LCL height means low level air in the updraft motion being saturated and consequently, higher temperature of saturated moist air. Therefore, the lower LCL heights contribute to increased total acoustic power radiated by a monopole source related to the heat production during the condensation of moisture. As a result, enhanced low frequency infrasound signals from supercell storm appear to be in strong correlation with an increased probability of tornadoes.

It is also known (Weisman and Klemp 1982; Rotunno and Klemp 1982; Rasmussen and Blanchard 1998; Rasmussen 2003; Thompson *et al.* 2003) that high values of supercell CAPE assist tornado formation. Indeed, high values of CAPE lead to an increase of the updraft persistence and thunderstorm activity and therefore increase probability of tornado formation. According to recent studies, when relatively larger CAPE occurs closer to the surface it could cause more intense low-level stretching of vertical vorticity required for low-level mesocyclone intensification and perhaps tornadogenesis (McCaul 1991; McCaul and Weisman 1996; Rasmussen 2003). Rasmussen (2003) found the 03 km above ground level (AGL) CAPE to be possibly important in discriminating between environments supportive of significant tornadoes and those that are not.

On the other hand, high values of CAPE mean high updraft velocity caused by large low-level accelerations and increased rms of turbulent velocities. According to Eq. (92) $N_q \sim M_t^4$. Therefore increasing rms of turbulent velocities results in strong enhancement of total acoustic power.

8. Conclusion

In this chapter we have reviewed properties of atmospheric infrasound, its characteristics and propagation features. We have reviewed the history of infrasound research, instrumentation and processing techniques for observation as well as benefits of atmospheric infrasonic wave study. Strong convective storms, the main geophysical infrasound source of interest in this chapter, have been considered and several infrasound sources related to supercell storms have been discussed. It was emphasized, that the significant infrasound in a passband from 0.5 to 2.5 Hz, with peak frequencies between 0.5 and 1 Hz, which is emitted substantially before ($0.5 - 1$ hrs) generation of tornadoes, does not appear to be related with tornado itself or other severe storm accompanying meteorological events, such as microbursts, lightning and sprites. Observations and numerical studies show correlation of infrasound generation mechanism with convective processes that precede tornado formation in supercell storms. Broad and smooth spectra of observed infrasound emission indicates that turbulence is the most promising sources of the radiation. Therefore, Lighthill's acoustic analogy, which represents the basis for understanding of the sound generation by turbulent flows, has been introduced. We have considered acoustic radiation from turbulent convection in the framework of a generalized acoustic analogy, taking into account effects of stratification, temperature fluctuations and moisture in the air. Several acoustic sources initiated by turbulent convection in a stratified, moist atmosphere have been studied: Lighthill's quadrupole source, dipole radiation related to temperature inhomogeneities, monopole sources related to nonstationary heat and mass production during the condensation of moisture, dipole source related to stratification, the dipole and quadrupole sources related to inhomogeneity of background velocity and monopole source related to variability of adiabatic index. Performed analysis shows that for low Mach number turbulent convection and typical parameters of strong convective storms, infrasound radiation should be dominated by a monopole acoustic source related to the nonstationary heat production during the condensation of moisture. The total power of this source is of order 10^7 watts, two orders higher than thermo-acoustic and Lighthill's quadrupole radiation power, and is in qualitative agreement with observations of strong convective storms (Bowman and Bedard 1971; Georges 1973; Georges and Greene 1975; Bedard and Georges 2000). We have performed detailed spectral analysis of the dominant monopole source of sound. We have also discussed the relationship between the acoustic power of this source and certain significant tornado forecast parameters. It has been shown that acoustic power of a monopole source related to the moisture of the air strongly depends on the same parameters that are the most promising in discriminating between nontornadic and tornadic supercells according to the recent study of tornadogenesis (Markowsky and Richardson 2008). Particularly, low LCL, which is known to favor significant tornadoes (Rasmussen and Blanchard 1998; Thompson *et al.* 2003) implies warmer air at the level of saturation. We have shown that the increase of temperature causes rapid enhancement of acoustic power. High values of CAPE (especially, occurring closer to the surface), which assist tornado formation (Weisman and Klemp 1982; Rotunno and Klemp 1982; Rasmussen 2003; Rasmussen and Blanchard 1998; Thompson *et al.* 2003), mean high updraft velocity and therefore, increased rms of turbulent velocities, which results in strong enhancement of total acoustic power. The study presented in this chapter gives physical explanation of the results obtained in the framework of recent studies which compare ISNeT output with occurrences of tornadoes (Bedard *et al.* 2004a) and correlate ISNeT signals with detailed radar output (Szoke *et al.* 2004). These studies show, that infrasound of a tornadic thunderstorm is much stronger than the infrasound of a nonsevere weather system. Consequently, data from infrasonic detection systems combined with the information from Doppler Radar observations could help to improve tornado forecast by determining potentially tornadic storms and reducing false alarms from non-tornadic supercells.

References

Abdullah, A. J. MWRv 1966, 94, 213–220.

Akhalkatsi, M.; Gogoberidze, G. QJRMS 2009, 135, 935–940.

Akhalkatsi, M.; Gogoberidze, G. QJRMS 2011, 137, 229–235.

Akhalkatsi, M.; Gogoberidze, G., and Morrison, P. J. (2004). Infrasound generation by turbulent convection. http://arxiv.org/PS_cache/astro-ph/pdf/0409/0409367v1.pdf

Bass, H. E. ASAJ 1980, 67, 1959–1966.

Bastin, F. ASAJ 1996, 99, 2444–2444.

Batchelor, G. K. An Introduction to Fluid Dynamics; ISBN 0-521-66396-2; Cambridge University Press: Cambridge, UK, 1967; pp 615.

Beasley, W. H.; Georges, T. M.; Evans M.W. JGR 1976, 81, 3133–3140.

Bedard, A. J.; Georges T. M. Phys. Today 2000, 53, 32–37.

Bedard, Jr., A. J., Bartram, B. W., Keane, A. N., Welsh, D. C., and Nishiyama, R. T. (2004a). The Infrasound Network (ISNet): Background, Design Details and Display Capability as an 88D Adjunct Tornado Detection Tool. http://www.esrl.noaa.gov/psd/programs/infrasound/isnet/pubs/paper_1.1.pdf

Bedard, A. J., Bartram, B. W., Entwistle, B., Golden, J., Hodanish, S., Jones, R. M., Nishiyama, R. T., Keane, A. N., Mooney, L., Nicholls, M., Szoke, E. J., Thaler, E., and Welsh, D. C. (2004b). Overview of the ISNet Data Set and Conclusions and Recommendations from a March 2004 Workshop to Review ISNet Data. http://www.esrl.noaa.gov/psd/programs/infrasound/isnet/pubs/p2.8.pdf

Bedard, A. J. MWRv 2005, 133, 241–263.

Bluestein, H. B. Synoptic-Dynamic Meteorology in Midlatitudes; ISBN 0195062671; Oxford University Press, NY, 1992; Vol. 1, pp 448.

Bolton, D. MWRv 1980, 108, 1046–1053.

Bowman, H. S.; Bedard, A. J. GeoJ 1971, 26, 215–242.

Buneman, O. PhRvL 1963, 10, 285–287.

Christie, D. R.; Campus, P. In Infrasound Monitoring for Atmospheric Studies; Le Pichon, A.; Blanc E.; Hauchecorne A.; Ed. 1st; ISBN 978-1-4020-9507-8; Springer Dordrecht Heidelberg: London, NY, 2010; pp 185–234.

Colonius, T.; Lele, S.K.; Moin P. JFM 1994, 260, 271–298.

Cook, R. K. Sound 1962, 1, 12–16.

Crighton, D. G.; Huerre, P. JFM 1990, 220, 355–368.

Curle, N. RSPSA 1955, 1187, 505–514.

Delyukov, A.; Didyk, L. IJB 1999, 43, 31–37.

Dessler, A. J. JGR 1973, 78, 1889–1896.

Donn, W. L.; Pfeffer, R. L.; Ewing, M. Sci 1963, 139, 307–317.

Donn, W. L.; Posmentier, E. S. JGR 1967, 72, 2053–2061.

Few, A. A. JGR 1985, 90, 6175–6180.

Ffowcs Williams, J. E.; Hall, L. H. JFM 1970, 40, 657–670.

Freund, J. B. AIAA Pap. 1999, 99–1858.

Freund, J. B. PhFl 2003, 15, 1788–1799.

Fujiwara, S. BCMO 1914, 2, 1–43.

Gavreau, V. Sci 1968, 4, 33–37.

Georges, T. M. RvGSP 1973, 11, 571–594.

Georges, T. M.; Greene, G. E. JApMe 1975, 17, 1303–1316.

Georges, T. M. NOAA Tech. Rep. ERL 380-WPL 1976, 49, pp 59.

Georges, T. M.; Beasley, W. H. JASA. 1977, 61, 28–34.

Gogoberidze, G.; Kahniashvili, T.; Kosowsky, A. PhRvD 2007, 76, 083002.

Goldreich, P.; Kumar, P. ApJ 1990, 363, 694–704.

Goldstein, M. E. Aeroacoustics; ISBN 0-07-023685-2; McGraw-Hill Inc.: McGraw Hill, NY, 1976; pp. 293.

Goldstein, M. E. AnRFM 1984, 16, 263–285.

Goldstein, M. E. JFM 2003, 488, 315–333.

Gossard, E. E.; Hooke, W. H. Waves in the atmosphere: Atmospheric infrasound and gravity waves - Their generation and propagation; ISBN 0444411968; Elsevier Science Ltd: Elsevier, NY, 1975; pp 472.

Gutin, L. Physik. Zeitschr. der Sowjetunion 1936, 9, 57–71.

Hardin, J. C.; Pope D. S. ASAJ 1989, 85, 1359–1362.

Haubrich, R. A. BSSA 1968, 58, 977–991.

Hickey, M. P.; Schubert, G.; Walterscheid, R. L. JGR 2001, 106, 21543–21548.

Hinze, J. O. Turbulence; ISBN 0070290377; McGraw-Hill Companies: McGraw Hill, NY, 1975; pp 790.

Howe, M. S. In Sound-flow interactions; Auregan, Y.; Maurel, A.; Pagneux, V.; Pinton, J. F.; Ed. 1st; ISSN 3540433325; Springer: Berlin, Germany, 2001; pp 308.

Jones, R. M., Emily S. G., and Bedard Jr., A. J. (2004). Infrasonic Atmospheric Propagation Studies Using a 3-D Ray Trace Model. `http://www.esrl.noaa.gov/psd/programs/infrasound/isnet/pubs/p2.9.pdf`

Jones, R. M.; Georges, T.M. 1976, 59, 765–779.

Kraichnan, R. H. PhFl 1964, 7, 1030–1048.

Kulichkov, S. N. IzAOP 1992, 28, 253–269.

Kulichkov, S. N. MAP 2004, 85, 47–60.

Lastovicka, J. 2006, 68, 479–497.

Lemon L. R.; Doswell III C. A. MWRv 1979, 107, 1184–1197.

Lighthill, M. J. RSPSA 1952, 211, 564–587.

Lighthill, M. J. RSPSA 1954, 222, 1–32.

Lilley, G. M. AGARD-CP 1974, 131, 13.1–13.12.

Lilley, G. M. ThCFD 1994, 6, 281–301.

Lindemann, F. A. Dobson, G. M. B. PRS 1922, 102, 411–437.

Liszka, L. Infrasound: a summary of 35 years of infrasound research; ISBN 978-91-977255-0-7; IRF Scientific Report 291, Swedish Institute of Space Physics: Ume, Sweden, 2008; pp 150.

Markowski, P. M.; Straka, J. M.; Rasmussen, E. N.; Blanchard, D. O. MWRv 1998, 126, 2959–2971.

Markowski, P. M.; Straka, J. M.; Rasmussen, E. N. MWRv 2002, 130, 1692–1721.

Markowski, P. M.; Straka, J. M.; Rasmussen, E. N. JAtS 2003, 60, 795–823.

Markowski, P. M.; Richardson, Y. P. AtmRe 2009, 93, 3–10.

McCaul, E. W. MWRv 1991, 119, 1954–1978.

McCaul, E. W.; Weisman, M. L. MWRv 1996, 124, 408–429.

Mitchell, B. E.; Lele, S. K.; Moin, P. AIAA Pap. 92–0374, 1992.

Mitchell, B. E.; Lele, S. K.; Moin, P. AIAAJ 1997, 35, 1574–1580.

Monin; A.S.; Yaglom A.M. Statistical Fluid Mechanics; ISBN 0262130629; The MIT Press: Cambridge, MA, 1975; pp 782.

Naugolnykh, K.; Rybak, S. JASA 2008, 124, 1–3.

Nicholls, M. E.; Pielke, R. A. QJRMS 1994a, 120, 305–332.

Nicholls, M. E.; Pielke, R. A. QJRMS 1994b, 120, 333–359.

Nicholls, M. E.; Pielke, R. A. JAtS 2000, 57, 3251–3271.

Nicholls, M. E.; Pielke, Sr.R. A., and Bedard, A. (2004). Preliminary Numerical Simulations of Infrasound Generation Processes by Severe Weather Using a Fully Compressible Numerical Model. `http://www.esrl.noaa.gov/psd/programs/infrasound/isnet/pubs/paper_8a3.pdf`

Ostashev, V. E.; Georges, T. M.; Clifford, S. F.; Goedecke, G. H. JASA 2001, 109, 2682–2692.

Panickar, P.; Srinivasan, K.; Raman G.; Juve D. PhFl 2005, 17, 096103.

Phillips, O. M. JFM 1960, 9, 1–28.

Pielke, R. A; Nicholls, M. E.; Bedard, A. J. EOSTr 1993, 74, 493.

Pierce, A. D.; Posey, J. W. GeoJI 1971, 26, 341–368.

Posmentier, E.S. GeoJ 1967, 13, 487–501.

Powell, A. JASA 1960, 32, 8, 982–990.

Proudman, I. RSPSA 1952, 214, 119–132.

Rasmussen, E. N.; Blanchard, D. O. Weather Forecast. 1998, 13, 4, 1148–1164.

Rasmussen, E. N. Weather Forecast. 2003, 18, 3, 530–535.

Rishbeth, H. 2006, 68, 469–478.

Rotunno, R.; Klemp, J. B. MWRv 1982, 110, 136–151.

Sarkar, S.; Hussaini, M.Y. ICASE Report 93-74, 1993; pp 30.

Schecter, D. A.; Nicholls, M. E.; Persing, J.; Bedard. A. J.; Pielke, Sr. R. A. JAtS 2008, 65, 3, 685–713.

Seror, C.; Sagaut, P.; Bailly, C.; Juve, D. PhFl 2001, 13, 476–487.

Shaw, W. N.; Dines, W. H. QJRMS 1904, 31, 39–52.

Sindelarova, T.; Buresova, D.; Chum, J.; Hruska, F. AdSpR 2009, 43, 1644–1651.

Stein, R. F. SoPh 1967, 2, 385–432.

Szoke, E. J.; Bedard, A. J.; Thaler, E., and Glancy, R. (2004). A comparison of inset data with radar data for tornadic and potentially tornadic storms in northeast colorado. `http://www.esrl.noaa.gov/psd/programs/infrasound/isnet/pubs/paper_1.2.pdf`

Tandy, V.; Lawrence, T. SPR 1998, 62, 360–364.

Tatom, F. B.; Knupp K. R.; Vitton S. J. JApMe 1995, 34, 572–582.

Thompson, R. L.; Edwards, R. Weather Forecast 2000, 15, 6, 682–699.

Thompson, R. L.; Edwards, R.; Hart, J. A.; Elmore, K. L.; Markowski, P. M. Weather Forecast 2003, 18, 1243–1261.

Weisman, M. L.; Klemp J. B. MWRv. 1982, 110, 504–520.

Whipple, F. J. W. QJRMS 1930, 56, 287–304.

Whipple, F. J. W. QJRMS 1935, 61, 285–308.

Whipple, F. J. W. QJRMS 1939, 65, 319–323.

Whitaker, R. W.; Mutschechner, J. P. The Design and Operation of Infrasonic Microphones; LA-I3257; Los Alamos National Laboratory, Los Alamos; NM; 1997.

Whitmire, J.; Sarkar, S. PhFl 2000, 12, 381–391.

Witkowska A., Brasseur J.G. and Juve D. AIAA Pap. 1995, 95–037.

In: Sound Waves: Propagation, Frequencies and Effects ISBN 978-1-61470-001-2
Editor: V. Abagnali and G. Fabbri

Chapter 7

ON THE NEUTRONS DIFFRACTION IN A CRYSTAL UNDER THE INFLUENCE OF A SOUND WAVE

A. G. Hayrapetyan[1,*], *K. K. Grigoryan*[2] *and R. G. Petrosyan*[2]
[1]Max-Planck-Institut für Kernphysik
Postfach 103980, D-69029 Heidelberg
Germany
[2]Yerevan State University
1 Alex Manoogian Str., 0025 Yerevan
Armenia

Abstract

In this chapter we consider neutron diffraction in a crystal under the influence of an external sound wave. We examine both the traveling and standing hypersonic waves' contribution in the neutron diffraction intensity, diffraction condition and the Debye–Waller factor (influence of thermal motion of atoms). The possibility of diffraction of both thermal and high-energy (short-wave) neutrons is shown in the process of multiphonon interaction of a neutron with a crystal and the field of a hypersonic wave. The difference between traveling and standing sound waves is discussed. The formation of a sublattice is illustrated in the process of neutron elastic scattering with respect to a standing hypersonic wave. The periodicity of a sublattice has the same order as the high-energy neutron wavelength, 0.01-0.1 Angströms. The analogy to the Kapitza–Dirac effect is considered for neutrons. The possible tuning of the Debye–Waller factor is examined as well. It is shown that the application of a hypersonic wave can either suppress or enhance the diffraction intensity. A realistic condition is obtained under which the negative influence of the thermal motion of atoms of which the matter is comprised is completely eliminated.

*E-mail address: armen@physi.uni-heidelberg.de, armen.hayr@gmail.com

1. Introduction

Scattering of neutrons and neutron diffraction are powerful modern tools for investigating the structure and properties of crystals, hence cause their wide use in the present-day physics. Using neutrons, it is possible to reveal the atomic structure of matter, to determine magnetic characteristics, to gain information on atomic thermal vibrations in liquids and crystals. The application of neutrons creates so many possibilities, sometimes perfectly unique, that despite considerable experimental difficulties, neutron research is growing rapidly. The overwhelming majority of the studies performed on reactors with use of neutron radiation deals at present with the physics of condensed matter.

Diffraction of particles is a fruitful technique for studies of different types of crystalline structures. However, it is well-known that X-ray diffraction scattering from inner electrons of atoms is used for the investigation of the atomic structure of matter consisting of large number of atoms. The intensity of such scattering from the atoms with small atomic numbers (for instance for the Hydrogen atom) is small and it is difficult to determine their positions. The electron diffraction is accepted as a productive method for structural analysis as well. Though this method can be extended for solid state surface investigations, the electromagnetic interaction between electrons and matter is stronger and the diffraction effects cannot be observed. These disadvantages are absent in the case of neutron diffractometry, the basic problem of which is the investigation of the atomic crystalline structure peculiarities of the matter.

In structural neutron diffractometry, a significant factor is that neutron interaction with nuclei has quite a different nature: neutrons barely interact with electrons but strongly interact with atomic nuclei. Meanwhile, due to the presence of an anomalous magnetic moment of the neutron, two types of forces exist which produce neutrons' interaction with atoms: these are the nuclear forces that define neutron interaction with an atomic nucleus, and the magnetic forces, that appear in the process of the interaction of the neutron magnetic moment of an atomic electron shell. The neutron diffraction in crystals due to the interaction with atomic nuclei which form a crystal is a fruitful method for atomic and molecular structure analysis of crystals. Due to the magnetic interaction with electron shell the neutron diffraction underlines the investigation of the crystal magnetic structure. Many books and articles relate to this topic: the classics such as [1, 2, 3, 4], and more recent such as [5, 6, 7] (see also references therein). In recent years the investigation of condensed media by means of neutrons scattering has been enlarged given that it is a fast developing branch of modern physics. Both experimental [8, 9, 10, 11, 12, 13, 14] and theoretical [14, 15, 16, 17, 18, 19, 20, 21] research has been performed, n.b. [15, 16]. In [15] the interaction of a neutron with a strong electromagnetic field is considered in the frames of the Dirac theory taking into account both the magnetic and electric anomalous moments of a neutron. In [16] the exact solution of the Schrödinger equation is obtained for the neutron moving in the field of a circularly polarized electromagnetic field. These results can be used in the many various types of scattering processes, for instance, in frames of non-stationary S-matrix theory which is a powerful method for investigation of the dynamics of a quantum system [22]; the neutron diffraction in crystals under the impact of external mechanical and thermal influences has been studied in [23, 24, 25, 26, 27, 19, 20]. For neutron scattering in a crystal under the influence of external fields, the S-matrix theory is developed in

[18, 19, 20, 21], where for the first time the possibility of short-wave neutron diffraction is illustrated.

This chapter is devoted to consideration of neutron diffraction in a crystal under the influence of external hypersonic waves (both traveling and standing), hereby we examine neutron-crystal-hypersonic wave interaction in details. Two new concepts are of special interest, *possibility of high-energy neutrons diffraction* and *control of Debye–Waller factor*. Taking a few steps ahead let us note that under the impact of external influences, neutrons can be diffracted into any angles, and the optimal values of the angles can be set. This could be another advantage of the application of an external fields.

In crystalline matter atoms are located in ordered arrangement. This fact radically changes the scattering picture. In the orthodox case, when the externa field is absent, it is significant that the characteristic distance of a structure (lattice period) and the neutron wavelength have the same order. In this case only so-called thermal neutrons give a diffraction pattern in the absence of an external fields. These are the neutrons with the wavelengths 10^{-8} cm and energy about 10^{-2} eV which corresponds to temperature ~ 100 K. This follows from the relation $\lambda = 0.287/\sqrt{E}$ between the wavelength and the energy of the neutron, where λ is measured in Angströms and the energy in electron-Volts. The scattering at longer wavelengths cannot reveal the atomic-scale details of a structure, while shorter wavelengths diffract by very small angles, which are difficult to record.

Nevertheless, it turns out that the shorter wavelengths give diffraction pattern when the external field exists. The external field has an efficient influence on neutron scattering (diffraction) intensity. For the first time the possibility of *short-wave* neutron diffraction is shown in [18], where neutron scattering is considered in the crystal under the influence of linearly polarized strong laser radiation. In [21] the short-wave neutron diffraction is discussed in a crystal under the influence of a standing laser wave. The solution is based on *multiphoton* interaction of the anomalous magnetic moment of a neutron within the field of laser radiation (Farri representation). Short-wave neutron diffraction occurs also when the crystal is subjected to the field of traveling (standing) hypersonic waves [19] ([20]). Note that in articles [19, 20] the neutron scattering under external actions is examined for the first time with taking into account the quantum nature of the acoustic field; the neutron-crystal-field interaction occurs via *multiphonon* channels. This is due to the consideration of displacements of nuclei from the equilibrium positions within the framework of secondary quantization.

The main advantage of the application of a sound wave comparable with that of a laser is that in this case the Debye–Waller factor can be changed and tuned, thus under the appropriate choice of parameters of the sound wave and the crystal the diffraction picture can be either increased or suppressed. The Debye–Waller factor describes the influence of nuclear thermal motion on diffraction intensity [28, 29]. The diffraction intensity can be decreased if atomic arrangement order is violated for some reason. One of the basic reasons of such a violation is nucleus thermal motion. A nucleus can be considered as a dot with a size about 10^{-13} cm. The thermal motion smears this dot over some region, the volume of which in some cases can coincide with atomic sizes. In contrary to the application of a laser field, the sound field gives a qualitatively new contribution in the Debye–Waller factor. This fact physically can be understood in the following way: electromagnetic waves are material on their own, and sound waves only exist within the matter in which they propagate - they are a propagation of alternate compressions and rarefactions with small amplitudes.

2. Neutron-Crystal Interaction Potential Under the Influence of a Sound Wave

The crystal is considered as a primary cubic lattice consisting of a single isotope element atoms with zero spin and located on the vertices of a cube. Such an approach allows one to examine neutron coherent scattering, which ultimately leads us to neutron diffraction. Neutrons interact with aggregation of atoms since their time of passing through a characteristic distance is comparable with the period of propagation of excitation occurring in the crystal under the simultaneous action of lattice oscillations and phonons of external sound waves. Neutron waves are added at the point of observation in accordance with the laws of interference.

The interaction of neutrons with a crystal subjected to the influence of a standing sound wave can be described via the Fermi *pseudopotential*

$$V(t) = -\frac{2\pi\hbar^2 A}{m}\sum_n \delta\left(\vec{r}-\vec{R}_n\right), \tag{1}$$

where A is the neutron scattering amplitude from the nucleus, m is the neutron mass. The position vector $\vec{R}_n$ determines the location of a nucleus near the site $\vec{n} = n_1\vec{a}_1 + n_2\vec{a}_2 + n_3\vec{a}_3$, where n_i are integers and $\vec{a}_i$ are the vectors of basic translations.

In the absence of the external sound wave the position vector has the form

$$\vec{R}_n = \vec{n} + \vec{\xi}_n\,, \tag{2}$$

where

$$\vec{\xi}_n = \sum_{S,\vec{q}}\sqrt{\frac{\hbar}{2MN\Omega_S(\vec{q})}}\,\vec{e}_S(\vec{q})\left[b_{qS}e^{i\vec{q}\vec{n}} + b^{\dagger}_{qS}e^{-i\vec{q}\vec{n}}\right] \tag{3}$$

is the harmonic shift operator of atomic displacement from the site $\vec{n}$ represented in terms of second quantization. In (3), M is the nucleus mass, N is the number of elementary cells in the crystal, $\Omega_S(\vec{q})$ is the thermal phonon frequency, $\vec{e}_S(\vec{q})$ is thermal phonon polarization vector, b_k and $b_k^{\dagger}$ are the canonical conjugate coordinates (phonon annihilation and creation operators respectively), satisfying Bose commutation relations $[b_k, b^{\dagger}_{k'}] = \delta_{kk'}$, $[b_k, b_{k'}] = 0$.

The external sound wave makes it possible both to illustrate the probability of high-energy neutron diffraction and essentially to eliminate problems connected with the nuclei thermal motion in the process of the observation of diffraction extrema. The sound wave is taken into account by introducing a new term in the expression (2).

When the traveling sound wave's field is turned on, the position vector (2) is replaced by

$$\vec{R}_n = \vec{n} + \vec{\xi}_n + \vec{\iota}, \tag{4}$$

where

$$\vec{\iota} = \vec{b}\sin\left(\vec{k}\vec{n} - \omega t\right) \tag{5}$$

with $\vec{b}$, $\vec{k}$ and ω being respectively the amplitude, wave vector and frequency of a traveling sound wave.

When a standing sound wave is turned on, the position vector (2) is replaced by

$$\vec{R}_n = \vec{n} + \vec{\xi}_n + \vec{\zeta}, \tag{6}$$

where

$$\vec{\zeta} = \vec{a}sin\left(\vec{\Lambda}\vec{n} - \Omega t\right) + \vec{a}sin\left(-\vec{\Lambda}\vec{n} - \Omega t\right). \tag{7}$$

Here $\vec{a}$, $\vec{\Lambda}$ and Ω are the standing sound wave amplitude, wave vector and frequency respectively. The vectors (4) and (6) are just the proposed forms of the vector $\vec{R}_n$ that allows consideration of the short-wave neutron diffraction and control the Debye–Waller factor.

3. S-Matrix Theory

The problem of neutron diffraction in crystals in the presence of an external sound wave is solved in the framework of the non-stationary S-matrix theory in the interaction representation where the neutron-phonon interaction is considered as a perturbation of the free motion of the neutron. The time dependence of the Hamiltonian determines the picture of the quantum dynamical behavior of the system during the scattering process. It gives an obvious description of the physical phenomena occurring in the microcosm. In this section we mainly follow the basic concepts of modern scattering theory (e.g.[22]). We are dealing with a three-particle system which consists of a neutron, crystal and sound wave (later on we will call it the neutron-phonon-phonon system).

The time-dependent Hamiltonian of the full neutron-phonon-phonon system is taken in the following form:

$$H(t) = H_0(t) + V(t), \tag{8}$$

where $H_0(t)$ is the Hamiltonian of the non-perturbed system (free neutron) and $V(t)$ is the perturbation Hamiltonian (neutron-crystal) which is the Fermi pseudopotential (1). In general both terms depend on time. Should one consider free neutron scattering on a crystal which is under the influence of an external sound wave the non-perturbed Hamiltonian has the form

$$H_0 = \frac{\hat{\vec{p}}^2}{2m}, \tag{9}$$

where $\hat{\vec{p}}$ is the free neutron momentum vector operator. And the perturbation Hamiltonian (1) includes both thermal (3) and external phonons ((5) or (7)).

For the calculation of the probability of the considered process, the evolution of the whole system is described by means of the evolution operator $U(t, t')$ in the case where the system is not conservative, i.e. the Hamiltonian of the system explicitly depends on time. In the Schrödinger representation, the time evolution operator of the whole dynamical system generates the final state $|\Psi(t)\rangle$ from the initial one $|\Psi(t')\rangle$ through the law

$$|\Psi(t)\rangle = U(t,t') \left|\Psi(t')\right\rangle . \tag{10}$$

Note that $|\Psi(t)\rangle$, being a representation of the state at some moment t, is totally defined by imposing $U(t,t')$ on $|\Psi(t')\rangle$ if the system is not subjected to a measurement in the (t,t') interval. The state vector $|\Psi(t)\rangle$ is postulated as a solution of the Schrödinger non-stationary equation

$$i\hbar\frac{\partial\,|\Psi(t)\rangle}{\partial t} = H\,(t)\,|\Psi(t)\rangle . \tag{11}$$

After substituting (10) into (11) one can find the following equation for the unitary operator $U\,(t,t')$

$$i\hbar\frac{\partial U\,(t,t')}{\partial t} = H\,(t)\,U\,\left(t,t'\right) , \tag{12}$$

with the initial condition

$$U\,\left(t',t'\right) = 1 . \tag{13}$$

Consequently, the problem is to construct the operator $U\,(t,t')$, preferably using exact solutions, which describes time evolution of dynamical states of the system by means of the equations (12) and (13) with the following properties:

$$U\,\left(t,t'\right)U^{\dagger}\,\left(t,t'\right) = U^{\dagger}\,\left(t,t'\right)U\,\left(t,t'\right) = 1 , \tag{14}$$

$$U^{\dagger}\,\left(t,t'\right) = U\,\left(t',t\right) = U^{-1}\,\left(t,t'\right) . \tag{15}$$

Here $U^{\dagger}\,(t,t')$ is the Hermitian conjugate operator. Condition (15) follows from the composition law

$$U\,\left(t,t'\right) = U\,\left(t,t''\right)U\,\left(t'',t'\right) . \tag{16}$$

In case of the absence of the scattering potential (1), the state vector of the free neutron is

$$|\psi\,(t)\rangle = e^{(i/\;)(\vec{P}\vec{r}-Et)} , \tag{17}$$

where $\vec{P}$ and E are the free neutron momentum vector and energy respectively. The state vector (17) satisfies the Schrödinger equation

$$i\hbar\frac{\partial\,|\psi(t)\rangle}{\partial t} = H_0\,(t)\,|\psi(t)\rangle . \tag{18}$$

The time behavior of the neutron is given by the unitary evolution operator $U_0\,(t',t')$ through the relation

$$|\psi(t)\rangle = U_0(t,t') \left|\psi(t')\right\rangle ; \tag{19}$$

and the unitary operator satisfies the equation

$$i\hbar\frac{\partial U_0\left(t,t'\right)}{\partial t}=H_0\left(t\right)U_0\left(t,t'\right)\,, \tag{20}$$

with the initial condition

$$U_0\left(t',t'\right)=1 \tag{21}$$

and possesses the same properties as (14)-(16). The free neutron time evolution operator has the form

$$U_0\left(t,t'\right)=e^{-\frac{i}{\hbar}E(t-t')}\,. \tag{22}$$

For the investigation of scattering phenomena it is convenient to turn to the so-called intermediate representation (interaction representation), where the state of the system is defined by means of the vector $|\Phi(t)\rangle$ which is generated from the whole system state vector $|\Psi(t)\rangle$ in the following way:

$$|\Phi(t)\rangle=U_0^{\dagger}(t,t')\,|\Psi(t)\rangle\,. \tag{23}$$

Substituting (23) into (11) and taking into account Eq. (20) one obtains the following equation for $|\Phi(t)\rangle$:

$$i\hbar\frac{\partial\,|\Phi(t)\rangle}{\partial t}=V_I\left(t\right)|\Phi(t)\rangle\,, \tag{24}$$

where the following notation is introduced:

$$V_I(t)\equiv U_0^{\dagger}(t,t')V(t)U_0(t,t')\,. \tag{25}$$

As it was done above, the time evolution of the system in the interaction representation is defined by means of the unitary operator $U_I(t,t')$ in the following way:

$$|\Phi(t)\rangle=U_I(t,t')\,\left|\Phi(t')\right\rangle\,. \tag{26}$$

The equation

$$i\hbar\frac{\partial U_I\left(t,t'\right)}{\partial t}=V_IU_I\left(t,t'\right) \tag{27}$$

can be obtained with the initial condition

$$U_I\left(t',t'\right)=1 \tag{28}$$

based on (24) and (26).

It is necessary to find the evolution operator $U_I(t,t')$ of the whole system in order to specify $U(t,t')$. For this purpose with the help of (21), (23) and (26) for subsequent calculations it is convenient to represent the state vector $|\Phi(t)\rangle$ in the form

$$|\Phi(t)\rangle=U_I(t,t')\,\left|\Psi(t')\right\rangle\,.$$

On the other hand (10) and (23) give

$$|\Phi(t)\rangle = U_0^\dagger(t,t')U(t,t')\,|\Psi(t')\rangle \,.$$

Matching the last two equations one has

$$U_I\left(t,t'\right) \equiv U_0^\dagger\left(t,t'\right) U\left(t,t'\right) , \tag{29}$$

which gives rise to the definition of $U(t,t')$ if the operator $U_0(t,t')$ is known. Note that the operator $U_I(t,t')$ in turn is subject to the same properties (14)-(16).

The solution of Eq. (27) with the corresponding initial condition (28) can be presented by means of the integral equation

$$U_I\left(t,t'\right) = 1 - (i\hbar)^{-1} \int\limits_{t'}^{t} V\left(\tau\right) U_I\left(\tau,t'\right) \mathbf{d}\tau, \,. \tag{30}$$

In the interaction representation the system's fundamental law of evolution is expressed by equations (27) and (28) or the integral equation (30).

Eq. (30) is solved iteratively. As a result the following expansion can be obtained:

$$U_I\left(t,t'\right) = \sum_{n=0}^{\infty} U_I^{(n)}\left(t,t'\right) , \tag{31}$$

where

$$U_I^{(0)}\left(t,t'\right) = 1, \quad U_I^{(n)}\left(t,t'\right) \equiv (i\hbar)^{-1} \int\limits_{t>\tau_n>\cdots>\tau_1>t'} \mathbf{d}\tau_n \cdots \mathbf{d}\tau_1 V_I\left(\tau_n\right) \cdots V_I\left(\tau_1\right). \tag{32}$$

Now , taking into account definitions (25) and (29) the following expression for $U(t,t')$ is obtained:

$$U\left(t,t'\right) = \sum_{n=0}^{\infty} U^{(n)}\left(t,t'\right) , \tag{33}$$

where

$$U^{(0)}\left(t,t'\right) = U_0\left(t,t'\right) ,$$

$$U^{(n)}\left(t,t'\right) = (i\hbar)^{-1} \int\limits_{t>\tau_n>\cdots>\tau_1>t'} \mathbf{d}\tau_n \cdots \mathbf{d}\tau_1 U_0\left(t,\tau_n\right) V\left(\tau_n\right) \times U_0\left(\tau_n,\tau_{n-1}\right) V\left(\tau_{n-1}\right) \cdots U_0\left(\tau_2,\tau_1\right) V\left(\tau_1\right) U_0\left(\tau_1,t'\right). \tag{34}$$

The expansions written above are power series in $V(t)$. Here $U_I^{(n)}$ and $U^{(n)}$ represent the n-th order contribution in the series (31) and (33) respectively. Thus, the time evolution operator $U(t,t')$ of the whole three-particle system is constructed. This is the key to the examination of dynamics of neutron scattering.

The time-dependent S-matrix theory is used for the consideration of three-particle interactions. The papers [18, 19] relate to neutral particles and the papers [30, 31] to charged particles, where specifically the electron-photon-phonon interaction is examined in a polar semiconductor under the influence of laser radiation. In Ref. [30] the problem is solved in the dipole approximation, and in contrary to this in Ref. [31] both position and time dependence of the external laser field is taken into account.

4. Diffraction Probability

In this section, we evaluate neutron diffraction probability both for traveling and standing sound waves in the first order contribution in the expansion (34). We turn to the calculation of the probability of the neutron-phonon-phonon interaction, i.e. consider the scattering of whole system from the initial state

$$|\Psi_i\rangle = e^{(i/\ )(\vec{P}_i\vec{r}-E_i t')} \prod_{S,q} |\nu_{Sq}\rangle \tag{35}$$

to a final state

$$|\Psi_f\rangle = e^{(i/\ )(\vec{P}_f\vec{r}-E_f t)} \prod_{S,q} |\nu_{Sq}\rangle \ . \tag{36}$$

Here the product presents the state of the crystal vibrations with ν_{Sq} phonons from the branch S and with the wave vector $\vec{q}$. Moreover, $\vec{P}_i$, $\vec{P}_f$ and E_i, E_f are the corresponding values of neutron momentum vector and energy in the initial and final states respectively. The choice of states in the form (35) and (36) means that the scattering of the neutron by the crystal is elastic, i.e. phonon excitation of the crystal does not occur.

In the expansion (34), the first order probability amplitude may be calculated by means of the following expression:

$$M_{fi} = \langle\Psi_f| U^{(1)}(t,t')|\Psi_i\rangle = (i\hbar)^{-1}\langle\Psi_f| \int_{t'}^{t} \mathbf{d}\tau U_0(t,\tau) V(\tau) U_0(\tau,t')|\Psi_i\rangle \ . \tag{37}$$

Inserting (1) and (22) in (37), the matrix element takes the form

$$M_{fi} = -\frac{2\pi\hbar A}{im}\int_{t'}^{t} \mathbf{d}\tau e^{(-i/\ )(E_i-E_f)\tau}$$
$$\times \int \mathbf{d}\vec{r} e^{(i/\ )(\vec{P}_i-\vec{P}_f)\vec{r}} \sum_{n}\prod_{S,q} \langle\nu_{Sq}|\,\delta(\vec{r}-\vec{R}_n)\,|\nu_{Sq}\rangle\,, \tag{38}$$

where the integration over position coordinates is performed over the volume occupied by the crystal.

Henceforth, as long as calculations for both traveling and standing sound waves are similar, we will consistently display them together. Integrating (38) over the position coordinates, using (4) for the traveling sound wave and introducing the notation

$$\vec{Q} \equiv (1/\hbar)(\vec{P}_i - \vec{P}_f) \tag{39}$$

one obtains

$$M_{fi} = -\frac{2\pi\hbar A}{im}\sum_n e^{i\vec{Q}\vec{n}}\prod_{S,q}\langle\nu_{Sq}|\, e^{i\vec{Q}\vec{\xi}_n}\,|\nu_{Sq}\rangle\int_{t'}^{t}\mathbf{d}\tau e^{(-i/\ )(E_i-E_f)\tau}e^{i\vec{Q}\vec{b}sin\left(\vec{k}\vec{n}-\omega\tau\right)}\,. \tag{40}$$

The matrix element in the case of the standing sound wave, using (6), reads as

$$M_{fi} = -\frac{2\pi\hbar A}{im}\sum_n e^{i\vec{Q}\vec{n}}\prod_{S,q}\langle\nu_{Sq}|\, e^{i\vec{Q}\vec{\xi}_n}\,|\nu_{Sq}\rangle$$
$$\times\int_{t'}^{t}\mathbf{d}\tau e^{(-i/\ )(E_i-E_f)\tau}e^{i\vec{Q}\vec{a}sin\left(\vec{\Lambda}\vec{n}-\Omega\tau\right)}e^{-i\vec{Q}\vec{a}sin\left(\vec{\Lambda}\vec{n}+\Omega\tau\right)}\,. \tag{41}$$

From here, using the expansion

$$e^{i\alpha sin\beta} = \sum_{s=-\infty}^{+\infty} J_s(\alpha)e^{is\beta}$$

of the exponent in terms of Bessel functions and integrating over τ with the notation

$$\Omega_s \equiv (1/\hbar)\left(E_i - E_f + s\hbar\omega\right)$$

and

$$\Omega_{s_1 s_2} \equiv (1/\hbar)\left(E_i - E_f + (s_1 - s_2)\hbar\Omega\right)\,,$$

one can see that the probability amplitudes take the forms

$$M_{fi} = -\frac{2\pi\hbar A}{m}\sum_{s=-\infty}^{+\infty} J_s(\vec{Q}\vec{b})\frac{e^{-i\Omega_s t} - e^{-i\Omega_s t'}}{\Omega_s}\sum_n e^{i(\vec{Q}+s\vec{k})\vec{n}}\prod_{S,q}\langle\nu_{Sq}|\, e^{i\vec{Q}\vec{\xi}_n}\,|\nu_{Sq}\rangle \tag{42}$$

and

$$M_{fi} = -\frac{2\pi\hbar A}{m}\sum_{s_1,s_2=-\infty}^{+\infty} J_{s_1}(\vec{Q}\vec{a})J_{s_2}(-\vec{Q}\vec{a})\frac{e^{-i\Omega_{s_1 s_2} t} - e^{-i\Omega_{s_1 s_2} t'}}{\Omega_{s_1 s_2}}$$
$$\times\sum_n e^{i(\vec{Q}+(s_1+s_2)\vec{\Lambda})\vec{n}}\prod_{S,q}\langle\nu_{Sq}|\, e^{i\vec{Q}\vec{\xi}_n}\,|\nu_{Sq}\rangle \tag{43}$$

for traveling and standing sound waves respectively.

The scattering probability is obtained when quantum and statistical double averaging is performed. First, in order to perform the quantum mechanical averaging, we take into account the Weil identity

$$e^{\alpha b_S+\beta b_S^{\dagger}}=e^{-\frac{1}{2}\alpha\beta}e^{\beta b_S^{\dagger}}e^{\alpha b_S}$$

with α and β being arbitrary numbers [22], which yields

$$\langle\nu_S|\,e^{\alpha b_S+\beta b_S^{\dagger}}\,|\nu_S\rangle=e^{\alpha\beta\left(\frac{1}{2}+\nu_S\right)},$$

from which we obtain the following forms for the probability amplitude:

$$M_{fi}=-\frac{2\pi\hbar A}{m}\sum_{s=-\infty}^{+\infty}J_s(\vec{Q}\vec{b})e^{-\nu}\frac{e^{-i\Omega_s t}-e^{-i\Omega_s t'}}{\Omega_s}\sum_n e^{i(\vec{Q}+s\vec{k})\vec{n}} \tag{44}$$

for the traveling sound wave, and

$$M_{fi}=-\frac{2\pi\hbar A}{m}\sum_{s_1,s_2=-\infty}^{+\infty}J_{s_1}(\vec{Q}\vec{a})J_{s_2}(-\vec{Q}\vec{a})e^{-\nu}\frac{e^{-i\Omega_{s_1s_2}t}-e^{-i\Omega_{s_1s_2}t'}}{\Omega_{s_1s_2}}\times\sum_n e^{i(\vec{Q}+(s_1+s_2)\vec{\Lambda})\vec{n}} \tag{45}$$

for the standing sound wave with the notation

$$\nu\equiv\sum_{S,\vec{q}}\frac{\hbar}{2MN\Omega_S(\vec{q})}\left(\vec{Q}\vec{e}_S(\vec{q})\right)^2\left(\frac{1}{2}+\nu_{Sq}\right).$$

The neutron scattering probability is

$$W_{\vec{P}_f,\vec{P}_i}=\lim_{\Delta t\to\infty}\frac{\partial\bar{R}}{\partial(\Delta t)}, \tag{46}$$

where $R=|M_{fi}|^2$ is the modulus squared of the probability amplitude, $\Delta t\equiv t-t'$ is the scattering period, and $\bar{R}$ is the statistical average of R.

Using (44) and (45) for $\bar{R}$ we obtain

$$\bar{R}=\frac{(2\pi)^2\hbar^2A^2}{m^2}\sum_{s,s'}J_s(\vec{Q}\vec{b})J_{s'}(\vec{Q}\vec{b})e^{-2w}\frac{e^{i(s-s')\omega t}}{\Omega_s\Omega_s'}\times\left(1-e^{-i\Omega_s\Delta t}-e^{i\Omega_{s'}\Delta t}+e^{-i(s-s')\omega\Delta t}\right)\left|\sum_n e^{i(\vec{Q}+s\vec{k})\vec{n}}\right|^2, \tag{47}$$

$$\bar{R}=\frac{(2\pi)^2\hbar^2A^2}{m^2}\sum_{s_1,s_2,s_1',s_2'}J_{s_1}(\vec{Q}\vec{a})J_{s_2}(-\vec{Q}\vec{a})J_{s_1'}(\vec{Q}\vec{a})J_{s_2'}(-\vec{Q}\vec{a})e^{-2w}\frac{e^{i(s_1-s_1'-(s_2-s_2'))\Omega t}}{\Omega_{s_1s_2}\Omega_{s_1's_2'}}\times\left(1-e^{-i\Omega_{s_1s_2}\Delta t}-e^{i\Omega_{s_1's_2'}\Delta t}+e^{-i(s_1-s_1'-(s_2-s_2'))\Omega\Delta t}\right)\left|\sum_n e^{i(\vec{Q}+(s_1+s_2)\vec{\Lambda})\vec{n}}\right|^2 \tag{48}$$

for traveling and standing sound waves respectively.

The statistical average here

$$w = \bar{\nu} = \sum_{S,\vec{q}} \frac{\hbar}{2MN\Omega_S(\vec{q})} \left(\vec{Q}\vec{e}_S(\vec{q})\right)^2 \left(\frac{1}{2} + \bar{\nu}_{Sq}\right),$$

with

$$\bar{\nu}_{Sq} = \left[\exp(\hbar\Omega_S(\vec{q})/(kT)) - 1\right]^{-1}$$

is the average number of bosons (thermal phonons). The averaging is performed according to Wick's theorem [32].

Expressions (47) and (48) must be simplified the reasons for which are outlined as follows. Since real processes occur in the finite space-time interval, and exponents $\exp\left[-i(s-s')\omega\Delta t\right]$ and $\exp\left[-i(s_1 - s_1' - (s_2 - s_2'))\Omega\Delta t\right]$ are strongly oscillating functions, naturally their maximal values can be obtained under the corresponding conditions $s = s'$ and $s_1 - s_1' = s_2 - s_2'$; the last one can in turn be identically written as $s_1 = s_1'$ and $s_2 = s_2'$. Based on this argument and using the representation $\pi\delta(X) = \lim\limits_{L\to\infty}[\sin(LX)/X]$ of the δ-function, we obtain scattering probabilities

$$W_{\vec{P}_f,\vec{P}_i} = \frac{(2\pi)^3\hbar^2A^2}{m^2} \sum_{s=-\infty}^{+\infty} e^{-2w} J_s^2(\vec{Q}\vec{b}) \left|\sum_n e^{i(\vec{Q}+s\vec{k})\vec{n}}\right|^2 \delta(\Omega_s), \tag{49}$$

$$W_{\vec{P}_f,\vec{P}_i} = \frac{(2\pi)^3\hbar^2A^2}{m^2} \sum_{s_1,s_2=-\infty}^{+\infty} e^{-2w} J_{s_1}^2(\vec{Q}\vec{a}) J_{s_2}^2(-\vec{Q}\vec{a}) \left|\sum_n e^{i(\vec{Q}+(s_1+s_2)\vec{\Lambda})\vec{n}}\right|^2 \delta(\Omega_{s_1s_2}). \tag{50}$$

Next, we consider that for large N the sum with respect to the vector $\vec{n}$ can be replaced by the rule $\sum\limits_n \cdots = (N/V)\int \cdots \mathbf{d}\vec{n}$, where the integration is performed in the volume corresponding to the first Brillouin zone, where V is the volume occupied by the crystal. We take into account the representation of the square of the δ-function as well: $\delta^2(\vec{r}) = \delta(\vec{r})/\left(v(2\pi)^3\right)$, where $v = \vec{a}_1[\vec{a}_2\vec{a}_3] = V/N$ is the elementary cell volume of the direct lattice. Thus, the squared modulus of the sum may be replaced by δ-function as follows:

$$\left|\sum_n e^{i(\vec{Q}+s)\vec{k})\vec{n}}\right|^2 = \frac{(2\pi)^3N}{V}\delta\left(\vec{Q} + s\vec{k} + \vec{g}\right) \quad \textit{for the traveling sound wave,}$$

$$\left|\sum_n e^{i(\vec{Q}+(s_1+s_2)\vec{\Lambda})\vec{n}}\right|^2 = \frac{(2\pi)^3N}{V}\delta\left(\vec{Q} + (s_1+s_2)\vec{\Lambda} + \vec{g}\right) \quad \textit{for the standing sound wave,}$$

where $\vec{g}$ is the reciprocal lattice vector defined by the expression $\vec{g} = g_1\vec{b}_1 + g_2\vec{b}_2 + g_3\vec{b}_3$ (g_i are integers, $\vec{b}_i$ are reciprocal lattice elementary vectors).

Finally, the expression

$$W_{\vec{P}_f,\vec{P}_i} = \frac{(2\pi)^6\hbar^4A^2N}{m^2V} \times \sum_{s=-\infty}^{+\infty} e^{-2w} J_s^2(\vec{Q}\vec{b})\delta\left(E_i - E_f + s\hbar\omega\right)\delta\left(\vec{P}_i - \vec{P}_f + s\hbar\vec{k} + \hbar\vec{g}\right) \tag{51}$$

for neutron diffraction in a crystal under the influence of a traveling sound wave is obtained, and subsequently the expression

$$W_{\vec{P}_f,\vec{P}_i} = \frac{(2\pi)^6\hbar^4 A^2 N}{m^2 V} \sum_{s_1,s_2=-\infty}^{+\infty} e^{-2w} J_{s_1}^2(\vec{Q}\vec{a}) J_{s_2}^2(-\vec{Q}\vec{a})$$
$$\times\, \delta\left(E_i - E_f + (s_1 - s_2)\,\hbar\Omega\right) \delta\left(\vec{P}_i - \vec{P}_f + (s_1 + s_2)\hbar\vec{\Lambda} + \hbar\vec{g}\right) \quad (52)$$

for neutron diffraction in a crystal under the influence of a standing sound wave.

The term $\exp(-2w)$ appearing in the last expression is the well-known Debye–Waller factor, a measure of the influence of thermal motion on observation of violations of lattice periodicity. The Debye–Waller factor can be tuned by means of energy and momentum conservation laws for the closed systems consisting of neutron, crystal and traveling sound wave (neutron, crystal and standing sound wave). Specifically, the δ-functions indicate these conservation laws. Being a resonance condition they illustrate neutron diffraction under the stimulated process of phonon emission (or absorption).

5. Analysis of the Results: Diffraction Condition and the Debye–Waller Factor

5.1. Diffraction condition

Neutron diffraction conditions are derived from the expressions (51) and (52) by means of energy and momentum conservation laws. We discuss influences of each type of wave on the neutron diffraction condition separately.

5.1.1. Influence of a traveling sound wave

As seen in formula (51), the energy and momentum conservation laws for the closed, neutron-crystal-sound wave system are written as

$$E_i - E_f + s\hbar\omega = 0, \quad (53)$$

$$\vec{P}_i - \vec{P}_f + s\hbar\vec{k} + \hbar\vec{g} = 0. \quad (54)$$

In expression (53) thermal phonons are absent because in derivation of formula (51) it was assumed that the energy spectrum of vibrations of the crystal lattice does not change: the initial $|\Psi_i\rangle$ and the final $|\Psi_f\rangle$ states are chosen in correspondence with this assumption.

The conservation laws (53) and (54) describe the process of inelastic scattering of neutrons with respect to the field. The value s is both positive and negative and indicates the number of acoustic phonons. Positive values correspond to absorption of phonons from the sound field, while negative values mean stimulated emission of acoustic-frequency phonons. Simultaneously the number of phonons decreases with the increase of the sound frequency. This is obvious, since in stimulated multiphonon processes the number of high-energy phonons is small.

It may be written, with taking into account conservation laws (53) and (54):

$$\cos\theta = \frac{1}{\sqrt{1+\frac{ms\omega\lambda_i^2}{2\pi^2}}}\left[1+\frac{ms\omega\lambda_i^2}{4\pi^2\hbar} - \frac{\lambda_i^2}{8\pi^2}\left(s^2k^2+g^2+2s\vec{k}\vec{g}\right)\right], \tag{55}$$

where $\lambda_i = 2\pi\hbar/P_i = 2\pi\hbar/\sqrt{2mE_i}$ is the de Broglie wavelength of the scattering neutron and θ is the scattering angle (the angle between the initial and final directions of the neutron momentum).

Expression (55) shows that in the absence of the field the well-known Bragg condition is obtained, $2d\sin\theta/2 = \lambda_n$, where $d = 2\pi/|\vec{g}|$ is the distance between the atomic planes (lattice parameter).

Based on the diffraction condition (55) the dependence of the diffraction angle both from phonon number and the initial neutron wavelength is established in figures 1-4. Curves are outlined for the lattice parameter $d \sim 10^{-8}$ cm, hypersonic frequency $\omega \sim 10^{10}$ Hz and wave vector $k \sim 10^5$ 1/cm, and for the positive number of phonons, also $\vec{k}$ is parallel to $\vec{g}$.

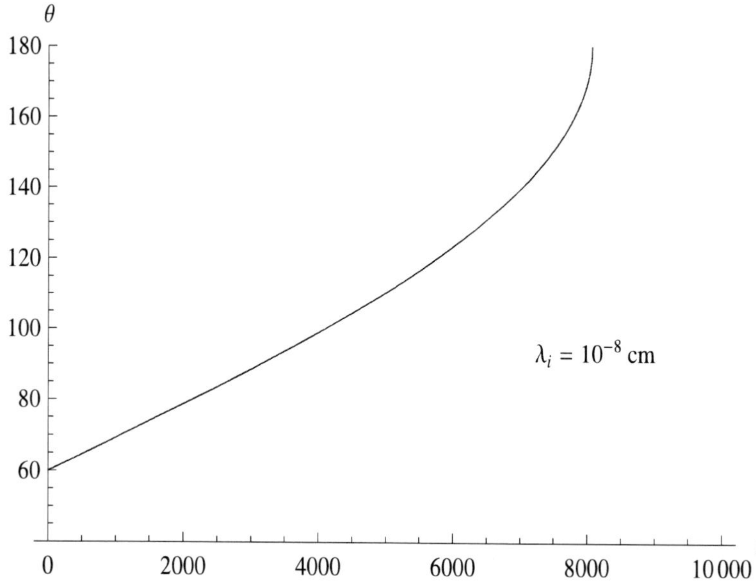

Figure 1. The thermal neutron diffraction angle vs. phonon number dependence for the lattice parameter $d \sim 10^{-8}$ cm, hypersonic frequency $\omega \sim 10^{10}$ Hz and wave vector $k \sim 10^5$ 1/cm. The diffraction angle is expressed in degrees.

The condition of non-negativity of the radicand gives that the number of phonons is limited. Figure 1 displays the diffraction angle of the thermal neutron depending on phonon number restricted by $s < 0.8 \cdot 10^3$. Figures 2 and 3 display the same dependence for the neutron with wavelengths and phonon number limitations $\lambda_i = 10^{-9}$ cm, $\lambda_i = 10^{-10}$ cm and $s < 1.4 \cdot 10^3$, $s < 10^6$ respectively. As seen with a decrease of a neutron wavelength the diffraction angle, which corresponds to a very few number of phonons absorbed by the crystal, is also decreasing. For the thermal neutrons it is $\sim 60^o$, for neutrons with wavelengths $\lambda_i = 10^{-9}$ cm and $\lambda_i = 10^{-10}$ cm the diffraction angle takes values between the intervals

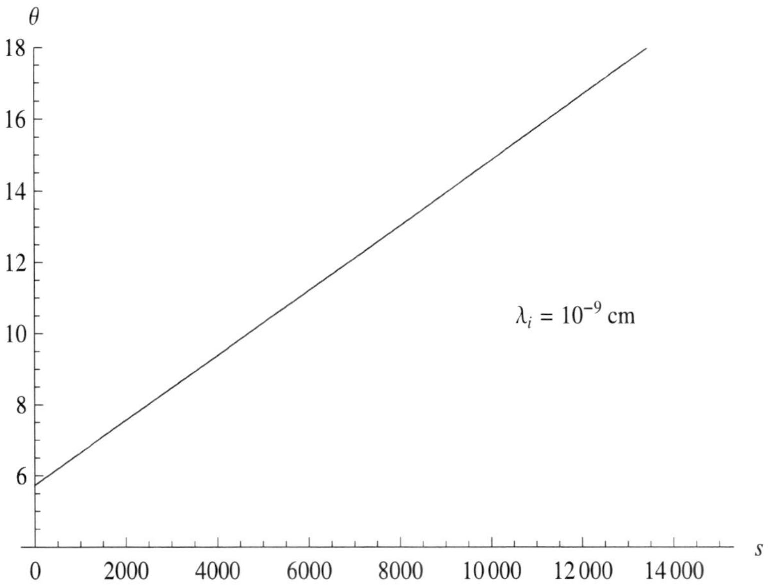

Figure 2. The dependence of short-wave (10^{-9} cm) neutron diffraction angle from phonon number for the lattice parameter $d \sim 10^{-8}$ cm, hypersonic frequency $\omega \sim 10^{10}$ Hz and wave vector $k \sim 10^5$ 1/cm. The diffraction angle is expressed in degrees.

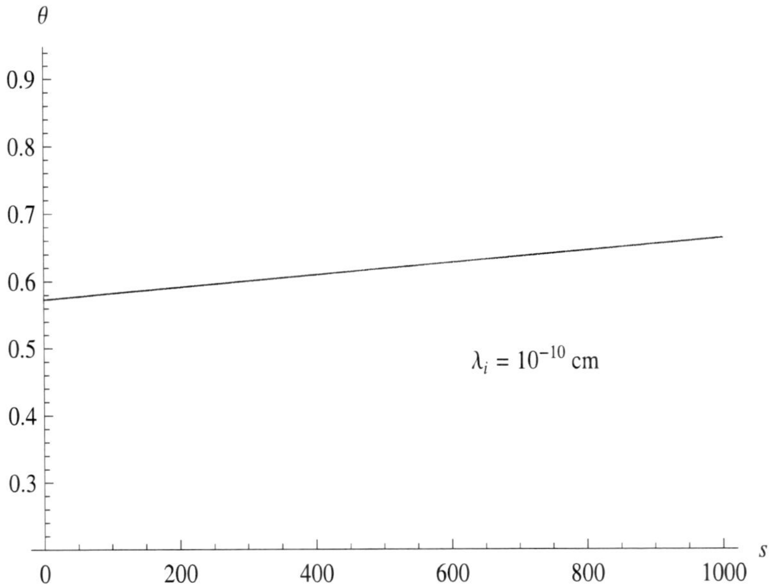

Figure 3. The dependence of short-wave (10^{-10} cm) neutron diffraction angle from phonon number for the lattice parameter $d \sim 10^{-8}$ cm, hypersonic frequency $\omega \sim 10^{10}$ Hz and wave vector $k \sim 10^5$ 1/cm. The diffraction angle is expressed in degrees.

$(5^o \div 6^o)$ and $(0.55^o \div 0.6^o)$ respectively. These angles are in terms that experimentally they can be easily achieved.

In Figure 4 the dependence of the diffraction angle from the neutron wavelength is displayed for three different values of phonons number, $s = 10^2$, 10^3 and 10^4 for the first (red), second (blue) and third (brown) curves respectively. As it seen as shorter the neutron wavelength the diffraction angle much smaller.

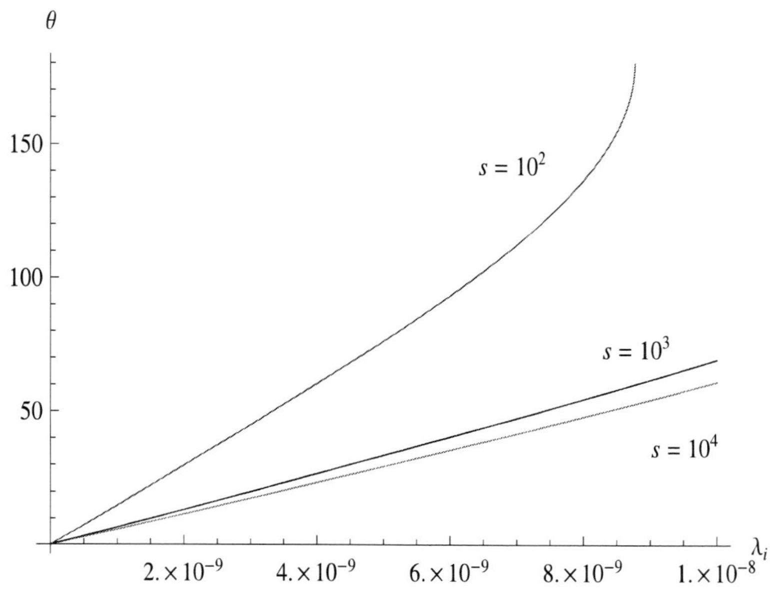

Figure 4. Diffraction angle vs. neutron wavelength dependence for the lattice parameter $d \sim 10^{-8}$ cm, hypersonic frequency $\omega \sim 10^{10}$ Hz and wave vector $k \sim 10^5$ 1/cm. The neutron wavelength is expressed in centimeters.

5.1.2. Influence of a standing sound wave

The energy and momentum conservation laws in the case of neutron elastic scattering with respect to a crystal under the influence of an external standing sound wave have forms

$$E_i - E_f + (s_1 - s_2)\,\hbar\Omega = 0\,, \tag{56}$$

$$\vec{P}_i - \vec{P}_f + (s_1 + s_2)\hbar\vec{\Lambda} + \hbar\vec{g} = 0. \tag{57}$$

The numbers s_1 and s_2 in (56) and (57) have both positive and negative values. They show the number of phonons in the external standing sound wave. Positive values correspond to the process of phonon absorption from the standing sound wave, negative values correspond to phonon emission. Illustratively, one can talk about s_1 phonons which propagate in one direction, and s_2 phonons which propagate in the opposite direction.

The conservation law (56) describes neutrons undergoing both elastic and inelastic scattering with respect to the field of the standing sound wave. First, we examine the process of neutron elastic scattering. From (56) it is evident that the phonon energy is not changed

$(E_{i(s_1=s_2)} = E_{f(s_1=s_2)})$ when $s_1 = s_2 \equiv s$, i.e. the same number of phonons participate in the absorption and emission processes. Under these conditions the momentum conservation law (57) takes the following form:

$$\vec{P}_i - \vec{P}_f + 2s\hbar\vec{\Lambda} + \hbar\vec{g} = 0\,, \tag{58}$$

which in turn can be written in the form

$$4d^2\sin^2\frac{\theta}{2} = \frac{1}{\pi^2}s^2\Lambda^2\lambda_n^2 d^2 + \lambda_n^2 + \frac{1}{\pi^2}s\vec{\Lambda}\vec{g}\lambda_n^2 d^2, \tag{59}$$

where λ_n is the neutron wavelength (the initial and final neutron wavelengths are the same for this case), θ is the scattering angle (the angle between the initial and final directions of the neutron momentum) and $d = 2\pi/\,|\vec{g}|$ is the interatomic plane distance. Note, that in this case the Bragg diffraction condition is obtained from (59) when the standing sound wave is absent.

Illustratively, the relation (59) can be written in the form

$$2d_{eff}\sin(\theta/2) = \lambda_n,$$

where

$$d_{eff} \equiv d/\sqrt{1 + s^2\Lambda^2 d^2/\pi^2 + s\vec{\Lambda}\vec{g}d^2/\pi^2} \tag{60}$$

is an effective period of a certain *sublattice* which is formed in the process of neutron interaction with the crystal and the standing sound wave.

The term $2s\hbar\vec{\Lambda}$ in the expression (58) can be interpreted in the following way. A standing wave can be considered as a set of phonons with momenta $\hbar\vec{\Lambda}$ and $-\hbar\vec{\Lambda}$, and the neutron scattering can be considered as the absorption of a phonon with momentum $\hbar\vec{\Lambda}$ and the stimulated emission of the phonon with the momentum $-\hbar\vec{\Lambda}$. As a result, the neutron momentum increases by $2s\hbar\vec{\Lambda}$ in addition to the part $\hbar\vec{g}$, and emitted into the θ angle without changing its value. This situation is very similar to the Kapitza–Dirac effect - the elastic scattering of an electron in the field of a standing light wave [33], this has been also discussed in [34, 35].

Now, let us consider the neutron's inelastic scattering. From the conservation laws (56) and (57) one has

$$\cos\theta = \frac{1}{\sqrt{1 + \frac{m(s_1-s_2)\Omega\lambda_i^2}{2\pi^2}}} \times \left[1 + \frac{m(s_1 - s_2)\Omega\lambda_i^2}{4\pi^2\hbar} - \frac{\lambda_i^2}{8\pi^2}\left((s_1+s_2)^2\Lambda^2 + g^2 + 2(s_1+s_2)\vec{\Lambda}\vec{g}\right)\right], \tag{61}$$

where λ_i is the de Broglie wavelength of the initial neutron, and the traditional Bragg diffraction condition is also obtained in the case of the absence of the standing sound wave. Analyzing the radicand we get the phonon number limitation: $s_1, s_2 < 10^5$.

Relations (60) and (61) allow us to theoretically predict values of certain physical quantities measured in the process of neutron diffraction. Particularly, they are important for the

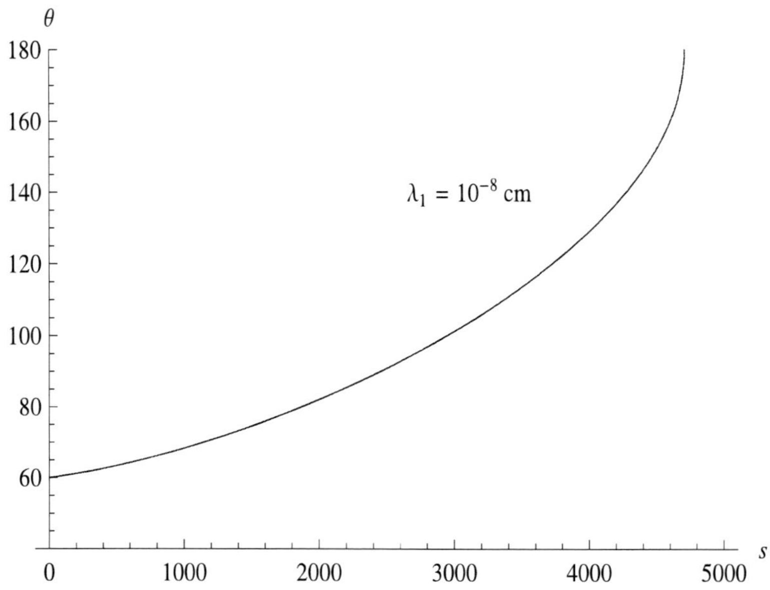

Figure 5. The thermal neutron diffraction angle vs. phonon number dependence for the lattice parameter $d \sim 10^{-8}$ cm, hypersonic frequency $\omega \sim 10^{10}$ Hz and wave vector $k \sim 10^5$ 1/cm. The diffraction angle is expressed in degrees, the sublattice has the order of 10^8 cm.

structural analysis of crystalline condensed matter. They can be used both for heavy and light nuclei matter for all scattering angles $\theta \neq 0$.

Here we bring the dependence of diffraction angle from phonon numbers and the neutron wavelength in case of neutrons elastic diffraction when the sublattice is formed. Curves are outlined for the lattice parameter $d \sim 10^{-8}$ cm, hypersonic frequency and wave vector $\Omega \sim 10^{10}$ Hz and $\Lambda \sim 10^5$ 1/cm respectively in case of phonons absorption by the crystal. As well as the vector $\vec{\Lambda}$ is parallel to $\vec{g}$. In Figures 5–7 the dependence of the diffraction angle from phonon number is displayed for the neutrons with the wavelengths 10^{-8} cm, 10^{-9} cm and 10^{-10} cm respectively. The corresponding limitations of the phonon number are $s < 5 \cdot 10^3$, $s < 6.3 \cdot 10^4$ and $s < 6.3 \cdot 10^5$. We see that the diffraction angle corresponding to an absorbtion of a very few phonons decreases with a decrease of the neutron wavelength. The angle is $\sim 60^o$, $\sim 4^o$ and $\sim 0.2^o \div 0.3^o$ for the sublattice parameter 10^{-8} cm, 10^{-9} cm and 10^{-10} cm respectively. Figure 8 displays the dependence of the neutron elastic diffraction angle from the neutron wavelength. The first (red), second (blue) and the third (brown) curves correspond to phonons number $s = 10^2$, 10^3 and 10^4 respectively. Finally note that conditions (59) and (61) can be used in experimental estimations of characteristic physical quantities, such as neutron energy, interatomic plane distances, both thermal and acoustic phonons energy and momentum.

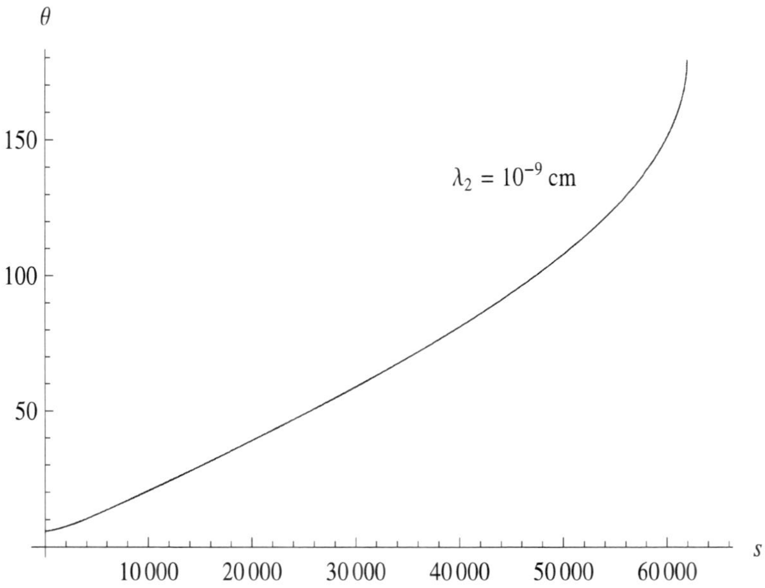

Figure 6. The 10^{-9} cm neutron diffraction angle vs. phonon number dependence for the lattice parameter $d \sim 10^{-8}$ cm, hypersonic frequency $\omega \sim 10^{10}$ Hz and wave vector $k \sim 10^5$ 1/cm. The diffraction angle is expressed in degrees, the sublattice has the order of 10^{-9} cm.

5.2. The Debye–Waller factor

Now, we turn to the examination of the Debye–Waller factor. Depending on the temperature and properties of the crystal, it can lead to the damping of the elastic coherent scattering for all scattering angles $\theta \neq 0$. The quantity w increases with an increase of the scattering angle, neutron energy and crystal temperature. When $T \to 0$ the function $\bar{\nu}_{Sq}$ approaches zero and the term $\exp(-2w)$ consequently takes its maximal value. In this case, when considering the neutrons scattering in a crystal in the absence of the external standing sound wave, the Debye–Waller factor $\exp(-2w) \sim 1$. A shift of the nuclei from the equilibrium position has essentially no effect on the coherent scattering intensity; nuclei become more inert, and neutrons scatter on the motionless crystal structure. In the process some difficulties appear connected with the structural analysis of light nuclei matter should the scattering intensity decrease. The consideration of neutron diffraction in such media with the presence of the standing sound wave can eliminate this problem, and vice versa, under a convenient choice of physical quantities, the diffraction intensity can be suppressed. All these scenarios are realized if the conservation laws are taken into account.

With use of formula (54) for the traveling sound wave the quantity w may be represented as

$$w = \sum_{S,\vec{q}} \frac{\hbar}{2MN\Omega_S(\vec{q})} \left((s\vec{k} + \vec{g})\vec{e}_S(\vec{q})\right)^2 \left(\frac{1}{2} + \bar{\nu}_{Sq}\right). \tag{62}$$

Using the momentum conservation law (58) in the case of elastic scattering the quantity w

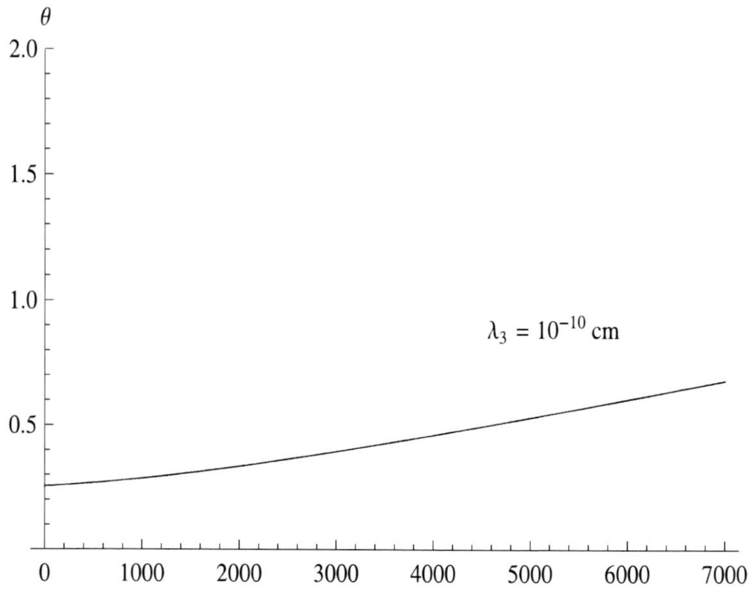

Figure 7. The 10^{-10} cm neutron diffraction angle vs. phonon number dependence for the lattice parameter $d \sim 10^{-8}$ cm, hypersonic frequency $\omega \sim 10^{10}$ Hz and wave vector $k \sim 10^5$ 1/cm. The diffraction angle is expressed in degrees, the sublattice has the order of 10^{-10} cm.

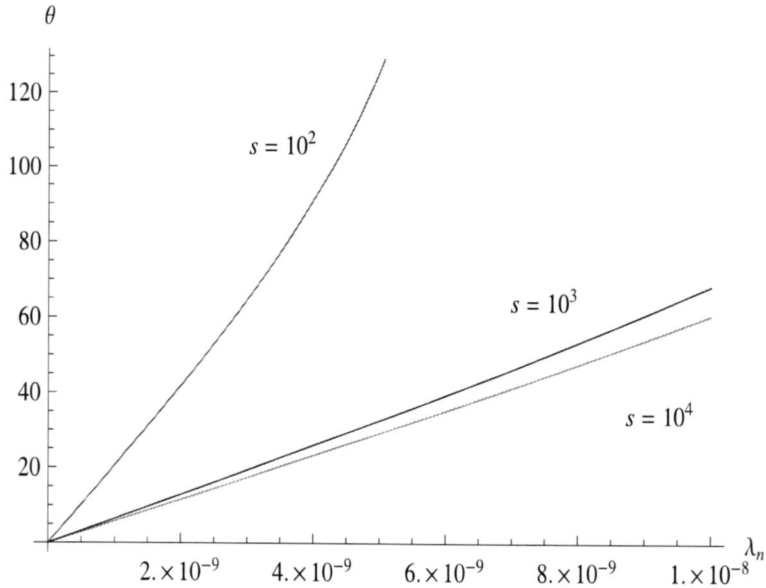

Figure 8. Scattering angle vs. neutron wavelength dependence for the lattice parameter $d \sim 10^{-8}$ cm, hypersonic frequency $\omega \sim 10^{10}$ Hz and wave vector $k \sim 10^5$ 1/cm. The neutron wavelength is expressed in centimeters.

can be written in the following form:

$$w = \sum_{S,\vec{q}} \frac{\hbar}{2MN\Omega_S(\vec{q})} \left((2s\vec{\Lambda} + \vec{g})\vec{e}_S(\vec{q}) \right)^2 \left(\frac{1}{2} + \bar{\nu}_{Sq} \right), \tag{63}$$

and using the momentum conservation law (57), in the case of inelastic scattering, w can be written in the form

$$w = \sum_{S,\vec{q}} \frac{\hbar}{2MN\Omega_S(\vec{q})} \left(\left((s_1 + s_2)\vec{\Lambda} + \vec{g} \right) \vec{e}_S(\vec{q}) \right)^2 \left(\frac{1}{2} + \bar{\nu}_{Sq} \right). \tag{64}$$

When $\vec{k} = 0$, $s = 0$ ($\vec{\Lambda} = 0$, $s_1 = s_2 = 0$), i.e. the sound waves are absent, the well-known expression for the Debye–Waller factor is obtained [29]. It becomes obvious from (62)-(64) that the Debye–Waller factor can be tuned under a correspondingly convenient choice of parameters for a given standing sound wave (number of phonons, frequency and wave vector). This can be achieved either by suppressing or increasing the neutron coherent scattering intensity even when the crystal temperature is not equal to zero ($T \neq 0$) and the medium consists of light nuclei atoms. It is also clear that the scattering intensity may either decrease or increase depending on the direction of the vectors $\vec{g}$ and $\vec{k}$ ($\vec{\Lambda}$). The Debye–Waller factor obtains its maximal value when $\vec{g} = -2s\vec{\Lambda}$, $\vec{g} = -(s_1 + s_2)\vec{\Lambda}$ and $\vec{g} = -s\vec{k}$ in the cases of elastic scattering with respect to the standing sound wave, inelastic scattering with respect to the standing sound wave and inelastic scattering with respect to the traveling sound wave respectively. These conditions can be realized in experiments with hypersonic waves with frequencies $\Omega \sim 10^8 - 10^{10}$ Hz.

Acknowledgments

The authors extend their sincerest gratitude to Mr. Sean McConnell for his proofreading of the text.

References

[1] W. Marshall, S. W. Lovesey, *Theory of Thermal Neutron Scattering* (Oxford: Univ. Press, 1971).

[2] G. E. Bacon, *Neutron Diffraction* (Oxford: Clarendon Press, 1975).

[3] I. I. Gurevich, L. V. Tarasov, *Fizika Neytronov Nizkikh Energij* (*Low Energy Neutron Physics*, Moscow, Nauka, 1965).

[4] Yu. Z. Nozik, R. P. Ozerov, K. Hennig, *Neytroni i Tverdoe Telo: Strukturnaya Neytronografiya, tom 1* (*Neutrons and Solids: Structural Neutron Diffractometry, vol. 1*); *Neytroni i Tverdoe Telo: Neytronografiya Magnetikov, tom 1* (*Neutrons and Solids: Neutron Diffractometry of Magnetics, vol. 2*); *Neytroni i Tverdoe Telo: Neytronnaya Spektroskopiya, tom 3* (*Neutrons and Solids: Neutron Spectroscopy, vol. 3*) (Moscow, Atomizdat, 1979).

[5] R. Golub, R. Gähler, T. Keller, Am. J. Phys., 62, 9, 779–788 (1994).

[6] V. K. Ignatovich, F. V. Ignatovich, Am. J. Phys., 71, 10, 1013–1024 (2003).

[7] V. K. Ignatovich, *Neytronnaya Optika* (*Neutron Optics*, Moscow, Fizmatlit, 2006).

[8] V. V. Fedorov, V. V. Voronin, NIMB, 201, 1, 230–242 (2003).

[9] S. H. Kilcoyne, Physica B: Cond. Matter, 350, 1-3, 91–97 (2004).

[10] T. Fukunaga, K. Itoh, S. Orimo, K. Aoki, Mat. Sci. Eng. B, 108, 1-2, 105–113 (2004).

[11] V. V. Fedorov, I. A. Kuznetsov, E. G. Lapin, S. Yu. Semenikhin, V. V. Voronin, NIMA, 593, 3, 505–509 (2008)

[12] T. F. Koetzle, P. M. B. Piccoli, A. J. Schultz, NIMA, 600, 1, 260–262 (2009).

[13] V. V. Fedorov, M. Jentschel, I. A. Kuznetsov, E. G. Lapin, E. Lelievre-Berna, V. Nesvizhevsky, A. Petoukhov, S. Yu. Semenikhin, T. Soldner, F. Tasset, V. V. Voronin, Yu. P. Braginetz, NIMA, 611, 2-3, 124–128 (2009).

[14] M. Agamalian, E. Iolin, L. Rusevich, C. J. Glinka, G. D. Wignall, Phys. Rev. Lett. 81, 3, 602–605 (1998).

[15] R. Simonovits, Phys. Lett. A, 199, 15–20 (1995).

[16] Q.-G. Lin, Phys. Lett. A, 342, 67–76 (2005).

[17] B. Sur, V. N. P. Anghel, R. B. Rogge, J. Katsaras, Phys. Rev. B 71, 014105 (2005).

[18] A. R. Mkrtchyan, A. G. Hayrapetyan, B. V. Khachatryan, R. G. Petrosyan, Phys. At. Nucl., 73, 3, 478–484 (2010).

[19] A. G. Hayrapetyan, J. Cont. Phys. (Arm. Ac. Sci.), 44, 4, 168–173 (2009).

[20] K. K. Grigoryan, A. G. Hayrapetyan, R. G. Petrosyan, NIMB, 268, 2366–2370 (2010).

[21] K. K. Grigoryan, A. G. Hayrapetyan, R. G. Petrosyan, NIMB, 268, 2539-2543 (2010).

[22] A. Messiah, *Quantum Mechanics* (Amsterdam: North Holland, vol. 1,2, 1970).

[23] P. Mikula, R. Michalec, B. Chalupa, L. Sedlakova, V. Petrzilka, Acta Cryst. A 31, 688–693 (1975).

[24] B. Chalupa, R. Michalec, L. Horalik, P. Mikula, Phys. Stat. Sol. (a) 97, 403–409 (1986).

[25] A. R. Mkrtchyan, R. G. Gabrielyan, H. A. Hunanyan, A. G. Beglaryan, Izvestiya Ak. Nauk Arm.SSR, Fizika (Proc. Ac. Sci. Arm.SSR, Physics), 21, 6, 313–316 (1986) (in Russian).

[26] R. Michalec, P. Mikula, M. Vrana, J. Kulda, B. Chalupa, L. Sedlakova, Physica B+C, 151, 113- 121 (1988).

[27] J. Kulda, M. Vrana, P. Mikula, Physica B+C, 151, 113- 121 (1988).

[28] C. Kittel, *Quantum Theory of Solids* (John Wiley and Sons, Inc., New York, London, 1963).

[29] A. S. Davydov, *Teoriya Tverdogo Tela* (*Theory of Solids*, Moscow, Nauka, 1976).

[30] W. Xu, J. Phys.: Condensed Matter 10, 6105–6120 (1998).

[31] A. R. Mkrtchyan, R. M. Avakyan, A. G. Hayrapetyan, B. V. Khachatryan, R. G. Petrosyan, Arm. J. Phys., 2, 4, 258–267 (2009).

[32] L. D. Landau, E. M. Lifshitz, L. P. Pitaevskii, *Electrodynamics of Continuous Media* (Elsevier Butterworth-Heinemann, 1984).

[33] P. L. Kapitza and P. A. M. Dirac, Proc. Cam. Philos. Soc. 29, 297 (1933).

[34] D. L. Freimund, K. Aflatooni, H. Batelaan, Nature 413, 142–143 (2001).

[35] H. Batelaan, Rev. Mod. Phys. 79, 929–941 (2007).

In: Sound Waves: Propagation, Frequencies and Effects ISBN 978-1-61470-001-2
Editor: V. Abagnali and G. Fabbri

Chapter 8

ON THE TRANSFORMATION OF SOUND WAVES IN NON-STATIONARY MEDIA

***A. G. Hayrapetyan*[1,*]*, K. K. Grigoryan*[2]*, R. G. Petrosyan*[2] *and B. V. Khachatryan*[2]**
[1]Max-Planck-Institut für Kernphysik
Postfach 103980, D-69029 Heidelberg
Germany
[2]Yerevan State University
1 Alex Manoogian Str., 0025 Yerevan
Armenia

Abstract

In this chapter the transformation of sound waves in the fluid dynamical approach is considered. In particular the reflection and transmission of sound waves for spatially homogeneous non-stationary media are considered. Time dependence of the properties of the medium (for example, mass density, sound velocity) is induced from the abruptness of its changes. Reflection and transmission coefficients for both sound wave amplitudes and energy fluxes are obtained. Quantitative relations between the reflection and transmission coefficients are adduced. It is shown that the sum of the energy flux reflection and transmission coefficients is greater than one and that the energy of a sound wave is not conserved, that is, exchange of the energy occurs between the wave and the medium. The non-conservation of the energy causes the sound wave to obtain a notable property: the transmitting wave carries an energy equal to the sum of the energies of the incident and reflected waves. The possibility of the amplification of sound waves is illustrated and a transformation of their frequencies is shown: the physical justification of these effects is also given. The problem is generalized for smooth variation of mass density and sound velocity and the transmission and reflection coefficients are obtained from the solution of generalized wave equation written for the potential of the velocity of the fluid.

*E-mail address: armen@physi.uni-heidelberg.de, armen.hayr@gmail.com

1. Introduction

Physical theory and experiment are based on the fundamental conservation laws of energy, momentum, angular momentum, as well as parity conservation law and selection rules etc., which are peculiar to the physics of microsystems. The energy, momentum and angular momentum conservation laws are the principal conservation laws in that they are connected with the properties of time and space (homogeneity and isotropy) [1].

The time homogeneity implies the energy conservation law. For a closed system the Lagrange function does not explicitly depend on time, i.e. the choice of the origin of time is entirely arbitrary. This fact implies the energy conservation law which is furthermore valid for open systems if these open systems are subjected to an external constant influence which does not depend on time.

The momentum conservation law is determined from the homogeneity of the space, this means that properties of the system are not changed under any translation in the space. As a consequence of the spatial isotropy, the angular momentum conservation law leaves the physical properties of a closed system the same under any rotation of the whole system. The angular momentum conservation law is also valid for open systems in external fields in a more restricted form. A projection of the angular momentum on any axis is conserved when an external field is symmetric with respect to that axis.

Conservation laws are not upheld under violations of the properties of space and time. Under these conditions, a physical insight can be gained through observation of certain physical effects. For instance transformation (reflection and refraction) of monochromatic plane electromagnetic (EM) waves from the interface of two spatially homogeneous media with different values of refractive index (or dielectric permittivity) illustrates the violation of the momentum conservation law for an EM wave [2]. Transformation of EM radiation in a medium with diverse types of spatial properties is of special interest [2, 3, 4], one of the simplest problems in this area is transition radiation (also diffraction radiation) of uniformly and rectilinearly moving charges which cross the interfaces of two or more homogeneous media.

In scientific literature, the transformation of sound waves and partially acoustic waves is considered from various aspects [5, 6, 7]. Sound waves propagating in media with spatial inhomogeneities undergo reflection and refraction, for instance the simplest problem is reflection and refraction of a sound wave from the interface of two homogeneous media described by two different values of mass density, pressure and sound velocity [5]. Note that in this case the energy of the system which consists of three waves (incident, reflected and refracted) is conserved, but the momentum is not, which is a consequence of the inhomogeneity of the space.

Consideration of physical phenomena in non-stationary media is of great interest [3, 4, 8, 9, 10, 11, 12], of most interest are ones considered in spatial homogeneous non-stationary media, the properties of which can be changed either abruptly or smoothly. From the mathematical point of view the situation is very similar to an analogous process in stationary spatial inhomogeneous media. Nevertheless, this type of phenomena has its own specificity. Moreover, due to the time inhomogeneity new effects have arisen: energy exchange between the medium and the waves, transformation of frequencies of the propagating waves in non-stationary media, generation of frequency difference etc. Though processes in non-stationary media have been considered since the 1970's [8, 9, 10, 11].

The first time that the transformation of EM waves (transition radiation) in a non-stationary medium was considered was in the work by Ginzburg [8]. Based on simple physical reasoning he proposed an original idea about a new type of transition radiation under the abrupt change of properties of the medium in time. Specifically, one defines that optical properties of a medium abruptly change at $t = 0$ such that for $t < 0$ the refractive index is $n = n_1$, and for $t > 0$ it is $n = n_2$. Such a situation arises, for instance, under the variation of pressure both in cases when a medium is subjected to an influence of a spatial homogeneous time-dependent electric field (the field in a condensate) or strong laser radiation when a medium is suddenly changed into plasma. Speaking generally, the field which depends on time is position-dependent as well. Nonetheless, having external field sources at our disposal, in principal, the homogeneity of the entire medium is achieved including along the trajectory of a charge. Besides Ginzburg's idea, enforced processes of acceleration of particles and enhancement of EM waves have been examined in [10, 11]. The idea in [8, 9] has been applied for sound waves in [12] for the first time.

In this chapter, we present general principles of fluid dynamics especially applied for description of sound waves based on [5]. Next, we consider the wave equation for sound waves propagating in a non-stationary medium based on [12]. In the end, we present transformation of sound waves in spatially homogenous non-stationary media with smoothly changing physical quantities. We analytically solve the generalized wave equation for the sound potential and derive the transmission and reflection coefficients as a particular case of the general solution.

2. Basic Equations

Sound waves generally are considered within the framework of fluid dynamics. The medium is considered as a *liquid* with a laminar flow; the viscosity, the turbulence and the gravitational force are neglected (large Reynolds number, $R \gg 1$, large Froude number, $F \gg 1$).

2.1. Basic equations of fluid dynamics

The effects on the state of a moving fluid can be described by functions which give the distribution of the fluid velocity $\vec{v} = \vec{v}\,(x, v, z, t)$ and of any two thermodynamic quantities pertaining to the fluid, for instance the pressure $P\,(x, y, z, t)$ and the mass density $\varrho\,(x, y, z, t)$. The state of a moving fluid is completely described by means of those five quantities which in general are functions of space and time, the spatial coordinates refer to a fixed point in space and not to specific particles of the fluid. The derivative over time of any physical quantity depending on spatial coordinates has to be expressed in terms of quantities referring to points fixed in space. Due to this the derivative for a certain physical quantity F is expressed as

$$\frac{dF}{dt} = \frac{\partial F}{\partial t} + (\vec{v}\vec{\nabla})F\,. \tag{1}$$

The conservation of the matter is expressed by the equation of continuity

$$\frac{\partial \varrho}{\partial t} + \vec{\nabla}\vec{j} = 0, \tag{2}$$

where $\vec{j} = \varrho\vec{v}$ is the mass flux density. It has the same direction as the fluid velocity and the magnitude is the mass of fluid flowing in unit time through unit area perpendicular to the velocity. There is another way of writing of the equation of continuity if one uses the expression (1) for the full derivative of the mass density

$$\frac{d\varrho}{dt} + \varrho(\vec{\nabla}\vec{v}) = 0. \tag{3}$$

The equation of motion of the fluid is described by means of Euler's equation

$$\frac{\partial \vec{v}}{\partial t} + (\vec{v}\vec{\nabla})\vec{v} = -\frac{\vec{\nabla}P}{\varrho}, \tag{4}$$

which with equation of continuity (2) form the fundamental equations of fluid dynamics.

In case of an ideal fluid, i.e. absent of energy dissipation and heat exchange between different parts of the fluid as well as between the fluid and any external matter adjoining it, the motion of the fluid remains adiabatic. Adiabatic motion of any moving particle within the fluid means the entropy is constant. This implies the continuity equation for entropy

$$\frac{\partial(\varrho s)}{\partial t} + \vec{\nabla}(\varrho s\vec{v}) = 0, \tag{5}$$

where s is the entropy per unit mass and $\varrho s\vec{v}$ is the entropy flux density. Eq. (5) can be simplified based on the following. As usually happens, if the entropy is constant throughout the volume of the fluid at some initial instant, it retains everywhere the same constant value at all times and for any subsequent motion of the fluid. In this case, Eq. (5) takes the form $s = const$. The motion of the fluid concerning this case is called isentropic.

2.2. Sound waves

Alternating compressions and rarefactions may propagate in a compressible fluid, otherwise known as sound wave. For the potential (non-vortical) motion it is convenient to introduce the potential of the fluid velocity $\vec{v}$

$$\vec{v} = \vec{\nabla}\varphi. \tag{6}$$

Since the oscillations and the fluid velocity $\vec{v}$ are small, the relative changes of mass density and pressure are also small. Thus, the variables ϱ and P can be expressed as

$$\varrho = \rho_0 + \rho\,, \quad P = p_0 + p\,, \tag{7}$$

where ρ_0 and p_0 are the constant equilibrium density and pressure, and ρ and p are their variations ($\rho \ll \rho_0, p \ll p_0$) and have the same order as the velocity $\vec{v}$ of fluid particles.

Neglecting small quantities of the second order, the continuity equation (2) and Euler's equation (4) can be written in the corresponding forms:

$$\frac{\partial \rho}{\partial t} + \rho_0 \vec{\nabla} \vec{v} = 0, \tag{8}$$

$$\rho_0 \frac{\partial \vec{v}}{\partial t} + \vec{\nabla} p = 0. \tag{9}$$

Using the concept of the velocity potential (6), Eqs. (8) and (9) take the following forms:

$$\frac{\partial \rho}{\partial t} + \rho_0 \Delta \varphi = 0, \tag{10}$$

$$p = -\rho_0 \frac{\partial \varphi}{\partial t}. \tag{11}$$

Eqs. (10) and (11) contain unknown functions p, ρ and φ. A new equation should be added in order to complete the set of required equations for the identification of these unknown functions. For isentropic motion, the pressure and density are connected linearly, their connection is given by means of the state equation

$$p = \left(\frac{\partial P}{\partial \rho_0} \right)_s \rho \, , \tag{12}$$

where

$$V = \sqrt{\left(\frac{\partial P}{\partial \rho_0} \right)_s} \tag{13}$$

is the velocity of sound, and $v \ll V$.

Using Eqs. (10), (11), (12) and (13), one obtains the following wave equation for the velocity potential:

$$\Delta \varphi - \frac{1}{V^2} \frac{\partial^2 \varphi}{\partial t^2} = 0 \, , \tag{14}$$

which describes the propagation of a sound wave in a fluid dynamical medium (this can be both liquid and gas).

Now, we consider the wave equation when the sound wave propagates in spatially homogeneous non-stationary media. The solution of Eq. (14) can be presented in the form of an expansion with respect to Fourier integral with regard to spatial coordinates

$$\varphi(\vec{r}, t) = \frac{1}{(2\pi)^{3/2}} \int e^{i\vec{K}\vec{r}} \varphi_{\vec{K}}(t) d\vec{K} \, , \tag{15}$$

where $\vec{K}$ is the wave vector of the sound wave.

Substituting the expansion (15) in the wave equation (14) one obtains the following equation for the Fourier-image:

$$\frac{d^2 \varphi_{\vec{K}}}{dt^2} + \Omega^2 \varphi_{\vec{K}} = 0 \, , \tag{16}$$

where Ω is the frequency of the sound wave.

Wave equation (16) has two linearly independent solutions

$$\varphi_{1\vec{K}} = a_+(\vec{K})e^{-i\Omega t}\,, \tag{17}$$

$$\varphi_{2\vec{K}} = a_-(\vec{K})e^{i\Omega t}\,. \tag{18}$$

The wave (17) propagates *normally* in time with a frequency Ω. From the mathematical point of view the wave (18) is necessary to guarantee the time matching conditions of a general solution

$$\varphi_{\vec{K}} = \varphi_{1\vec{K}} + \varphi_{2\vec{K}} \tag{19}$$

of the wave equation (16). From the physical point of view, according to Feynman and Stueckelberg [13], the solution (18) can be interpreted in the following way. The exponent in the solution (17) characterizes a wave with positive energy and *normal* flow of the time. If one rewrites it in the way $\exp(-i\Omega t) = \exp[-i(-\Omega)(-t)]$, then it can be interpreted as a wave which propagates backward in time with negative energy. Thus, the wave (18) characterizes a wave propagating backward in time but with positive energy, $\exp(i\Omega t) = \exp[-i\Omega(-t)]$.

Given that the velocity potential φ is an observable quantity the essence of $a_-(\vec{K})$ can be represented by the condition $\varphi_{\vec{K}} = \varphi^*_{-\vec{K}}$ which in turn gives

$$a_-(\vec{K}) = -a_+^*(-\vec{K})\,.$$

One may recognize that $a_-(\vec{K})$ describes a wave which flows *normally* in time opposite to the direction of $\vec{K}$.

3. Transformation of Sound Waves in Non-Stationary Media with Abruptly Changing Parameters

The approximation of the abrupt change of the properties of the medium, which models real processes, is allowed based on the mathematical convenience from one side of the problem. From the physical side drastic changes of the properties of the medium occur if their duration τ is much less than the characteristic frequency Ω of the sound wave, $\tau \ll 2\pi/\Omega$,

Let us consider a spatially homogeneous medium and assume that its properties drastically change at some point in time ($t = 0$); let for $t < 0$ fluid density and the sound velocity take the values ρ_1, V_1, and for $t > 0$ the corresponding values ρ_2, V_2.

We write the potentials of *incident*, *reflected* and *transmitted* waves in the following form (hereinafter these concepts relate to sound waves' propagation, reflection and transmission *in time*):

$$\varphi_1 = A_1 e^{i(\vec{K}_1\vec{r} - \Omega_1 t)}\,, \tag{20}$$

$$\varphi_2' = A_2' e^{i\left(\vec{K}_2'\vec{r} - \Omega_2' t\right)}, \tag{21}$$

$$\varphi_2 = A_2 e^{i\left(\vec{K}_2\vec{r} - \Omega_2 t\right)}, \tag{22}$$

where A_1, A_2', A_2 (complex in general) are the amplitudes of incident, reflected, transmitted waves respectively, $\vec{K}_1$, $\vec{K}_2'$, $\vec{K}_2$ and Ω_1, Ω_2', Ω_2 are the corresponding wave vectors and frequencies. If the longitudinal sound wave propagates in the medium with the density ρ_1 for $t < 0$, the drastic change of properties of the medium at the moment $t = 0$ generates two waves in the medium with the density ρ_2 for $t > 0$: reflected with frequency Ω_2' and transmited with Ω_2. To satisfy the boundary conditions it is necessary to assume that a minimum of three waves exist. This explains the presence of the three waves (20)-(22).

A general physical argument supports the choice of potentials in the forms (20)-(22). Since the space is homogeneous, wave vectors are equal ($\vec{K}_1 = \vec{K}_2' = \vec{K}_2 = const$) as a consequence of the momentum conservation. This fact is visually interpreted from the quantum point of view: the momentum of the monochromatic wave is connected with the wave vector by the relation $\vec{P} = \hbar\vec{K}$, where $\hbar$ is Planck's constant. Hence, based on the dispersion relation for sound waves, $\Omega \equiv \vec{K}\vec{V}$, relations are straightforwardly established between the frequencies of the incident, reflected and transmitted waves:

$$\frac{\Omega_1}{V_1} = -\frac{\Omega_2'}{V_2} = \frac{\Omega_2}{V_2}, \tag{23}$$

i.e. the transformation of the frequency of the sound wave occurs (here we have taken into account that the coordinate axis is aligned along the direction of the propagation of the wave, and they are antiparallel for the reflected wave). The relations (23) can be interpreted in the following way. Since time and energy are canonically conjugate quantities and the problem under consideration is non-stationary, the energy in the longitudinal sound wave is not conserved. From the quantum point of view, this means that the frequencies of the propagating waves are different because of the relation between the energy and frequency in quantum physics, $E = \hbar\Omega$. A sign "-" for Ω_2' for a reflected wave causes the time-dependence $e^{i\Omega_2 t}$. While time flows normally for incident ($e^{-i\Omega_1 t}$) and transmitted waves ($e^{-i\Omega_2 t}$) (along the positive direction of the time axis), it flows backward for the reflected one (along the negative direction of the time axis).

Let us turn now to obtaining quantitative relations between the amplitudes and intensities of incident, reflected and transmitted waves. We proceed from the fact that the pressure and the fluid velocity in the sound wave are continuous quantities. Let us make the corresponding values in both media equal at the moment $t = 0$:

$$\Omega_1\rho_1 A_1 e^{i\vec{K}_1\vec{r}} = \Omega_2\rho_2\left(-A_2' e^{i\vec{K}_2'\vec{r}} + A_2 e^{i\vec{K}_2\vec{r}}\right), \tag{24}$$

$$A_1\vec{K}_1 e^{i\vec{K}_1\vec{r}} = A_2'\vec{K}_2' e^{i\vec{K}_2'\vec{r}} + A_2\vec{K}_2 e^{i\vec{K}_2\vec{r}}. \tag{25}$$

Conditions (24) and (25) must be satisfied in the entire space. Because exponents are linearly independent functions, (24) and (25) are true for all values of $\vec{r}$ (x, y, z) at the moment $t = 0$ when $\vec{K}_1 = \vec{K}_2' = \vec{K}_2$. We have already obtained this condition above based on physical reasoning. Thus, we finally get the following system of equalities for amplitudes:

$$A_1 - A_2' = A_2\,, \tag{26}$$

$$\frac{\rho_1 V_1}{\rho_2 V_2} A_1 + A_2' = A_2\,, \tag{27}$$

with the help of which the amplitude and the energy flux reflection and transmission coefficients are easily evaluated. We obtain

$$R_a = \frac{A_2'}{A_1} = \frac{\rho_2 V_2 - \rho_1 V_1}{2\rho_2 V_2} \tag{28}$$

for the amplitude reflection coefficient and

$$T_a = \frac{A_2}{A_1} = \frac{\rho_2 V_2 + \rho_1 V_1}{2\rho_2 V_2} \tag{29}$$

for the amplitude transmission coefficient.

The energy flux reflection coefficient R_e is defined as the ratio of the time average energy flux densities in the reflected and incident waves, and the energy flux transmission coefficient T_e is defined as the ratio of the time average energy flux densities in the transmitted and incident waves. Since the energy flux density in a plane wave is $V\rho v^2$ we have

$$R_e = \frac{V_2 \rho_2 \overline{v_2'^{\,2}}}{V_1 \rho_1 \overline{v_1^2}} = \frac{V_2 \rho_2 \left|A_2'\right|^2}{V_1 \rho_1 \left|A_1\right|^2}$$

and

$$T_e = \frac{V_2 \rho_2 \overline{v_2^2}}{V_1 \rho_1 \overline{v_1^2}} = \frac{V_2 \rho_2 \left|A_2\right|^2}{V_1 \rho_1 \left|A_1\right|^2}$$

for the energy flux reflection and transmission coefficients respectively.

The relation between the amplitude and energy flux coefficients is given by the expressions

$$R_e = \frac{V_2 \rho_2}{V_1 \rho_1} \left|R_a\right|^2, T_e = \frac{V_2 \rho_2}{V_1 \rho_1} \left|T_a\right|^2 .$$

Finally, we obtain the following forms of the energy reflection and transmission coefficients

$$R_e = \frac{\left(\rho_2 V_2 - \rho_1 V_1\right)^2}{4 V_1 V_2 \rho_1 \rho_2} \tag{30}$$

and

$$T_e = \frac{\left(\rho_2 V_2 + \rho_1 V_1\right)^2}{4 V_1 V_2 \rho_1 \rho_2} \tag{31}$$

respectively.

Note that in the case $V_1 = V_2$, $\rho_1 = \rho_2$ one has $R_a = 0$, $T_a = 1$, $R_e = 0$, $T_e = 1$, i.e. only the initial wave propagates through the homogeneous stationary medium.

The following relation follows from (30) and (31):

$$R_e + T_e = \frac{1}{2}\left[\frac{\rho_1 V_1}{\rho_2 V_2} + \frac{\rho_2 V_2}{\rho_1 V_1}\right] \geq 1 . \tag{32}$$

which implies the non-conservation of sound wave energy. In order to agree with the energy conservation law, it is sensible to assume that an exchange of energy occurs between the wave and the medium. This result is quite natural, since the energy conservation law is a consequence of time homogeneity, whereas for the wave time is inhomogeneous. Note that when $V_1 = V_2$, $\rho_1 = \rho_2$ (i.e. time is homogeneous) the energy is conserved.

A notable property of the sound wave in the non-stationary media follows from (30) and (31)

$$T_e - R_e = 1 , \tag{33}$$

which means that as a result of an exchange of energy between the sound wave and the medium, the transmitted wave carries an energy equal to the sum of the energies of the incident and reflected waves. Note that EM waves do not obtain such a property [12].

The energy flux density vector of a plane sound wave is defined as [5]

$$\vec{q} = V \rho v^2 \vec{n} ,$$

where $\vec{n}$ is the unit vector along the direction of propagation of the wave. Since we examine a spatially homogeneous medium we consider the modulus of the energy flux density vector. The variation of the modulus of the energy density flux vector in a sound wave, which is defined as a difference between the time average energy flux density of the incident wave ($\bar{q}_i$) and the sum of the energy flux densities for the reflected ($\bar{q}_r$) and transmitted waves ($\bar{q}_t$), can be written in the following way:

$$\Delta \bar{q} = \bar{q}_i - (\bar{q}_r + \bar{q}_t) . \tag{34}$$

For each wave we have

$$\bar{q}_i = \frac{K_1^2}{2} V_1 \rho_1 |A_1|^2 , \quad \bar{q}_r = \frac{K_2'^2}{2} V_2 \rho_2 |A_2'|^2 , \quad \bar{q}_t = \frac{K_2^2}{2} V_2 \rho_2 |A_2|^2 .$$

Inserting the last expressions into (34) one obtains

$$\Delta \bar{q} = V_1 \overline{W}_1 \left[1 - \frac{1}{2}\left(\frac{\rho_1 V_1}{\rho_2 V_2} + \frac{\rho_2 V_2}{\rho_1 V_1}\right)\right] , \tag{35}$$

where $\overline{W}_1 = \rho_1 \overline{v_1^2}$ is the time average of the energy density in the incident sound wave. Note that $\Delta \bar{q} \leq 0$ for all values of ρ_1, V_1, ρ_2, V_2. This fact proves that the sound wave takes energy from the non-stationary medium. This in turn means that a sound amplification takes place: the transmitted wave is reinforced because of the relation $T_e = 1 + R_e$. This amplification was naturally expected since the sound wave is caused by the medium, it is a result of the alternate compressions and rarefactions in the medium.

In a particular case when $\rho_1 = \rho_2$, $V_1 = V_2$, the variation of the energy flux density does not occur ($\Delta \bar{q} = 0$).

4. Transformation of Sound Waves in Non-Stationary Media with Smoothly Changing Parameters

A more realistic examination of transformation phenomena requires the consideration of the time-dependent physical parameters that change smoothly with time. In order to proceed with the consideration of smoothly changing parameters, let us write the continuity equation (3) expressed in terms of pressure. For this purpose we use the state equation (12). Combining Eqs. (3) and (12) one has

$$\frac{\partial P}{\partial t} + (\vec{v}\vec{\nabla})P + \varrho V^2(\vec{\nabla}\vec{v}) = 0\,, \tag{36}$$

where the form (1) is used for the derivative of pressure.

Taking into account the perturbation (7) both for density and pressure, we rewrite Eq. (36) applying it for sound waves

$$\frac{\partial p}{\partial t} + \rho_0 V^2(\vec{\nabla}\vec{v}) = 0\,,$$

which with Eq. (11) implies

$$\frac{\partial^2 \varphi}{\partial t^2} + \frac{d\ln\rho_0}{dt}\frac{\partial\varphi}{\partial t} - V^2\Delta\varphi = 0\,. \tag{37}$$

Here the equilibrium density is no longer constant, it depends on time $\rho_0 = \rho_0(t)$. This equation is the generalization of the ordinary wave equation (14) should the sound wave propagate through a non-stationary medium.

Taking the Fourier-transformation (15) and considering only its kernel and subject, due to the homogeneity of the space, the following second order differential equation can be found for the Fourier-image of the velocity potential

$$\frac{d^2\varphi_{\vec{K}}}{dt^2} + \frac{d\ln\rho_0}{dt}\frac{d\varphi_{\vec{K}}}{dt} + K^2V^2\varphi_{\vec{K}} = 0\,. \tag{38}$$

Introducing the transition time interval τ and a dimensionless time $\eta = t/\tau$, Eq. (38) obtains the following form:

$$\varphi''_{\vec{K}} + (\ln\rho_0)'\varphi'_{\vec{K}} + K^2\tau^2V^2\varphi_{\vec{K}} = 0\,, \tag{39}$$

where the prime stands for the derivation with respect to η.

In contrary to EM waves where the abrupt change of only one parameter (refractive index) is required [8, 9, 10, 11, 12], for sound waves two independent parameters appear in the *boundary* conditions (24) and (25), mass density and sound velocity. Generalizing and replacing the abrupt change of these parameters with smooth variation, we assume that in the same period τ the sound velocity changes faster than the mass density. Under this assumption the energy flux reflection (30) and transmission (31) coefficients obtain the form

$$R_e = \frac{(V_2 - V_1)^2}{4V_1V_2}\,,\quad T_e = \frac{(V_2 + V_1)^2}{4V_1V_2}\,. \tag{40}$$

As far as the real processes occur in finite space and time intervals, the change of the sound velocity and the mass density are naturally described by the following smoothly changing functions, which can be modeled by sigmoidal functions, for instance,

$$V = \sqrt{V_2^2 + \frac{V_1^2 - V_2^2}{1 + e^\eta}}\,, \tag{41}$$

$$\rho_0 = \frac{\rho_1 \rho_2}{\sqrt{\rho_1^2 + \frac{\rho_2^2 - \rho_1^2}{1 + e^\eta}}}\,, \tag{42}$$

thus, when $\tau \to 0$, then $V \to V_1$, $\rho_0 \to \rho_1$ and $V \to V_2$, $\rho_0 \to \rho_2$ for $t < 0$ and $t > 0$ respectively.

Eq. (39) can be solved analytically. Introducing a new function $u(\eta)$ in the form

$$u = \sqrt{\rho_0}\varphi_{\vec{K}} \tag{43}$$

and representing the derivation of the logarithm in the way $(\ln \rho_0)' \approx \tau\lambda/\rho_0$ ($\Delta\rho \equiv \lambda\tau$ is an increase of equilibrium mass density during the transition period τ, and $\lambda = const$), Eq. (39) obtains the form

$$u'' + f(\eta)u = 0\,, \tag{44}$$

where

$$f = \frac{\lambda^2\tau^2}{4\rho_0^2} + K^2V^2\tau^2\,. \tag{45}$$

The smooth change of sound velocity (41) and mass density (42) gives rise to a smooth change of the f function which can be presented in the form

$$f = f_2 + \frac{f_1 - f_2}{1 + e^\eta} \tag{46}$$

with

$$f_1 = \frac{\lambda^2\tau^2}{4\rho_1^2} + K^2V_1^2\tau^2\,, \quad f_2 = \frac{\lambda^2\tau^2}{4\rho_2^2} + K^2V_2^2\tau^2\,. \tag{47}$$

Inserting (46) into Eq. (44) one obtains the solution of the equation (44) which is expressed by means of hypergeometric functions [14]. Finally, the velocity potential obtains the form

$$\varphi = \frac{C}{(2\pi)^{3/2}\sqrt{\rho_0}} e^{i\vec{K}\vec{r}} \left(-e^\eta\right)^\nu F\left(\alpha, \beta, \gamma, -e^\eta\right)\,, \quad \eta = \frac{t}{\tau} \tag{48}$$

with

$$C = const\,, \ \alpha = \nu - i\sqrt{f_2}\,, \ \beta = \nu + i\sqrt{f_2}\,, \ \gamma = 2\nu + 1\,, \ \nu^2 = -f_1\,. \tag{49}$$

The velocity potential (48) contains the explicit form of all three waves (20)-(22) which can be obtained in the asymptotical case. When $t \to -\infty$, from (48) one has

$$\varphi \to \frac{C}{(2\pi)^{3/2}\sqrt{\rho_1}} e^{i\vec{K}\vec{r}} e^{-i\frac{t\sqrt{f_1}}{\tau}} . \tag{50}$$

The exponent $\exp\left(-it\sqrt{f_1}/\tau\right)$ in (50) can be replaced with $\exp\left(-i\Omega_1 t\right)$, since $\lambda \to 0$ when $\tau \to 0$, faster than τ. Indeed, it becomes obvious if the derivative of the density (which is proportional to λ) with respect to η is written in the following way:

$$\lambda \approx \frac{\rho_1^2\rho_2^2\left(\rho_2^2-\rho_1^2\right)}{2\tau\rho_0\left(\rho_2^2 e^{-t/2\tau}+\rho_1^2 e^{t/2\tau}\right)^2} \to 0\,, \quad \textit{when} \quad \tau \to 0 \quad \textit{for all} \quad t\,.$$

Comparing (50) with (20) the integration constant can be defined

$$C = (2\pi)^{3/2}\sqrt{\rho_1}A_1 \,. \tag{51}$$

In the case when $t \to +\infty$ the argument of the hypergeometric function in (48) is divergent. Thus one has to switch to a convergent variable $-e^{-\eta}$. Using a well-known expression for the hypergeometric functions [14] the velocity potential in the region $t > 0$ can be expressed in the form

$$\begin{aligned}\varphi = A_1\left(-e^{\eta}\right)^{\nu}\sqrt{\frac{\rho_1}{\rho_0}}e^{i\vec{K}\vec{r}}&\left[\frac{\Gamma(\gamma)\Gamma(\beta-\alpha)}{\Gamma(\beta)\Gamma(\gamma-\alpha)}e^{-\alpha\eta}F(\alpha,\alpha+1-\gamma,\alpha+1-\beta,-e^{-\eta})\right.\\ &\left.+\frac{\Gamma(\gamma)\Gamma(\alpha-\beta)}{\Gamma(\alpha)\Gamma(\gamma-\beta)}e^{-\beta\eta}F(\beta,\beta+1-\gamma,\beta+1-\alpha,-e^{-\eta})\right], \end{aligned}\tag{52}$$

which in the asymptotic case yields the superposition of reflected (21) and transmitted (22) waves

$$\varphi \to A_1\sqrt{\frac{\rho_1}{\rho_2}}e^{i\vec{K}\vec{r}}\left[\frac{\Gamma(\gamma)\Gamma(\beta-\alpha)}{\Gamma(\beta)\Gamma(\gamma-\alpha)}e^{i\Omega_2 t}+\frac{\Gamma(\gamma)\Gamma(\alpha-\beta)}{\Gamma(\alpha)\Gamma(\gamma-\beta)}e^{-i\Omega_2 t}\right]. \tag{53}$$

From the other side, in the case of the abrupt change of the parameters, from (21), (22) and (28), (29) one has

$$\varphi(\tau \to 0\,,\; t > 0) = R_a A_1 e^{i\vec{K}\vec{r}} e^{-i\Omega_2' t} + T_a A_1 e^{i\vec{K}\vec{r}} e^{-i\Omega_2 t}\,.$$

From the last two forms one obtains the amplitude reflection and transmission coefficients as

$$R_a = \sqrt{\frac{\rho_1}{\rho_2}}\frac{\Gamma(\gamma)\Gamma(\beta-\alpha)}{\Gamma(\beta)\Gamma(\gamma-\alpha)}, \tag{54}$$

$$T_a = \sqrt{\frac{\rho_1}{\rho_2}}\frac{\Gamma(\gamma)\Gamma(\alpha-\beta)}{\Gamma(\alpha)\Gamma(\gamma-\beta)}. \tag{55}$$

Straightforwardly, one can show that the expressions

$$R_e = \frac{V_2}{V_1}\left|\frac{\Gamma(\gamma)\Gamma(\beta-\alpha)}{\Gamma(\beta)\Gamma(\gamma-\alpha)}\right|^2 , \tag{56}$$

$$T_e = \frac{V_2}{V_1}\left|\frac{\Gamma(\gamma)\Gamma(\alpha-\beta)}{\Gamma(\alpha)\Gamma(\gamma-\beta)}\right|^2 \tag{57}$$

for energy flux reflection and transmission coefficients follow from expressions (54) and (55) in the case when the properties of the medium are changed smoothly.

Using the properties

$$\Gamma(\vartheta+1) = \vartheta\Gamma(\vartheta) ,$$

$$|\Gamma(i\theta)|^2 = \Gamma(i\theta)\Gamma(-i\theta) = \frac{\pi}{\theta\sinh(\pi\theta)}$$

for the Gamma-function, we obtain

$$R_e = \frac{\sinh^2(\pi K(V_2-V_1)\tau)}{\sinh(2\pi K V_1\tau)\sinh(2\pi K V_2\tau)} \tag{58}$$

and

$$T_e = \frac{\sinh^2(\pi K(V_2+V_1)\tau)}{\sinh(2\pi K V_1\tau)\sinh(2\pi K V_2\tau)} \tag{59}$$

for the energy flux reflection and transmission coefficients respectively. These expressions straightforwardly lead us to the energy non-conservation in the sound wave

$$R_e + T_e = \frac{\tanh(\pi K V_2\tau)}{2\tanh(\pi K V_1\tau)} + \frac{\tanh(\pi K V_1\tau)}{2\tanh(\pi K V_2\tau)} \geq 1 \tag{60}$$

and amplification of the transmitted sound wave

$$T_e - R_e = 1 . \tag{61}$$

Especially note that the properties (60) and (61), which gain the sound wave, have general nature since they are valid for all finite transition time τ. Finally, expanding the hyperbolic functions in (58) and (59), we obtain the coefficients (40) under the condition of abrupt change of the sound velocity.

5. Conclusion

We have considered transformation (reflection and transmission) of sound waves in non-stationary media both in cases when properties of the medium change abruptly and smoothly in time.

We have derived an equation for the velocity potential should the properties of a non-stationary medium smoothly change in time. We express the solution of the derived equation (44) in terms of hypergeometric functions and obtain both amplitude and energy flux reflection and transmission coefficients in the case of smoothly changing parameters and as a particular case, the energy flux reflection and transmission coefficients are obtained when the transition period tends to zero, i.e. the change of parameters become abrupt.

The inhomogeneity of time implies sound take an energy from the medium in contrary to the case of EM waves which either take or give a part of their energy to the medium [12]. This is connected with the fact that the EM waves are material on their own, and the sound waves are caused by the medium.

As a matter of principle, the obtained results are applicable for all media. But it is important to underline the case of plasma for the following three reasons; firstly the richness of physical processes is involved in plasma (different types of waves may propagate in plasma: EM as well as sound). Secondly the characteristic frequencies in plasma vary in such volume that the frequency, corresponding to a characteristic period of change of the properties of the medium, has the same order as frequencies of processes in plasma. In contrast to EM waves, frequencies of which are in the minimal case about 10^{15} Hz in the optical range, the advantage of sound waves is their easy experimental realization as their oscillation with inherent frequency equal to 10^{10} Hz. Thirdly the sound wave influences the parameters of plasma, and vice versa. Such systems are often called acoustoplasma. A few interesting phenomena appear when the acoustic waves propagate in plasma (amplification of acoustic waves, decrease of temperature of gas under constant pressure and current in the discharge tube etc [15]).

The obtained results are also applicable in astrophysics. Heavenly bodies consist of plasma, the properties of which significantly differ from the laboratory plasmas. Of specific interest is the case of superdense heavenly bodies whose characteristic period of change of the state of the medium is very small: the plasma is in a strong non-stationary state.

Acknowledgments

The authors extend their sincerest gratitude to Mr. Sean McConnell for his proofreading of the text.

References

[1] L. D. Landau, E. M. Lifshitz, *Course of Theoretical Physics, vol. 1, Mechanics* (Butterworth-Heineman, 2003).

[2] L. D. Landau, E. M. Lifshitz, L. P. Pitaevskii, *Course of Theoretical Physics, vol. 8, Electrodynamics of Continuous Media* (Elsevier Butterworth-Heinemann, 1984).

[3] V. L. Ginzburg *Theoretical physics and astrophysics* (Pergamon Press, 1979).

[4] V. L. Ginzburg, V. N. Tsytovich *Transition Radiation and Transition Scattering* (Taylor & Francis, 1990).

[5] L. D. Landau, E. M. Lifshitz, *Course of Theoretical Physics, vol. 6, Fluid Dynamics* (Butterworth-Heineman, 2000).

[6] R. B. Bird, W. E. Stewart, E. N. Lightfoot, *Transport Phenomena* (John Wiley & Sons, Inc., 2007).

[7] G. Ben-Dor, *Shock Wave Reflection Phenomena* (Springer-Verlag Berlin Heidelberg, 2007).

[8] V. L. Ginzburg, Radiofizika 16, 512 (1973).

[9] V. L. Ginzburg, V. N. Tsytovich, Zh. Eksp. Teor. Fiz. 65, 132 (1973).

[10] H. K. Avetisyan, A. K. Avetisyan, R. G. Petrosyan, *Proc. Int. Symp. on High Energy Electron Transition Radiation* (Yerevan, 1977).

[11] H. K. Avetisyan, A. K. Avetisyan, R. G. Petrosyan, Zh. Eksp. Teor. Fiz. 75, 382 (1978).

[12] A. R. Mkrtchyan, A. G. Hayrapetyan, B. V. Khachatryan, R. G. Petrosyan, Mod. Phys. Lett. B 24, 1951 (2010).

[13] R. P. Feynman, S. Weinberg, *The 1986 Dirac memorial lectures* (Cambridge University Press, 1987).

[14] I. S. Gradshtein, I. M. Ryzhik, *Table of Integrals, Series and Products* (Academic Press, 2000).

[15] G. A. Galechyan, A. R. Mkrtchyan, *Acoustoplasma* (Yerevan, 2005).

In: Sound Waves: Propagation, Frequencies and Effects
Editor: V. Abagnali and G. Fabbri
ISBN: 978-1-61470-001-2

Chapter 9

TOMOGRAPHY TECHNIQUE TO SYNOPTIC MAPPING OF OCEAN MESO-SCALE FIELD

T. V. Ramana Murty, **Y. Sadhuram and B.Sridevi***
National Institute of Oceanography,Regional Centre, Visakhapatnam, India

ABSTRACT

Ocean Acoustic Tomography (OAT) is a remote-sensing technique for the collection of synoptic data from the ocean interior pertaining to density and meso-scale ocean flow fields that has been studied by many scientists. Dynamic ocean processes could be observed by measuring the change in the travel time of acoustic signals transmitted over a number of ocean paths. The travel time is subjected to change by thermal anomalies and inhomogeneity along the acoustic ray paths. An estimate of these quantities can be calculated by using Inversion Techniques. In this chapter the stochastic inverse method of estimating the sound velocity perturbations field from information on travel times of sound pulses through a real ocean has been discussed. A simulation experiment on forward and inverse problems for the observed sound velocity perturbation field has been discussed (in the vertical slice) keeping the single source- receiver configuration (at the channel axis depth) in the SOFAR channel, under peculiar characteristics viz, depth-limited environment and weak gradient waters of the northern Indian Ocean. For the formulation of the Stochastic Inverse, both vertical and horizontal structure of the ocean has been modeled using empirical orthogonal modes.

In this chapter, we report (a) acoustic characteristics of a sub-surface cold core eddy observed (below the mixed layer between depths of 50 and 300m, with a diameter of about 200 km having temperature drop of 5^0 C at the center) in the Bay of Bengal during south-west monsoon period (21-29 July1984) and explore the possibility to reconstruct the acoustic profile of the eddy by Stochastic Inverse Technique and (b) results from a medium range, short duration acoustic tomography experiment for mapping of the synoptic ocean thermal field in the Arabian Sea during summer (2-12 May 1993). This OAT experiment has many components: preparation of available data on ocean parameters relevant to OAT studies, acoustic model simulation in reference ocean and relevant mathematics, optimal estimation of sound velocity perturbations prevailing at the

* Corresponding author: E-mail address: tvrmurty@nio.org (T.V.Ramana Murty).

time of observations and finally the interpretation of results for oceanographic applications.

Keywords: Tomography, cold core eddy, Bay of Bengal, Forward problem, Inverse problem, EOF, Stochastic Inverse, SOFAR channel, acoustic modeling.

INTRODUCTION

Physical variability occurs in the ocean with space (scales of millimeters to planetary) and time (from seconds to geological times) due to external and internal forcing (Robinson and Ding Lee, 1994). Physical oceanographers are interested to study the variability of ocean environment (of different spatial scales: global – basin – subbasin – meso-scale – submeso-scale – fine and microstructure) rather than its mean, especially due to meso-scale eddies. There has been considerable interest in recent years, in meso-scale ocean flow fields of the order of 100 kms and time scale of the order of 100 days, since they contain major portion of the total kinetic energy of the ocean circulation (99% versus 1% in large scale)(MODE group, 1978). By analogy to the atmosphere (spatial scale: 1000 km versus 1000 km, time scale: 3 days), the general circulation may be regarded as climatology; the meso-scale variability is the ocean weather. As these meso-scale eddies transport and disperse chemicals, dissolved substances, particulate matter, nutrients, heat and small organisms in the sea (Robinson, 1983, Prasanna Kumar et al, 2004), observational methods/tools covering synoptic scales are essential. Unlike atmospheric weather the meso-scale variability of ocean is difficult to observe. Eddies are always found in all the oceans and in all the seasons with different spatial, temporal scales and energy levels (MODE group, 1978; Richman et al., 1977). The scales of meso-scale eddy are: a few hundred kilometers in the tropics, of the order of a hundred kilometers in the subtropics, and a few tens of kilometers in the high latitudes (Chelton et al., 1998). Meso-scale eddies are known to exist in the Bay of Bengal since way back from 1957 till to date and were reported earlier (Ramasastry and Balarama Murty, 1957; Rao and Sastry, 1981; Legeckis, 1987; Babu et.al., 1991; Babu et.al., 2003; Madhusudhanan and James, 2003). Eddies in the Bay of Bengal are influenced by the unique seasonal reversing monsoons and transient cyclonic storms (Swallow, 1983). Oceanic eddies are circulating water bodies which are either cyclonic (cold water and low sea level at the center) or anticyclonic (warm water and high sea level at the center).

The conventional method of ship board observations of this meso-scale space-time variability of the ocean field requires simultaneous deployment of several full time vessels, which is very expensive. Though these ship board measurements permit small scale resolution in vertical structure of ocean field at places where the instruments are lowered, they offered little scope for synoptic measurements over a large region. This problem of synoptic observation can be solved to some extent, by the deployment of telemetering data buoys moored at large number of places. This type of measurement can provide good time resolution, but gives poor spatial resolution and they need considerable engineering and logistic efforts in launching and retrieving of these buoys. Another method of data collection is through the satellite remote sensing techniques. This method yields synoptic data on ocean surface features, but fails to yield information below the sea surface because of the incapability of the electromagnetic radiation used in this technique to penetrate ocean waters.

Of late, acoustic monitoring of oceanic features is becoming more and more relevant in understanding the energies and dynamics of meso-scale phenomena. It is well known that in the ocean, sound (in contrast to light) is the only viable means for communication and range finding and also synoptically probing of above oceanic features. Sound can fulfill all the three requirements: (1) penetrative range is more, (2) high resolving power, and (3) high velocity of transmission. Sound velocity in the ocean is most sensitive to temperature and pressure effects, and decreasing temperature with depth produces a decrease of sound velocity with depth in the upper ocean while the increasing pressure eventually more than balances this effect, resulting in a sound velocity minimum (channel axis) at about 1 km (example, northern Indian ocean). This configuration provides a wave guide (also called SOFAR - Sound Fixing and ranging channel). In the language of ray optics, the ray paths oscillate about the sound axis (Fig.1). It has been proved that the existence of SOFAR channel in the ocean makes possible propagation of sound pulses over long distances without many losses (Ewing and Worzel, 1948). It is this property of the ocean which led to the development of acoustic remote sensing techniques viz., Ocean Acoustic Tomography (OAT) (Munk and Wunsch, 1979) and more recently the Acoustic Thermometry (Munk and Forbes, 1989; Spiesberger and Metzger, 1991) to monitor the global warming. This technique is analogous to the medical procedure called Computer Aided Tomography (CAT), which yields a two dimensional display of the anatomy of the part of the human body under investigation. This satellite remote sensing and the OAT together form a complete system to provide synoptic data in real time over large regions of ocean.

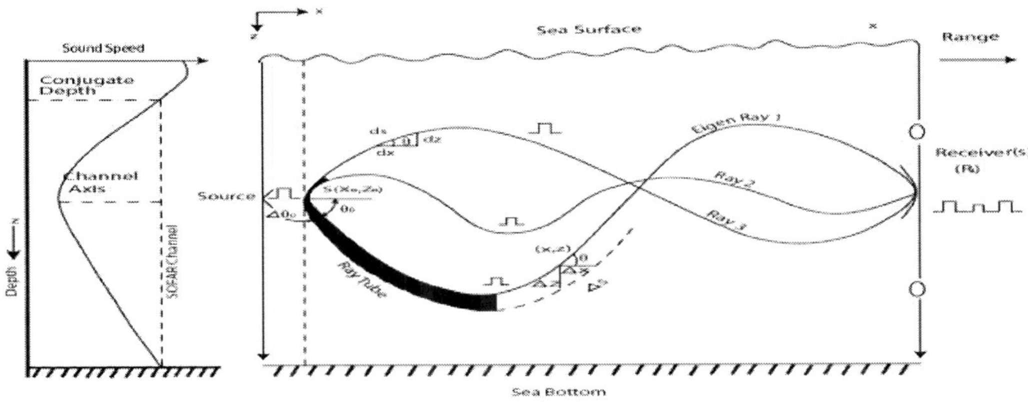

Figure 1. Schematic presentation of acoustic rays from source to receiver.

OAT has special advantage that, one can get the snapshots of small scale to large scale phenomena in the ocean, which is so difficult to fetch by conventional shipboard and mooring techniques. The velocity of sound in the ocean is as fast as 1.5 km/sec which permits the reconstruction of synoptic fields of data, which is equivalent to all that of one-point measurement stations, while N tomography stations produce the data number of ${}^{N}C_2$. With the increasing number of the stations, the spatial resolution for the acoustic tomography becomes much better when compared with the conventional method. In OAT the entire data is path-averaged one all along the ray and no point-measurement data are acquired. Another advantage of the acoustic tomography is that it is one of the remote sensing techniques to explore the ocean interior, where the direct observation is difficult due to ship traffic, fishing grounds, military training areas, etc, that become targets to be measured continuously by the tomography stations arrayed at the periphery of the observational region (Park, 2001).

The OAT technique essentially attempts to estimate this state of the ocean at any given time using the travel time measurements of sound pulses between low frequency sound sources and receivers distributed at suitable locations in the ocean by solving Forward and Inverse problems. In this technique we assume a reference ocean stratified in the vertical direction and trace the sound ray paths from a source to a receiver. For this purpose, the climatologically mean state of the ocean could be used as the reference state. The tracing of the sound ray paths and estimation of sound pulse travel time and its intensity at the receiver is referred to as the Forward Problem.

The travel times for different ray paths measured in a real ocean will differ from the computed travel times for the reference ocean. This will be due to the perturbations in the prevailing thermal structure and the consequent perturbations in the sound velocity profile of the real ocean over that of the reference ocean.

From the differences between the measured and the computed travel times and with the information on environmental noise if available, an estimate of the sound speed perturbation in the ocean along the ray paths may be obtained. The procedure used for the optimal estimation of the sound speed perturbation from this travel time information is called the Inverse problem. The inverse problem forms an important first step in the interpretation of the travel time data collected by tomographic experiments, which would be ultimately used for the description of the ocean flow field.

In India, OAT studies were initiated at the National Institute of Oceanography (NIO) during mid nineteen eighties under the project "Development of Acoustic techniques for Remote sensing of Oceans" and were confined to simulation aspects to the Arabian Sea and the Bay of Bengal (Ramana Murty et al, 1989, 1990, 1992; Prasanna Kumar et al, 1988, 1994; Somayajulu et al, 1993, 1994) utilizing historical and insitu data sets. The papers describe the acoustic propagational characteristics and tomography studies of Northern Indian Ocean wherein maps of channel parameters such as channel depth (depth of minimum sound velocity), axial sound velocity (sound velocity at the channel depth), conjugate depth (depth below the sea surface at which the sound velocity equals the near bottom values) on acoustic propagation were generated. The depth of sound channel axis varies between 1450 - 1850 m in the Arabian Sea while it is 1100-1750 m in the Bay of Bengal.

In general, the depth of the channel axis increases towards the northern latitudes in the Arabian Sea, while it decreases in the Bay of Bengal. The sound velocity profile in the Arabian Sea shows the presence of large gradients above the axis of SOFAR channel in contrast to that in the Bay of Bengal and relatively higher sound velocity values at any given depth indicating a strong waveguide. Coming to the conjugate depth, it shows variation from 75 - 300 m in the Arabian Sea and 100 - 300 m in the Bay of Bengal. Both the seas are the depth limited nature of the profile, i.e. surface sound velocity exceeds the near bottom values. This has an important implication in the sound propagation in the SOFAR channel. Acoustic rays in an ocean with depth limited profile will propagate as surface refracted, bottom reflected (RBR) rays.

As a result, the effective sound channel lies much below the sea surface. Under these circumstances and the presence of peculiar characteristics (viz., depth limited, weak wave guide etc.), we have detailed in the present chapter, the complete frame work of simulation including methodology from beginning to end and also demonstrated, how OAT principle (utilizing stochastic inverse technique) can work for a single source-receiver configuration in the SOFAR channel, to infer/map the observed typical subsurface cold core eddy. Here, dimension of the observed eddy is very much smaller as compared with the SOFAR propagation range.

It is interesting fact to note that the eddy is sandwitched between the sub surface and the conjugate depth which can not be identified by the satellites. Hence OAT is the alternative approach to identify this kind of sub surface meso scale ediies. Even if we keep source-receiver configuration in deep SOFAR channel, emitted sound pulses directed with steep eigen angle rays, which can enter the location of shallow subsurface cold core eddy during their propagation and in turn retain the information on refractivity of media in their memory and finally will present at the receiver end in terms of travel times and amplitudes. Ultimately, this information will be used in the Inversion as input to infer the location of that eddy. However under present context, OAT's principle cannot be applied to monitor the subsurface warm core eddies, which are located above the conjugate depth wherein propagated sound energy cannot be trapped and thereby no information is available at the receiver end and in this case, one cannot identify the location of that eddy. Anyhow, for prospective point of view with respect to the biological implications, monitoring the location of cold core eddies of northern Indian Ocean are essential and of an important issue because of their feasibility to infer them. But, both play an important role with respect to their own dynamic processing mechanisms to understand the physical aspects of monsoons, cyclones, depressions etc., of the Northern Indian Ocean and that is lead to have a need of scientific thought to achieve knowledge on the monitoring of them especially subsurface warm core eddy in the synoptic nature and thereby in this case became a major issue that is left to the scientific community.

The concept of OAT, triggered off a number of investigations towards the development of an operational tomography system in terms of travel times and amplitudes system (Spiesberger et al, 1980; Munk and Wunsch, 1982; Cornuelle et al, 1985; Worcester et al, 1985). Worcester and Cornuelle (1991) provided a lucid review of the developments towards this direction.

Having carried out a number of simulation experiments and inversion, reconstructing the sound velocity anomaly adequately, the next step was to conduct an experiment for a short duration to test models developed and to see how close one could reconstruct the sound velocity and temperature anomaly from the measured data. In view of this, acoustic tomography experiment was conducted in the eastern Arabian Sea during 2 – 12 May, 1993 by keeping the transceivers near the sound channel axis.

In this chapter, we present (a) acoustic characteristics of a sub-surface cold core eddy observed (below the mixed layer between depths of 50 and 300m, with a diameter of about 200 km having temperature drop of 5° C at the center) in the Bay of Bengal during south-west monsoon season and explore possibility to reconstruct the acoustic profile of the eddy by Stochastic Inverse Technique and (b) results from a medium range, short duration acoustic tomography experiment for mapping of synoptic ocean thermal field in the Arabian Sea during summer 1993.

1. Mathematical Formulation and Modeling

1.1. Ocean Model

Oceanic Variability and its Effect on Acoustics

The Ocean fluctuates in space and time and any dynamical process that alters the mean ocean state affect the sound propagation, through perturbations in the index of refraction field. This provides a method to probe the ocean process whose time scale is much longer than any acoustic period. The spatial inhomogenities produce significant scattering whose characteristics depend on acoustic wave length, the length scale and amplitude of the inhomogenities and the propagation range (Sastry et.al., 1984).

Oceanic changes relevant to underwater acoustics are classified into mean and fluctuating components (De Santo, 1979) with the following general model of the sound velocity equation

$$C(r,t)= C_o(z) + \delta C_1(r) + \delta C_2(r,t) \qquad \text{..........} \quad (1)$$

z - depth, t - time, r - three dimensional position having characteristics: 1) a mean vertical profile $C_o(z)$ that represents local climatology ($C_0 \sim$ 1500 ms^{-1}), 2) a meso – scale component ($\delta C_1(r)$) (fronts, eddies) -- deterministic with respect to the acoustic time scale ($\delta C_1 / C_0 \sim 10^{-2}$), and 3) a time-varying component ($\delta C_2(r,t)$) (i.e small scale time varying component with respect to the acoustic time scale, $\delta C_2 / C_0 \sim 10^{-4}$) that represents small scale fluctuations caused by Internal Waves (IW) of fine and microstructure.

$\delta C_1(r)$ components:

1.1a. Synoptic (Meso-Scale) Eddies

Meso–scale eddies are most frequently observed near intensive frontal currents (Gulf Stream etc), which have the shape of rings and are formed as a result of the separation of great meanders from the main flow. Synoptic eddies are also found in the open ocean. The most energetic variability phenomena in the oceans are expected in the meso-scale (MODE group, 1978). The parameters of synoptic eddies vary over a wide range. The diameter of the eddy may vary from 25-500 km, the velocity of water 30-150 cm/sec and the speed of the eddy itself upto 10 cm/sec. Sound velocity has a complicated structure within in the eddy zone. Ocean eddies cause significant variations in the velocity field primarily because of their anomalous temperature structure. Deterministically treated meso-scale effects are primary concern of this chapter.

1.1b. Large Scale Currents and Frontal Zones

Boundaries of large scale currents (Gulf Stream, Somali Current etc.) represent frontal zones separating water masses with essentially different physical characteristics. Temperature, salinity, density and sound speed undergo strong variations within these frontal zones. The variations (fluctuations) in the acoustic propagation can be explained with the variations of the profile along the path

$\delta C_2(r,t)$ components:

1.2a. Internal Waves

Considerable fluctuations of the intensity and phase of sound waves arise in the presence of internal waves, formed at the boundary of two layers having different densities. Their amplitudes are greatest at the top of the thermocline and decrease with depth. These internal waves are generated by surface atmospheric disturbances such as the passage of storms or the flow of water over a rough terrain from few seconds to several hours. The upper limit of the period is the inertial period and the lower limit is the period of stability oscillation. Fluctuations will be found in the amplitude and phase of low frequency sound propagated through internal waves in the ocean. Phase and amplitude fluctuations are the functions of time, space and acoustic frequency.

1.2b. Vertical Fine Structure

The properties of the ocean water such as salinity, temperature and the characteristic flows as observed by continuous recording instruments show small scale discontinuities. They remain constant within certain layers and change rapidly in passing from one layer to another. The thickness of these layers varies from tens of centimeters to tens of meters; the horizontal extent may reach upto tens of kilometers. Sometimes the vertical profiles of temperature, salinity and density look like regular step like structures. This introduces the possibility of "split-beam effect" at the bottom of the layers that has a positive gradient of sound velocity. Thus, there is a divergence of energy at the bottom of these layers part of the energy being refracted upward and part downward i.e. the layering effects significantly when rays vortex within the layers. The thickness of the layers comprising the deep stair case is several orders of magnitude larger than those observed in the thermocline. The result of this feature is that the rays can remain under the influence of each layer for a greater portion of their path.

1.2c. Small Scale Turbulence

Oceanic turbulence has a fairly wide spectrum of scales. Small scale turbulence with spatial scales from a few centimeters to dozens of meters is very important for acoustic propagation related to acoustic tomography. Due to the direct influence of the atmosphere, turbulence is permanently observed in the upper mixed layer of the ocean. At large depths it becomes apparent in the form of separate "patches" in which the fluctuations of physical parameters of the medium are observed. These fluctuations lead to the fluctuations in the sound velocity field.

In OAT we assume a reference ocean stratified in the vertical direction, and trace the sound ray paths from a source to a receiver (eigenrays). While solving forward problem (section 1.3) to simulate predicted travel time (T) utilizing acoustic model, the local climatological (or insitu) mean state of the ocean ($C_o(z)$) could be used as the reference state (initial condition) based on the Fermat's principle. Next step is, on the basis of Fermat's principle (that sound propagates along paths which extremize the travel time for given sound velocity field) and careful analysis of IWs effects of $\delta C_2(r,t)$, one should predict that the paths should be stable, so that changes due to the evolution of the ocean meso-scale of $\delta C_1(r)$ would be resolvable at receiver end and those will be used in the inverse problem (section 1.4). The temporal resolution required to distinguish multipath arrival at receiver is 50 ms; the precision required to measure meso-scale perturbations is estimated at 25 ms (Munk and Wunsch, 1979). In order to attain this, the selection of appropriate acoustic model and their operational frequency are very much important factors to map the meso-scale variability in real time. For the present OAT's simulation and experimental set up, we have considered the ray model and low operational frequency of 400 Hz (as compare to depth of the ocean in area of interest). At 400 Hz, sound in the ocean has wavelengths of about 3.75 m, small when compared with typical scales for the sound velocity structure of either the basic climatological ($C_o(z)$) state or the meso-scale fluctuations ($\delta C_1(r)$), but large compared to vertical microstructure and most fine structure ($\delta C_2(r,t)$) (Gregg and M.G. Briscoe, 1979). The slow variation of interesting structures when compared with the sound wavelengths allows a simplification of the acoustic wave equation called the geometrical optics approximation, using concept of acoustic rays which has been described in the Appendix-A. At typical sonar operating frequencies, the main cause of these fluctuations ($\delta C_2(r,t)$) are IWs on the $C(r,t)$, which produce both sound velocity gradients and current shear at scales of the order of meters. These features are comparable with the acoustic wavelength, and tend to scatter the sound, blurring the simple ray paths calculated for large-scale refraction into ensemble of micero-multipaths which change with IWs. These shifting paths interfere with one another, producing variations in overall travel time for the path and significant changes in the intensity of the received sound (Cornuelle, 1983). In the present theoretical frame work, mathematical approximations are considered to remove IW effects on $C(r,t)$. Anyhow, for propagation over long ranges (ranges exceeding several hundred kilometers), it has been noted that sound pulse arrivals at receiver end sometimes become so smeared together due to fluctuations of IWs that they cannot be used for travel-time measurements (Simmen et.al, 1997). But short – time changes in amplitude and phase of $\delta C_2(r,t)$ on $C(r,t)$ transmission cannot be neglected (Flatte, et al., 1979; Ramana Murty et.al., 2007a; Stefanie, 2006; Sridevi, et.al 2010, 2011).

1.3. Forward Problem

Sound waves, which propagate from the source to receiver, are influenced by the sound velocity $C(x,z)$ and current velocity $u(x,z)$ in the intervening water (Fig. 1). This is called the forward problem and the detailed properties of received signals can be calculated by solving the sound wave equation with the given values of C and u (Appendix-A). Understanding the forward problem is so important to solve optimally, the inverse problem because received acoustic signals are to be identified for making the accurate and successful inversion. For the OAT studies, the forward problem is solved by the computer simulation in which priori sound velocities from the observations are used. The received signals can be identified by comparing the simulated (predicted) results with observed ones (experimental data).

In the OAT studies the observable data are travel times and intensities of received signals. The latter are out of scope of this study, while the former are used in the present inverse problem. Early and late arrival rays take different ray paths and thus experience different sound velocity and current velocity fields (Park, 2001).

The travel times of sound between two acoustic transceivers along the reciprocal ray paths can be formulated as (in 2-D form)

$$T^{\pm} = \oint_{\Gamma^{\pm}} \frac{ds}{C(x,z) + \mathrm{u}(x,z).\mathrm{n}} \qquad \text{.......... .. (2)}$$

where $\pm$ represents the positive/negative direction shooting angle from one transceiver to another, n the unit vector tangent to the ray path and s the arc length measured along the ray path. $\Gamma^{\pm}$ are the positive and negative paths refracted by $C(x,z)$ and $u(x,z).n$. The ray paths are generally not coincident for two ways in the reciprocal transmission, but we can approximate $\Gamma^{+} \approx \Gamma^{-} \equiv \Gamma_0$ when $u.n/c << 1$. The mean travel time perturbation δT and the differential travel time ΔT are expressed by

$$\delta T = \frac{1}{2}(T^{+} + T^{-}) - T_0 \approx -\oint_{\Gamma_0} \frac{\delta C(x,z)}{C_0(z)^2} ds \qquad (3)$$

$$\Delta T = \frac{1}{2}(T^{+} - T^{-}) \approx -\oint_{\Gamma_0} \frac{u(x,z).n}{C_0(z)^2} ds \qquad (4)$$

Where T_0 the travel time along the ray path is Γ_0 associated with a reference sound speed $C_0(z)$ and $\delta C(x,z)$ $(= C(x,z) - C_0(z))$ is the sound speed perturbation. We also assume $\delta C / C << 1$. These are the linearized formulations for the acoustic travel times obtained in the oceanic environment with C and u. The velocity term in the denominator of the integrand (i.e., , $u(x,z).n)$*)*, in Eqn. (2) accounts for changes in the apparent velocity of sound due to current, provided local shear can be ignored (Hamilton et. al., 1980). In the present simulation experiment Eqn. (4) is not considered because currents are ignored in the ray tracing as the magnitude of the current shear in the ocean is typically $O(10^{-4})$ when compared with sound velocity shear which is of $O(10^{-2})$.

Eqn. (3) is integral equation with the unknown variable δC. This equation constitutes acoustic forward problem: continuous δC field is used to predict the data δT_i. The corresponding inverse problem takes data and estimates the continuous field i.e. δC. This is an infinitely under-determined problem because, values of δC must be specified at all *x* and *z* points using only a finite amount of data δT_i. So the continuous field δC will therefore be parameterized with a finite number of discrete parameters. An ocean model specifying this parameterization is supposed to be oceanographically meaningful. This has three advantages: a) the parameterization can be very efficient, b) the inverse solution becomes better determined because we are injecting our oceanographic prejudices concerning permeable structure of the fields (reducing the degree of freedom) and lastly, the model data can tell us whether the Oceanographic model is consistent with observed data or not. There are two convenient model representations being used, which have been routinely exploited by the physical oceanographic community, viz., the Empirical Orthogonal Functions (EOF's) and the Quasi-Geostropic (QG) ocean modes (Pedlosky, 1980; Ramana Murty and Mahadevan, 1995). In this chapter, we consider EOF model studies (Appendix - B).

1.4. Stochastic Inverse

The least square method, damped least square method and Gause-Markov method are introduced as suitable inversion techniques for the ocean acoustic tomography (Cornuelle, 1983; Park, 2001). The Gause-Markov method (Liebelt, 1967) has been used as the familiar technique. However, in the case of lack of statistical information, the least squares method may be an alternative to be selected. When the least squares method has problems related to smaller eigenvalues for coefficient matrix (data kernel), it is to be replaced by the damped least squares method (Menk, 1989, Aki and Richards, 1980 and Park, 2001). Prasanna Kumar et.al. (1999) simulated tomographic travel time anomaly inversion of a cold core eddy in the Bay of Bengal, under range - independent environment utilizing deterministic generalized inverse approach. Since the real ocean is of range - dependent nature, the same data of all eddy profiles are used in the above paper revisited in the present OAT's simulation study to explore possible means to reconstruct the eddy, which is above the conjugate depth. Deterministic inverse procedure doesn't permit a continuous representation of unknown field, so we demonstrate how the stochastic inverse technique (Cornuelle, 1983) works for the objective mapping of meso-scale eddy field, wherein partially sampled ray paths information is available.

In the stochastic inverse method both the data *(d)* and the unknown field $\delta C\ (x, z)$ are assumed random variables having known mean and covariance functions. The best estimate $\delta\hat{C}(x,z)$ of the model parameters $\delta C\ (x, z)\ (= C\ (x, z) - C_0 = D)$ can be expressed in the form

$$\delta\hat{C}(x,z) = \Sigma a_i^{\,1}(x,z)\delta T_i = A(x,z)d \qquad \text{.......... .... (5)}$$

such that the ensemble average of the square of the difference between the estimate of the unknown field $\delta\hat{C}\ (x, z)$ and the true field ($\delta C\ (x, z)$) at each point in the medium is a minimum, i.e., $a_i^{\,'}s$ are determined such that the square of the deviation (ensemble) is a minimum at each point (x, z) . Using the Gauss-Markov theorem (Liebelt, 1967), a_i*'s* are given by

$$A(x,z) = \left\langle \delta C(x,z)\delta T_i^{\ T} \right\rangle \left[\left\langle \delta T_i \delta T_j^{\ T} \right\rangle \right]^{-1} \qquad \text{..........} \quad \text{...} (6)$$

where $\left\langle \delta C(x,z)\delta T_i^{\ T} \right\rangle$ and $\left\langle \delta T_i \delta T_j^{\ T} \right\rangle$ are the model-data and data-data covariance matrices. These are evaluated using the linearized direct problem,

$$\delta T_i = -\int_{\Gamma_{i0}} \frac{\delta C(x,z)}{C_0^2(z)} ds + \varepsilon_i \qquad \text{.........} (7)$$

where ε_i represents noise due to smaller scale phenomena and measurement errors, assumed to have zero ensemble average and to be uncorrelated with the meso-scale sound velocity fluctuations. Using Eq. (7) the model-data and data-data covariance matrices can be expressed as

$$< \delta C(x,z) d^T > = < \delta C(x,z) \delta T_i^{\ T} >$$

$$= \int_{\Gamma_{0_i}} \frac{\langle \delta C(x,z) \delta C(x',z') \rangle}{C_0^2(x',z')} ds' \qquad \text{..........} (8)$$

$$= \int_{\Gamma_{0_i}} \frac{\left[\left(\sum_i^p a_i(x) u_i(z) \right) \left[\sum_{i=1}^p a_i(x') u(z') \right] \right]}{C_0^2(x',z')} \frac{dx'}{\cos\theta'}$$

$$< dd^T > = < \delta T_i \delta T_j^T >$$

$$= \int_{\Gamma_{0_j}} \frac{dx''}{\cos\theta''} \int_{\Gamma_{0_i}} \frac{\left[\left(\sum_i^p a_i(x') u_i(z') \right) \left[\sum_{i=1}^p a_i(x'') u(z'') \right] \right]}{C_0^{\ 2}(x',z') C_0^{\ 2}(x'',z'')} \frac{dx'}{\cos\theta'} \qquad (9)$$

where $\gamma_{ij} = < \varepsilon_i \varepsilon_j >$ is the error covariance matrix $(x',z'), \theta'$ and $(x'',z''), \theta''$ are the coordinates and angles of the *i-th* and *j-th* along rays respectively. In eqns.(8) and (9), the integrations are carried out along the eigenrays(Γ_{o_i} & Γ_{o_j}). In the above equations, $u_i(z)$ are the vertical EOF modes (column vectors of *U*) of covariance matrix

$D = U\Gamma V^T$ (SVD form) and horizontal modal amplitudes $a_i(x)$ are obtained (Appendix-B) as follows:

$$a_i(x) = \sum_{i=1}^{p} \delta C(x, z_n) u_i(z_n) \qquad \text{.......... ... (10)}$$

i = mode index, n = layer index, p = factor index ($\leq$ r = rank of D

For each point in a given region, the coefficients $a_i^{'}(x, z)$ are determined from the eigenrays and operated on the travel time perturbations to objectively estimate the model parameter ($\delta\hat{C}(x, z)$) . This procedure permits a continuous representation of the unknown field. The data-data and model-data covariance matrices are determined for the reference sound velocity distribution with a suitably selected source-receiver configuration. Using these matrices the sound velocity field ($\delta\hat{C}(x, z)$) of the real ocean is obtained for the given travel time perturbations (δT) .

The required components of OAT are discussed below and they are:1) preparation of available data 2) acoustic model simulation of forward problem in reference ocean, 3) optimal estimation of sound velocity perturbations from travel time perturbations (inverse problem) and finally 4) the interpretation of results for oceanographic applications.

2. Data and Simulation

The hydrographic survey was carried out on board ORV Sagar Kanya during 21-29 July 1984, and occupied a total number of 42 CTD stations near the western boundary of North Bay of Bengal (Figure 2). Sound velocity has been computed from the CTD data following Chen and Millerio (1977) at standard depths (Levitus, 1982).

3. Results and Discussion

3.1. Sound Velocity Field

In order to examine the variability of sound-velocity in the presence of a cold core eddy, the sound velocity structure $C(x, z)$ across the eddy (track parallel to the coast shown in Figure 2) is presented along with the vertical profiles at stations #23, #33, #40, #56 and #61 (Figs. 3 & 4). All the profiles are similar in shape having a well defined deep sound channel and a smaller duct near the surface (Figure 3). The sonic layer depth (depth of near-surface sound-velocity maximum) changed abruptly by about 20 m across the eddy. At the surface the sound velocity varied by 8 ms^{-1} with a decreasing trend from south to north. This variation is brought about by the presence of low saline cold water in northern Bay of Bengal (associated with freshwater discharge), reducing the sound velocity and comparatively high saline warm water in the southern side giving rise to higher sound velocity. Below the surface, the eddy is clearly discernible from the up tilt of 1500-1520 ms^{-1} sound velocity isopleths (Figure 4).

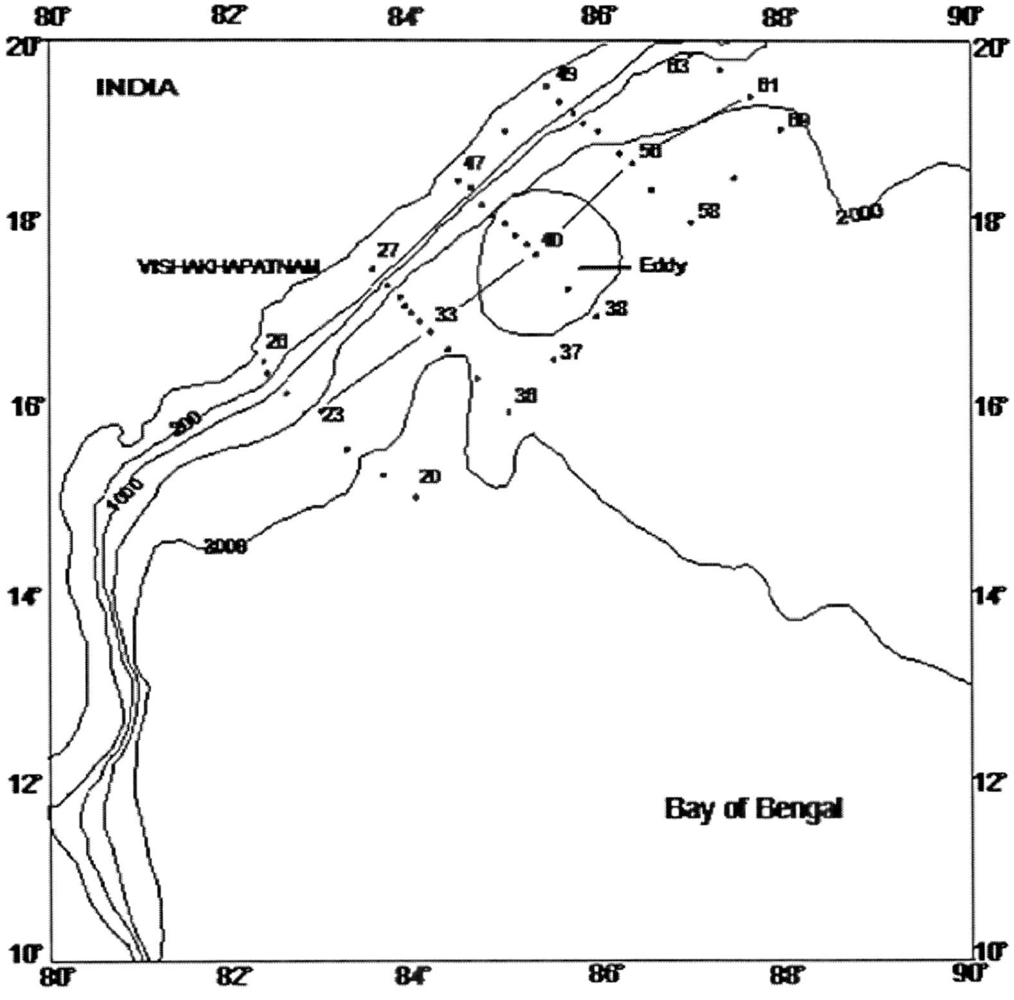

Figure 2. Sub surface cold core eddy (Center at stn 40) along with the station locations for CTD data collection.

The presence of eddy brings about a reduction in the sound-velocity. At 60 m depth, the sound velocity reduces by 4 ms^{-1} across the eddy. A maximum reduction occurs at 120-140 m layer, at the centre of the eddy (#40) where the sound-velocity drops up to 14 ms^{-1}. Below 300 m depth the variability in sound-velocity is less than 2 ms^{-1}. The effect of cold core eddy is to reduce the ambient sound-velocity by approximately 10 ms^{-1}, associated with this the depth of channel axis and axial velocity were also undergone changes (Table-1).

Table 1. Sound channel parameters

St No.	Latitude (^{0}N)		Longitude (^{0}E)		Sound channel depth (m)	Sound velocity at channel depth (m/sec)	Conjugate Depth (m)	Sound velocity at conjugate depth (m/sec)
	Deg	Min	Deg	Min				
23	15	59.84	82	54.94	1350	1493.04	250	1508.46
33	17	00.26	83	59.80	1400	1492.00	200	1506.09
40	17	40.12	85	19.76	1500	1491.11	125	1507.52
56	18	40.05	86	17.85	1750	1493.02	150	1518.48
61	19	21.98	87	39.96	1750	1493.21	150	1516.45
Reference Profile	-		-		1500	1493.07	200	1507.51

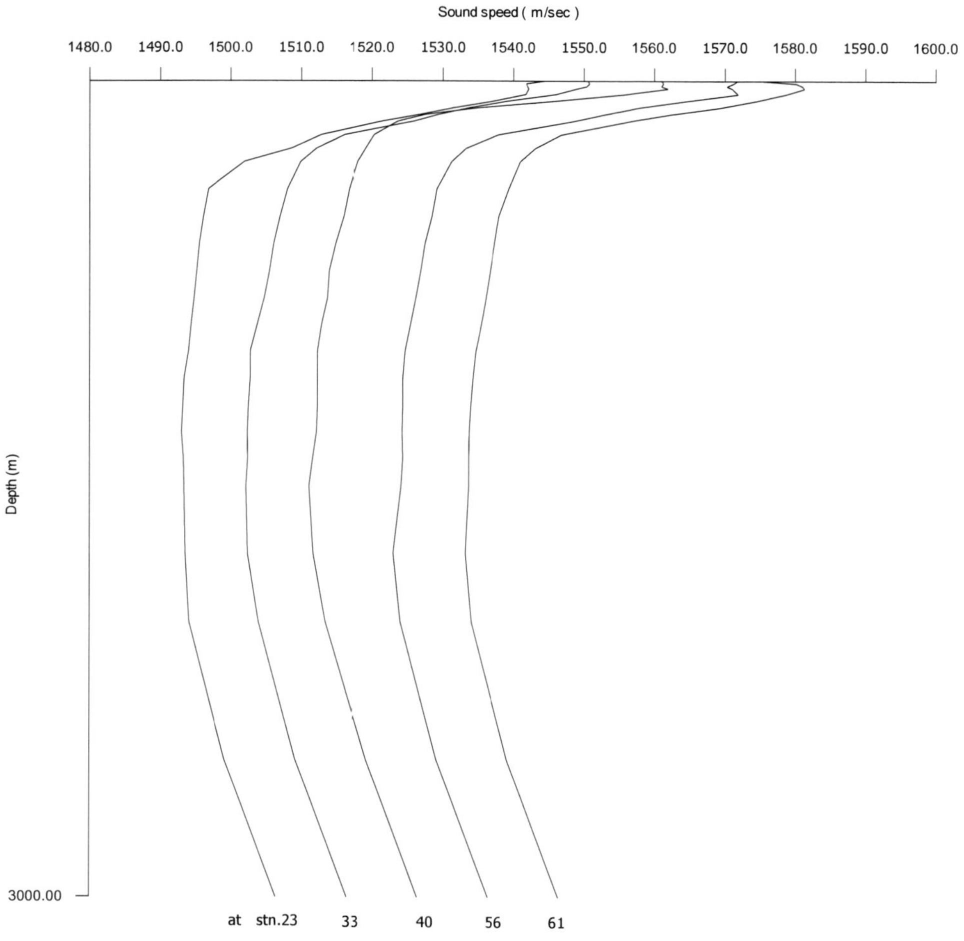

Figure 3. Sound velocity profiles along parallel track (Fig. 2). (The consequent profiles are drawn by incrementing 10m/sec in horizontal direction).

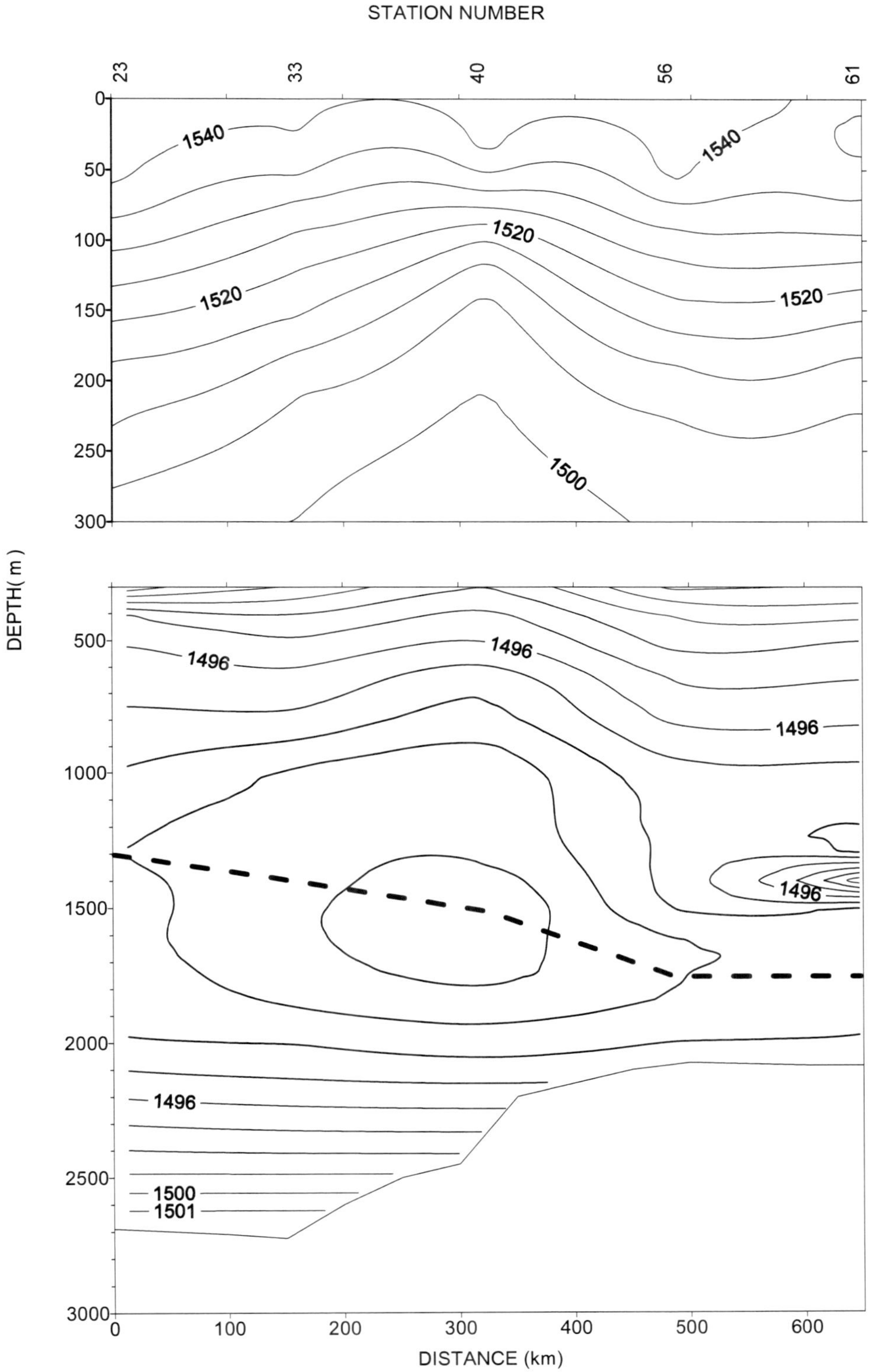

Figure 4. Vertical sound velocity (m/sec) structure across the eddy parallel to the Coast (------- indicates the depth of SOFAR channel axis).

In general, the depth of deep sound channel axis increases from south to north along the track parallel to the coast. However, below the cold core eddy, the channel axis depth remains constant (1600 m). Accordingly the sound velocity at the channel axis below the eddy reduces by about 1.5 ms^{-1}. The conjugate depth is decreasing from south to the center of the eddy and then onwards again it starts increasing. This indicates that cold core cyclonic eddy, uplift isopleths by 75m with reference to normal conditions (Table-1). The importance of monitoring such meso-scale cyclonic eddies and their role of such processes in enhancing surface ocean productivity has been extensively studied in the past (Denman and Dower, 2001). After utilization of OAT principle and stochastic inverse, we later prove that it is possible to monitor such cyclonic eddies to some extent even under depth-limited environment of tropical waters. Unlike cyclonic eddies, in the case of anticyclonic eddies, (i.e., have downward-tilting isopycnal surfaces in the interior) leading to little expected change in productivity. However this phenomenon has important role in studying the dynamics of interior water. Any how there is no limitation to monitor the anticyclonic eddies, below the conjugate depth, using OAT principle, but in the present study, it is not the case. In our simulation we have used the knowledge of the sound velocity perturbation itself to compute the covariance.

In order to use the stochastic formalism, it is necessary to define a mean state of $C_0(z)$ for the sound-velocity and the expected covariance around this basic state. Because we are interested in deriving reliable snapshots of the evolution of the sound-velocity anomalies $\delta C(x,z)(= C(x,z) - C_0 = D)$ due to meso-scale dynamics, we are more interested in the minimum variance properties of the estimator. For this reason, the basic state only is specified reasonably close to the true state to avoid problems due to linearization. It means that any archived estimate of the local mean sound-velocity is adequate to use as basic state, so that the closer the assumed mean state to the true mean, the smaller the variance around the mean will increase the effectiveness of the estimator.

For the initial estimate to the numerical experiment, following a simple average $C_0(z)$ be written as

$$C_0(z_k) = \frac{1}{n}\sum_{j=1}^{n} C_j^1(x_j, z_k) \qquad (11)$$

n is no of profiles,

where sound velocity profiles $C^1(x,z)$ that are computed from Levitus hydrographic data along the same track (shown in Fig.2) are chosen to the basic state. The basic state is taken to be stationary and horizontally homogeneous, (i.e., $C'(x,z,t) = C_0(z)$), both for simplicity and also the data available to date are inadequate to support any assumptions to the contrary.

The estimate of covariance for the sound-velocity anomaly $C(x,z) - C_0(Z) = D$ is also derived from the observed eddy profiles and Levitus archived data (of the same track)

(Fig.2), and then used in the forward problem to calculate the expected data-data covariance matrix. The decomposition into vertical modes $\{u_i(z)\}$ with horizontally varying amplitudes $\{a_i(x)\}$ is described in the Appendix-B. This model is used throughout the inversions to follow:

$$\delta C(x,z) = \sum_{i=1}^{r} u_i(z) v_i(x) l_i^2 \qquad \text{.......... ...(12)}$$

r = rank of D

The modes chosen as basis (i.e. depth - dependent set $s(z)$; range – dependent set $s(r)$ that are defined in Appendix-B) are the Empirical Orthogonal Functions(EOF) of sound velocity variation shown in Figure2. It is also more logical to use the historical averaged sound velocity profile as a reference, rather than the average of the observed eddy profiles. In the actual experiment, basis supposed to be chosen before the data available by actually conducting the experiment, so that the model would be independent of the traditional measurements made during the experiment. However traditional measurements made are useful for validation.

In deriving the EOF modes, we have tacitly assumed flat bottom geometry for the ocean, for which the modes are complete and mutually orthogonal set. In the present study area, ocean bottom is not actually flat (Fig.2) and so this EOF expansion cannot be made, and therefore one needs to look for an alternatives. To circumvent this difficulty, we employ the artifice of "false bottom" located at the deepest bathymetry point (Robinson and Ding Lee, 1994), to fill the values of sound velocity at unobserved deeper depths. While employing the "false bottom" method, it doesn't affect the top layer, as sound velocity variability with reference to subsurface eddy is negligible.

3.2. EOF Modes

This method has been widely used in many types of distributions in many studies. The geometrical interpretation of this method has been briefly explained by Wiggins (1972). It is important to point out that EOF may or may not have physical significance or interpretation but depends certainly on the data examined. However, various researchers from different disciplines employed the technique for different purposes. Stidd (1967) used eigenvectors to represent the seasonal variation of rainfall over Nevada and found that the first three eigenvectors, in order of importance have features in common with the three natural cycles of precipitation. Winant et al. (1975) and Aubray (1978) used this technique to beach data and attributed the first three eigenvectors to beach profiles, mean seasonal cycles and high frequency random oscillations respectively. Ramana Murty et al. (1986) used this method to grain-size data to identify the different energy environments along central west coast of India. Hua et. al., (1985) interpreted that first two modes represent barotropic and baroclnic modes of sound velocity structure. Cornuelle (1983) used this technique and derived EOF modes from historical hydrographic data to construct the total data – data covariance matrix for formalization of stochastic inverse. He explained that EOF's allow variance in the upper layers of the ocean, presumably due to seasonal effects.

Computational procedure of vertical and horizontal EOF's of perturbation matrix D of order $(m\,x\,n)$ and their horizontal modal amplitude are detailed in Appendix-B for two cases $m<n$ and $m>n$. In the present case we have $n = 5$, vertical profiles (Fig.2) each containing m=28 horizontal layers and then D^T will be 5 × 28 matrix. Each column of the matrix will be a profile of sound velocity perturbations from the back ground $C_0(z)$ and comes under the case $m>n$. The computed eigenvalues (l_i) arranged in the descending order in the diagonal matrix of Γ and percentage contribution of each eigenfunction (eigenmode) using closeness approach (Eqn. B6) are shown in the Table-2.

Table 2. Eigenvalues of sound velocity perturbation matrix $\delta C(x,z)$

No.	Eigenvalue	Closeness ratio (%)
1	5.396	60.42
2	3.010	18.78
3	2.331	11.27
4	2.130	9.42
5	0.218	0.09

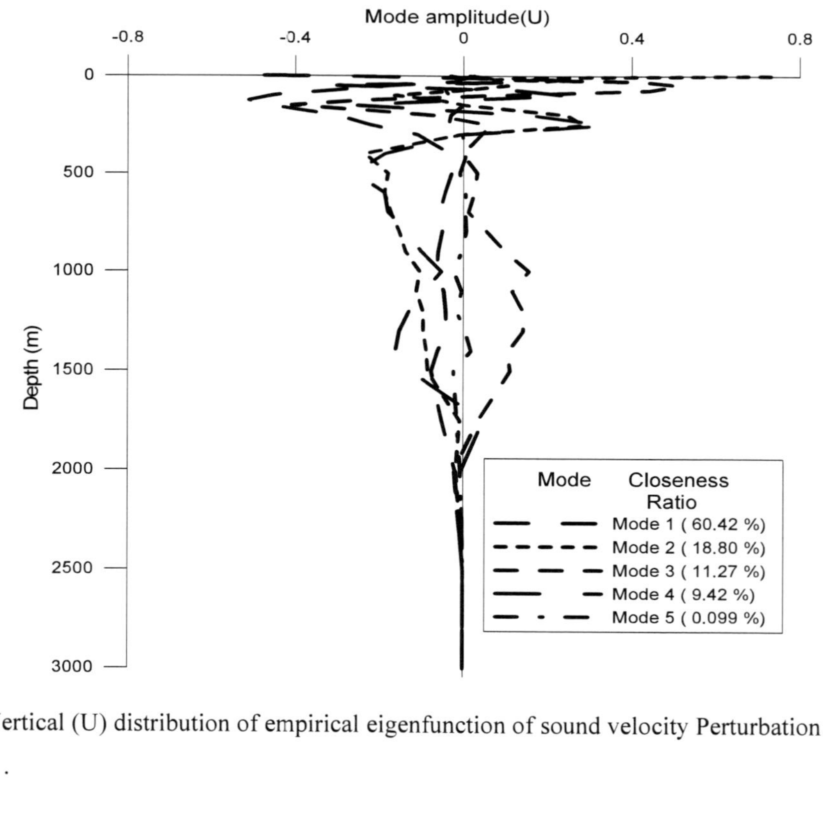

Figure 5. Vertical (U) distribution of empirical eigenfunction of sound velocity Perturbation $\delta C(x,z)$.

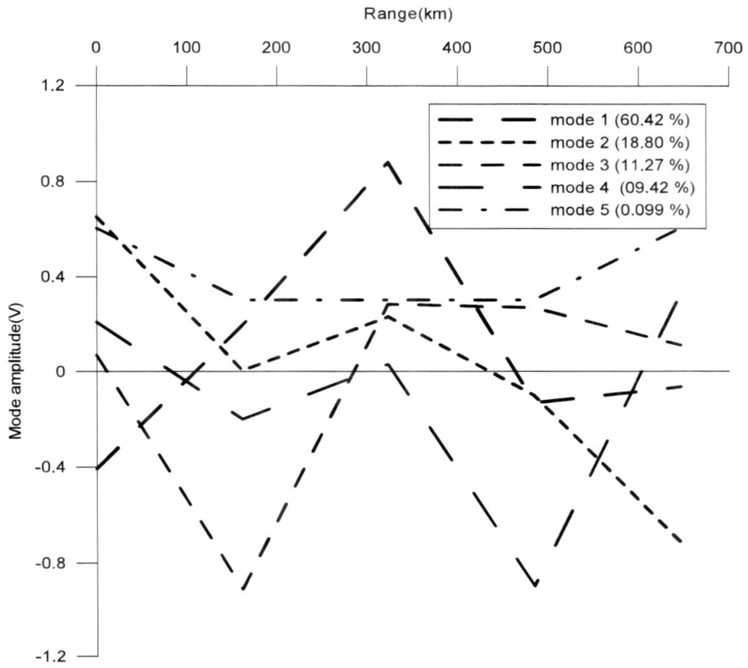

Figure 6. Horizontal/Spatial (V) distribution of empirical eigenfunction of sound velocity perturbation $\delta C(x,z)$.

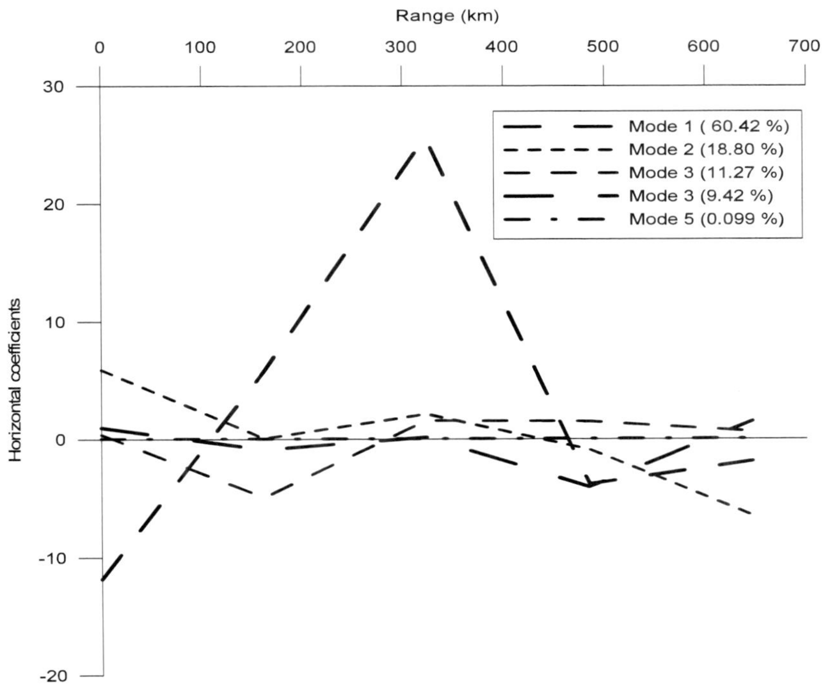

Figure 7. Horizontal modal coefficients $(a_i(x))$.

Table 3. Horizontal model coefficients $(a_i(x))$

Mode No. (i)	Station No.				
	St. 23	St. 33	St. 40	St. 56	St. 61
	$a_i(1)$	$a_i(2)$	$a_i(3)$	$a_i(4)$	$a_i(5)$
1	-11.85	5.70	25.62	-3.83	-1.90
2	5.88	0.039	2.09	-0.92	-6.49
3	0.38	-4.95	1.54	1.46	0.60
4	0.94	-0.91	0.14	-4.08	1.48
5	0.03	0.014	0.014	0.014	0.03

The computed depth - dependent eigenmodes of $s(z)$ (i.e., u_i, i=1,n), range - dependent eigenmodes of $s(r)$ (i.e., v_i, i=1,n) and horizontal modal amplitudes $(a_i(x), i=1,n)$ (using eqn. B7) are shown in Figs.(5-7) and Table-3. Column vectors u_i (i=1... 5) of $s(z)$ of corresponding eigenvalues l_i (measure of variance D accounted for mode i) having 28 elements and each element has a value indicating amplitude which explains sharing contribution of it's variance mode i. The vectors u in the $s(z)$ represents the trend or direction of depth-spatial common features contained in perturbation matrix D. Its corresponding eigenvalues l_i represents energy level (Wiggins, 1972). In a similar manner the components of vectors in set $s(r)$ can also be explained.

The percentage contribution of individual components of $s(z)$ and $s(r)$ are computed through closeness ratio approach (using Eqns. B7 & B8) and are shown in Tables 4 and 5. From these tables, the first four eigenfunctions of $s(z)$ and $s(r)$ account for more than 98% of variance, relatively dominated by the first function alone (60.42%). The first depth - dependent eigenfunction u_1 of $s(z))$ contribute 60.42% of the total distribution showing maximum variance between the depths 50 – 300m (vertical width of eddy) and is less at its top and below layers of the ocean (Fig.4 & Table-4). Maximum dominant contribution (modal value) of u_1 is 9.46% (Table-4) of total distribution (61.42%) at depth 125m indicating conjugate depth (Table-1) of sound velocity of cold core eddy. The first range – dependent eigen function v_1 of $s(r)$ contribution is 60.42% of total distribution showing maximum variance at stations #23 and #40 and relatively less contribution at the other stations (Fig.5 & Table – 5). Maximum dominant contribution (modal value) of v_1, is 31.65% (Table-5) of total distribution (61.42%) at station #40 indicating centre of the eddy. All the depth-dependent eigenmodes of $s(z)$ have higher values in the top vertical layers of the ocean and somewhat less in the middle layers and no variation below 2000 m depth. From Figs.6&7, one can infer that the shape of eigenmodes is similar, but in Fig.7, $\{a_i(x)\}$ are amplified by the energy levels of times v_i, (Fig.6 & Table-3). A great advantage of this representation is that the EOFs are the most efficient representation available in the sense that one needs least number of modes to account for a given percentage of the variance between the actual field and its representation. For a real ocean wave number spectrum, the modes sum can be truncated since the higher modes will contain little energy. As the ocean does have a red spectrum, this reduces the number of modes needed and, the more important number of $\{a_i(x)\}$ one has to store. The following resolution matrices tell us which combination of the data is well resolved (Fukumori, 1989).

Table 4. Percentage contribution of individual component data space $s(z)$

Depth(m)	u_1%	u_2%	u_3%	u_4%	u_5%
0	-0.548	3.851	0.053	-1.055	0.002
10	-0.087	0.704	0.18	-0.084	0.001
20	-0.304	0.383	0.251	0.521	-0.001
30	-0.133	0.418	1.37	0.411	-0.001
50	-3.163	0.556	1.662	-0.709	0
75	-5.772	-0.039	1.434	0.296	-0.003
100	-8.411	-0.866	0.208	0.551	-0.002
125	-9.464	-0.295	-0.764	-0.059	0
150	-8.641	0.033	-1.329	-0.579	0
200	-5.618	1.303	-0.227	0.357	-0.002
250	-4.082	1.512	0.211	0.76	-0.002
300	-2.012	-0.041	0.139	-0.014	0
400	-0.542	-1.235	-0.004	-0.416	0.001
500	-0.503	-0.946	0.107	-0.579	-0.001
600	-0.789	-0.991	0.085	-0.417	0
700	-1.02	-0.911	0.041	-0.398	0
800	-0.914	-0.804	0.184	-0.294	0
900	-1.098	-0.728	0.32	-0.224	0
1000	-1.155	-0.551	0.507	-0.115	-0.001
1100	-0.846	-0.591	0.379	-0.202	0
1200	-0.763	-0.505	0.463	-0.288	-0.001
1300	-0.74	-0.501	0.462	-0.34	0
1400	-1.12	-0.461	0.348	-0.359	0.001
1500	-1.398	-0.444	0.36	-0.279	-0.001
1750	-0.94	-0.044	0.113	0.104	0.066
2000	-0.357	-0.089	-0.068	-0.01	0
2500	0	0	0	0	0
3000	0	0	0	0	0
Total	60.42	18.8	11.27	9.42	0.09

Table 5. Percentage contribution of individual components of model space $s(r)$

St.No	v_1%	v_2%	V_3%	v_4%	v_5%
23	-14.637	7.171	0.475	1.182	0.026
33	7.053	0.047	-6.247	-1.134	0.013
40	31.645	2.55	1.944	0.176	0.013
56	-4.733	-1.125	1.846	-5.084	0.013
61	-2.352	-7.906	0.758	1.845	0.026
Total	60.42	18.8	11.27	9.42	0.09

3.3. Data Resolution Matrix $(U_p U_p^T)$

Table 6. Diagonal elements of data resolution $(u_p u_p^T)$ for each depth for factor $p = 1$ to 5

Depth (m)	1 Mode	2 Modes	3 Modes	4 Modes	5 Modes
0	0	0.53	0.54	0.76	0.76
10	0	0.02	0.02	0.02	0.02
20	0	0.01	0.01	0.07	0.07
30	0	0.01	0.18	0.22	0.22
50	0.03	0.04	0.3	0.4	0.4
75	0.1	0.1	0.29	0.31	0.31
100	0.21	0.21	0.24	0.3	0.3
125	0.26	0.26	0.32	0.32	0.32
150	0.22	0.22	0.39	0.45	0.45
200	0.09	0.15	0.16	0.18	0.18
250	0.05	0.13	0.14	0.25	0.25
300	0.01	0.01	0.01	0.01	0.01
400	0	0.06	0.06	0.09	0.09
500	0	0.03	0.03	0.1	0.1
600	0	0.04	0.04	0.07	0.07
700	0	0.03	0.03	0.07	0.07
800	0	0.03	0.03	0.05	0.05
900	0	0.02	0.03	0.04	0.04
1000	0	0.01	0.04	0.04	0.04
1100	0	0.01	0.03	0.04	0.04
1200	0	0.01	0.03	0.05	0.05
1300	0	0.01	0.03	0.05	0.05
1400	0	0.01	0.02	0.05	0.05
1500	0.01	0.01	0.02	0.04	0.04
1750	0	0	0	0.01	1
2000	0	0	0	0	0
2500	0	0	0	0	0
3000	0	0	0	0	0

The data resolution matrix is an identification of the information density of the data kernel, i.e. it indicates which data contribute independent information to the solution. The computed diagonal elements of the data resolution matrix in the $s(z)$ base are shown in Table 6. Higher values are found at depths 75 -300 m with one mode $(p = 1)$ with contribution of 60.42%. Resolution has improved at depths between 200 – 250 m and also deeper depths considering 2 modes (p=2) with contribution of 79.22%, but no gain at depths between 77 – 150 m. Considering 4 modes $(p = 4)$, resolution improved, but not much up to depth of 30m. For modes p=5, it is observed that there is no change in the resolution from $(p = 4)$

3.4. Model Resolution Matrix $(V_p V_p^T)$

Model parameter resolution is an indication of how perfectly and independently each model parameter is determined. The diagonal elements of model parameter resolution matrix (Table 7) give a clue how well the individual model parameters in the $s(r)$ base are resolved. A value of unity indicates perfectly resolved parameter, where as a smaller value means inadequate resolution. From Table 7, it can be seen that parameter resolution is found to be higher at two stations namely Stn #40 (center of eddy) & #23, due to first mode $(p = 1)$ itself with contribution of 31.64% and 14.64% respectively (Table-5). Considering the four modes $(p = 4)$, resolution at stations #33, 40 & 56 has improved well with 99% contribution, but relatively less improved at Stn #23 & 61 respectively.

Table 7. Diagonal elements of model resolution $(v_p v_p^T)$ for each station for factor p = 1 to 5

Range (km)	1 Mode	2 Modes	3 Modes	4 Modes	5 Modes
0.00	0.17	0.59	0.59	0.64	1.00
161.62	0.04	0.04	0.87	0.91	1.00
323.23	0.77	0.83	0.91	0.91	1.00
484.83	0.02	0.03	0.10	0.91	1.00
646.49	0.00	0.52	0.53	0.64	1.00

3.5. Eigen Rays

In order to understand the effect of cold core eddy on acoustic propagation, a range independent acoustic ray tracing programme is utilized (Ramana Murty et. al., 2005). Appendix-B gives the details of the ray model used for the computation. The ray tracing is carried out using the mean (left panel of Fig.8) as well as the eddy profiles (Fig.3) for a range of 650 km, with the same source - receiver configuration kept at a depth of 1500 m in the SOFAR channel. The nature of mean profile (left panel in Fig.8) is depth limited since its bottom velocity is less than the surface velocity. The conjugate depth of this profile is 200m (column 8 of Table -1). So in the present acoustic simulation experiment under such a depth limited environment, emitted sound pulses never enter into the top 200m layer from any source to receiver kept in the SOFAR channel (Table 1). Eigenrays between the source and receiver for unperturbed ocean are shown in Figure 8 (right panel) that also reveals the same conclusion. In order to know the effect of sub-surface cold core eddy on the propagation, ray paths are plotted (Figs. 9a & 9b) at the neighboring eddy location under unperturbed and perturbed conditions. In Fig.9b, eigenrays are trapped and uplifted 75m due to up tilt of 1500-1520 ms^{-1} sound-speed isopleths (Fig.4) in the region of cold core eddy as compared to unperturbed condition as shown in the Fig.9a. Here cyclonic eddies played a significant role in the interaction with the surface waters and in turn acoustic field, by raising the thermocline locally. It is evident from Fig.9b that further scanning of traced rays (in contrast to unperturbed condition, Fig.9a) has taken place in the water column above the conjugate depth. Due to the displacement of eigenrays by cold core eddy, there are delays in travel times (with reference to mean) are found (column 6 of Table – 8 & Figs.10 & 11).

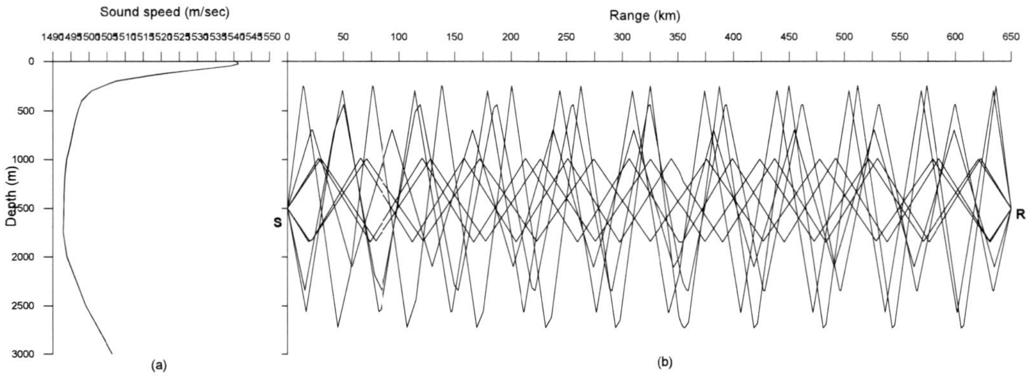

Figure 8. (a). Mean sound velocity profile $C_o(z)$ constructed from Levitus Climatological data along parallel track (Fig. 2) (b). Eigenray plot for a source and receiver located at 1500m depth over a range of 650km.

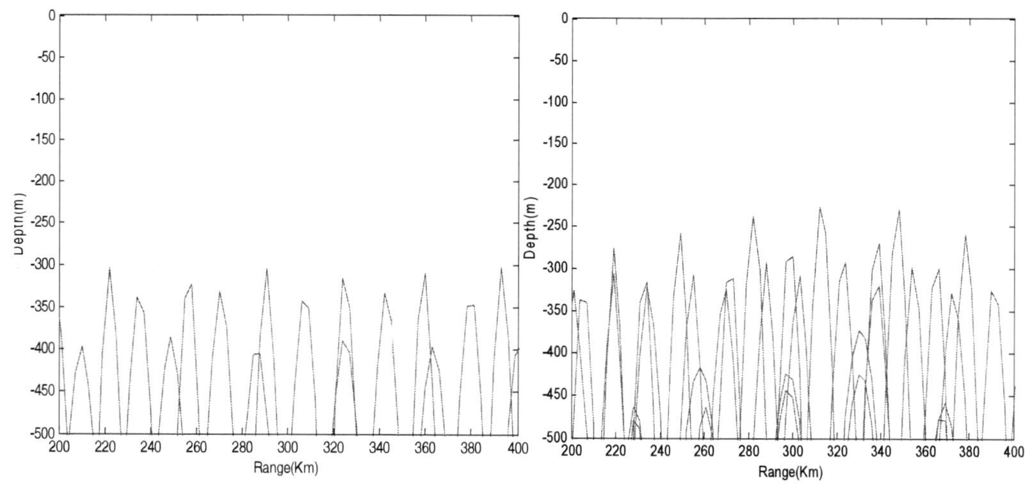

(a) (b)

Figure 9. (a) Ray paths in the reference ocean $(C_0(z))$ at cold core eddy location. (b) Ray paths in the presence of cold core eddy location.

To perform inversion of the travel time data, each arrival must be associated with a particular ray path corresponding to the reference as well as to the eddy profile. This identification is achieved by comparing the simulated arrivals for both mean and eddy in the time domain (Fig.10 & Table-8). Table-8 presents the characteristics of each of the 8 eigenrays. The column wise ray nomenclature in Table -8 is detailed as a footnote. The columns 7 to 12 of Table 8 explains number of upper (p^+) and lower (p^-) ray turnings, number of top (p_t^+) and bottom (p_b^-) ray bonks, and total number of ray turnings and bonks respectively. This information can be used as the ray identifier.

Table 8. Acoustic ray parameters for source depth = 1500 m, receiver depth = 1500 m and range =650 km

S. No.	θ_0 (deg)	θ_r (deg)	T (sec)		Travel time anomaly (sec) (δT^M)	P^+	p^-	p^+_t	p^-_b	$p^+ + p^-$	$p^+_t + p^-_b$	-db	
			Reference	Presence of Eddy								Reference	Presence of Eddy
1	6.52	-6.52	435.8272	435.8610	-0.03380	11	11	0	0	22	0	108.3	112.2
2	4.15	-4.16	435.5750	435.6500	-0.07500	10	10	0	0	20	0	106.9	116.4
3	3.08	3.09	435.5202	435.5493	-0.02910	9	10	0	0	19	0	119.3	118.7
4	0.86	0.86	435.4674	435.3148	0.15260	7	8	0	0	15	0	113.6	116.4
5	0.76	-0.80	435.4582	435.3060	0.15220	7	7	0	0	14	0	114.3	123.4
6	-0.71	0.71	435.4539	435.2991	0.15480	7	7	0	0	14	0	114.1	120.8
7	-4.66	4.69	435.5539	435.6272	-0.07330	10	10	0	0	20	0	109.8	116.3
8	-5.75	-5.81	435.6700	435.7369	-0.05790	11	10	0	0	21	0	117.1	114.9

Θ0 - ray launching angle at source, Θr - ray angle at receiver, T – travel time, p+ - number of upper turning points, p- - number of lower turning points, p_t^+ – number of top bonks, p_b^+ – number of bottom bonks, , d Θ – acoustic intensity loss.

There are four classes of rays 1) R/R (purely refracted both above and beneath the axis, $p_t^+ = p_b^- = 0$) 2) SR/R (surface reflected and refracted, $p_t^+ \neq 0, p_b^- = 0$), 3) R/BR (refracted and bottom reflected, $p_t^+ = 0, p_b^- \neq 0$) 4) SR/BR (surface and bottom reflected, $p_t^+ \neq 0, p_b^- \neq 0$). It is interesting to note that in the Bay of Bengal, long-range acoustical transmission paths are mostly refracted/bottom-reflected, *(* R/BR *)*, and refracted/refracted, *(* R/R *)*. In the present studies, we have considered the R/R category rays among all possible above said classes to avoid bottom losses (Table 8).

3.6. Ray Arrival Pattern

The travel times of eigenrays are determined by two factors viz., the ray path length and the sound-velocity along the path. The flat rays which have small initial angles that travels along shorter paths comparatively, but in lower sound-velocity region. On the other hand, the steep rays, which have larger initial angles, travel in faster sound-velocity regions but over greater path lengths. In general, the path lengths of steep rays in the top layer are less as compared to deeper layers. The combined effect of path length and sound-velocity on the ray travel times shown as a function of launch angle in Figure 10.

The predicted arrival pattern depicts the typical characteristics of the Bay of Bengal sound propagation namely the early arrival clusters of near – axial flat angle rays ($+0.85^0$ to -0.85^0) and the later arrivals of deep-turning rays *(R/R)*. This type of arrival pattern arises due to the weak acoustical wave guide nature of the Bay of Bengal, contradictory to general pattern of other waters (Prasanna Kumar et.al, 1994).

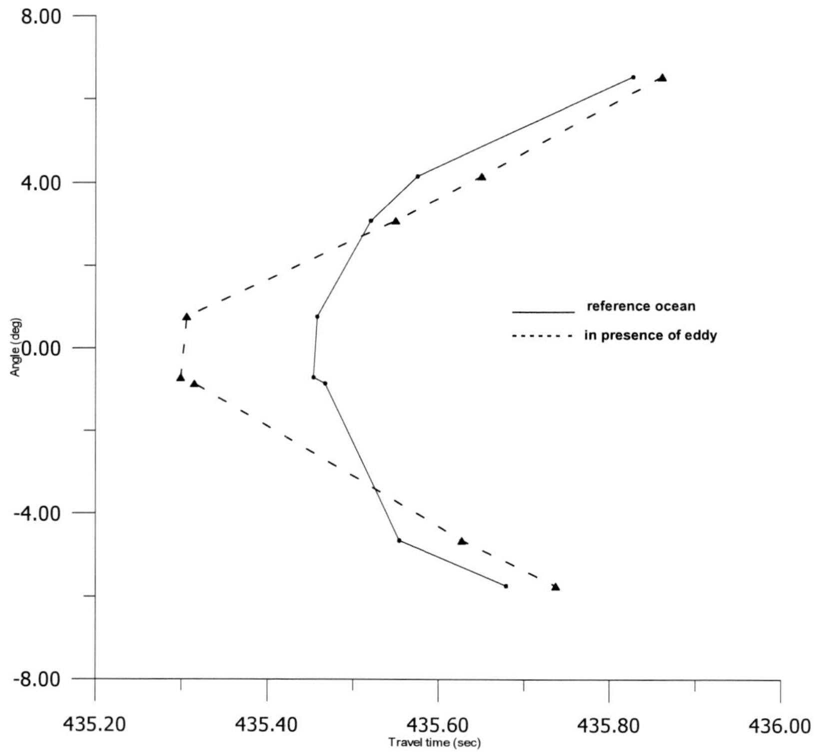

Figure 10. Travel time versus eigen ray angle for the reference and eddy profiles.

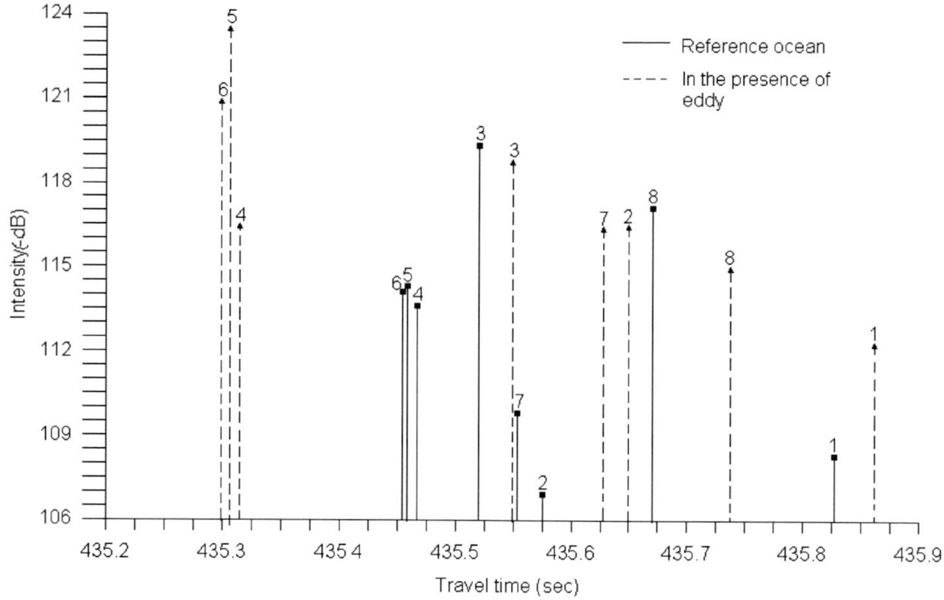

Figure 11. Comparison of predicted arrival pattern obtained from reference ocean $(c_0(z))$ and eddy profiles. The arrivals are labled and identified in Table 8.

It is discernible from the diagram that although the arrival angle (Fig.10) remains the same to a large extent for both the profiles, the arrivals are delayed by about 20 – 75 ms (milliseconds) in the presence of the eddy for step angle rays (column 6 of Table 8). The travel times of flat angle rays also showed delays due to channel depth variation (400m from south to north, as shown in Fig.4), and also weak acoustic wave guide nature (Table 8). In the present simulation, one can expect that the travel time perturbations of steep angle rays which travel through eddy are supposed to show more as compared to flat angle rays, for a long range of 650 km. However, slowness of steep angle rays in the present case are due to less transgressed in the top layer (Table 8).

3.7. Acoustic Intensity

The computed intensity of eigenrays (using the ray tube method) at the receiver end (see Appendix-A for details) shows the loss associated with steep angle rays through the average field is about 112.28dB, while that through the eddy field is about 115.7dB (Fig 11 & columns of 13 & 14 of Table - 8). The flat angle rays ($+0.85^0$ to -0.85^0), which do not sample the eddy; undergo average intensity loss of 114.05dB and 119.25dB for unperturbed and perturbed conditions respectively (Fig 11 & Table 8).

3.8. Inversion

Building the Estimates

For each point in a given region, the coefficients $a_i^1(x,z)$ (in Eq.5) are determined from the eigenrays and operated on the travel time perturbations(*d*) to estimate the model parameter $\delta\hat{C}(x,z)$ objectively. This procedure permits a continuous representation of the unknown field. The data-data and model-data covariance matrices are determined for the reference sound velocity distribution with a suitably selected source-receiver configuration. Using these matrices the estimated sound velocity field $\delta\hat{C}(x,z)$ of the real ocean $\delta C(x,z)$ is obtained for the given travel time perturbations (δT).

Once the model covariance (the unknown field - in this case, the sound velocity anomaly) is obtained, the model-data covariance $(<\delta C(x,z)d^T>)$ of order *(M x 1)* and data-data covariance matrices $(<dd^T>)$ of order (*M x M*) can be constructed as described in the above theory (Eqs.6-10). In the present example, model-data covariance matrix $(<\delta C(x,z)d^T>)$ is of order *(M x 1)* where *M*=8 is the number of identified eigenrays (Table-8). For example, computation of each element in matrix $(<\delta C(x,z)d^T>)$ is generated by evaluating the single integration (Eq.8) along the *i-th* identified eigenray (columns 7-12 of Table 8)). The integrand in the Eq.8 consists of a point (x,z) in the vertical domain (where model parameter estimation $\delta\hat{C}(x,z)$ to be computed), refractive index, vertical (U) and horizontal $(a_i(x))$ i.e., EOF modes. While performing integration, the cubic spline interpolation is carried out to evaluate vertical functional values of (U) of $s(z)$ distribution and horizontal modal amplitudes $(a_i(x))$ *of* $s(r)$ for all modes along *i-th* eigenray co-ordinates $(x^i(s), z^i(s))$ (where *s* is the distance along ray) (Appendix -A). The model-data covariance is constructed for mapping 5 points in the horizontal (at the station locations of the 5 CTD casts in the survey). This is carried out for easy comparisons between the estimates of sound velocity from assumed acoustic data and those calculated from the CTD stations. Note that the changes to the output of the estimator affect only the rows of the model-data covariance matrix. Similarly data-data covariance matrix $(<dd^T>)$ can be computed.

In the present example, data-data covariance matrix $(<dd^T>)$ is of the order (M x M) where *M*=8 is the number of identified eigenrays (Table -8). For example, computation for each element in matrix $(<dd^T>)$ is generated by evaluating the double integration (Eq.9) along both the *i-th* and *j-th* identified eigenrays (Table -8).

The data-data covariance matrix is fixed by the data available in the experiment or assumed simulation, and therefore does not change when a new output is desired. When any particular field or distribution of mapping points are desired, one needs only to compute the appropriate model-data covariance matrix and then multiply it by the inverse of the data-data covariance matrix, which is computed once and saved. Here Generalized Inverse Operator (GIO) (i.e. $A_p^{-1} = V_p \Gamma_p^{-1} U_p^T$) of the data-data covariance matrix $(<dd^T>)^{-1}$ is built using SVD employing eigenvalues technique (Jackson, 1972). Here *p* indicates the number of factors used in SVD, which is less than or equal of the rank of the matrix ($(<dd^T>)$ (i.e., *p* $\leq$ *M, where M* is the number of eigenrays). Also, U_p, and V_p are the column eigenvectors of the matrix $(<dd^T>)$ of U and V. Γ_p is the diagonal matrix of the matrix $(<dd^T>)$. The elements of diagonal matrix (eigenvalues) arranged in descending order and corresponding closeness ratio is presented in the Table -9. The SVD provides a simple frame – work for determining how well the inverted model parameters fit the data, and how closely the model parameter estimates the true values. The larger eigenvalues determine the large scale (gross) features of the problem, while the smaller eigenvalues correspond to small scale or higher frequency features. Then Eq.6 becomes.

$$A(x,z) = < \delta C(x,z) \delta T_i^T > A_g^{-1} \qquad \text{.......... .....} \quad (13)$$

Using this matrix $A(x,z)$ the estimated sound velocity anomaly field $(\delta\hat{C}(x,z))$ (Eq.5) of the real ocean $(\delta C(x,z))$ for different factors *p* (=1 to 5), is obtained for the given travel time perturbations (δT) of Table 8. Fig.12 gives the comparison between the tomographically reconstructed sub-surface cold core eddy field with that of observed ones (right of lower panel in Fig.12). Even 3 factors (contribution of 86%, Table – 9) are more than sufficient to objectively resolve the sub-surface eddy (gross features), where as for fine structures, it is needed at least five factors.

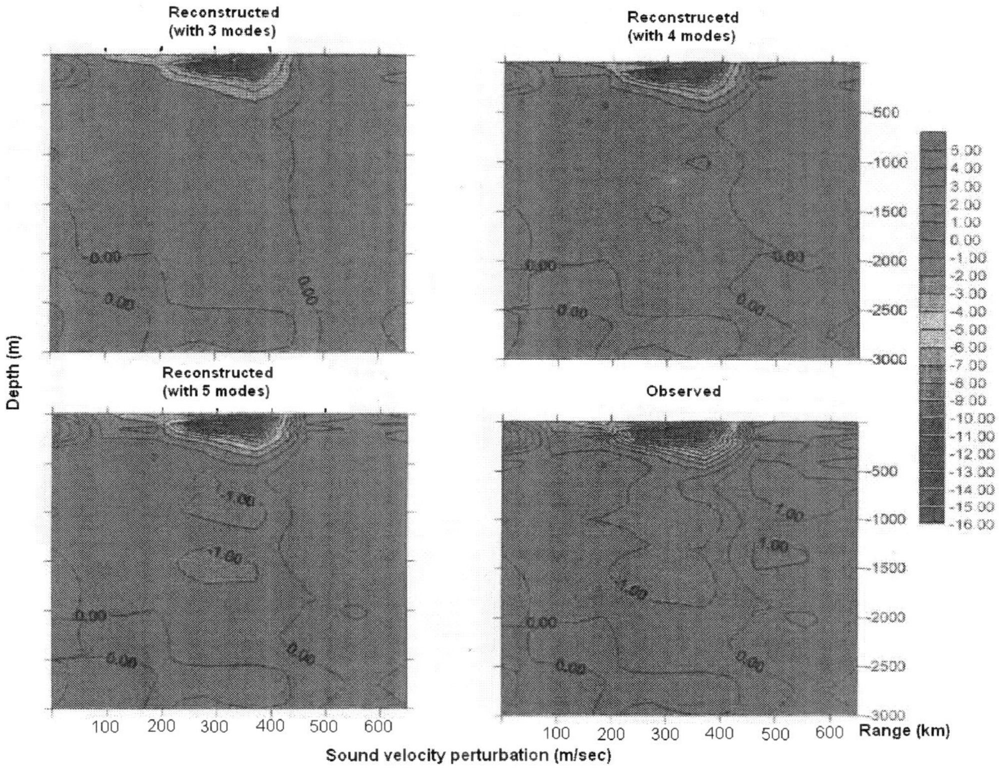

Figure 12. Tomographic inversion of a cold core eddy (21 – 29 July, 1984) in the Bay of Bengal.

Present, OAT's simulation experiment results reveal that one can objectively map /monitor the even small meso-scale sub-surface eddies by stochastic inverse technique keeping source and receiver in the SOFAR channel separated by probably for larger distances also under such a depth-limited environment of Bay of Bengal.

Table 9. Eigenvalues of $[< dd^T >_p < dd^T >_p^T]$

No.	Eigenvalue	Closeness Ratio (%)
1	0.709	44.97
2	0.610	34.19
3	0.283	7.15
4	0.267	6.38
5	0.227	4.62
6	0.154	2.11
7	0.071	0.45
8	0.037	0.13

For the assessment of the validity of the layering in the model used for inversion, it is essential to look at the model and data resolution matrices, but in the present case, both are the same. Here after, we call them resolution matrix$[<dd^T>_p<dd^T>_p^T]$. The diagonal elements of this matrix for different factors (p=1 to 8) are presented in the Table 10. A value of unity indicates contribution of information independent of the other observations. Among all the rays, the most independent information is contained by 6.52° ray. Table 10 therefore enables one to infer which of rays present poor information resolution (resulting from sampling nearly the same ocean region). Flat angle rays show maximum values in the vicinity of the sound velocity layer(Table-10, sl.no.4) . This is due to the nature of the SOFAR channel where ray bending takes place towards the sound velocity minimum focusing an envelope.

Table 10. Diagonal elements of resolution of data matrix $[<dd^T>_p<dd^T>_p^T]$ for each angle for factor p = 1 to 8

S.No	Eigen Ray angle (θ_0^0)	1 Mode	2 Modes	3 Modes	4 Modes	5 Modes	6 Modes	7 Modes	8 Modes
1	6.52	0.48	0.98	0.99	0.99	0.99	1	1	1
2	4.15	0	0	0.01	0.02	0.05	0.12	0.46	0.7
3	3.08	0	0	0.03	0.04	0.05	0.12	0.53	0.89
4	0.86	0	0	0.33	0.34	0.77	0.95	0.99	1
5	0.76	0.01	0.01	0.15	0.36	0.42	0.87	1	1
6	-0.71	0.34	0.75	0.8	0.96	0.96	0.99	1	1
7	-4.66	0	0	0	0	0.01	0.03	0.04	0.41
8	-5.75	0.01	0.01	0.39	0.43	0.86	0.97	1	1

Having carried out a number of simulation experiments and inversion, reconstructing the sound velocity anomaly adequately, the next step was to conduct an experiment for a short duration to test modes developed and to see how close one could reconstruct the sound velocity and temperature anomaly from the measured data. In view of this, acoustic tomography experiment was conducted in the eastern Arabian Sea (Prasanna Kumar et.al., 1999; Ramana Murty et.al., 2007b; Acoustic Group, 1993). For more details on OAT's experiment (viz., the deployment of mooring acoustic system, acoustic data collection, method of ray identification etc., interpretation of results with observed one), one may refer the paper by Prasanna Kumar et.al (1999) and are not covered in the present chapter.

4. Objective Mapping of Temperature Field by Stochastic Inverse Method Using Acoustic Tomography Experimental Data of Eastern Arabian Sea

An acoustic transmission experiment was conducted in the eastern Arabian Sea along 12.5°N latitude for a duration of ten days (2-12 May, 1993), with two transceiver systems(TR1 and TR2) having low central frequency 400Hz(100Hz bandwidth) deployed on deep sea moorings, separated by a range of 270.92 km(Figure 13).

Hourly reciprocal transmissions were carried out with time lag of 30 minutes between each direction. The received signals were cross-correlated with the replica of the source signal during the post processing, to generate amplitude peaks corresponding to the multipath arrivals (Fig.14). While transmissions were going on, a hydrographic survey was simultaneously carried using a Sea-Bird CTD system.

Vertical profiles of temperature and salinity, at half degree spacing were taken at 16 locations (Fig.13), for the *in situ* information on acoustic profile and the prevailing hydrographic conditions during the experiment. Range-independent and dependent ray tracing was carried out with Levitus Climatological mean sound velocity $C_0(z)$ and in situ sound velocity $C_0(x,z)$ to compute predicted (simulated) eigenray(Fig. 15) arrivals(Figure 16).

A sequence of the predicted ray arrivals in the Arabian sea is as follows: purely refracted rays with emergence angle between ±5-7^0 (with respect to horizontal) reach the receiver first, followed by lower angle near axial rays, and the steep rays with emergence angles±8-9^0 : the refracted- bottom reflected rays with emergence angle $\pm 10^0$ are the final arrivals. A close examination of the ray arrival time and the corresponding intensity of various arrivals revealed that the arrival patterns are stable in time. In view of this, the ray identification is carried out by correlating the axial arrival peaks of the measured travel time (Fig.16) with that of the predicted ones. A typical ray identification is shown in (Fig.16). It was found that during the 4 days transmissions, out of the 14 eigenrays, 9 rays were stable in time (Table 11), enabled reconstruction of temperature anomaly from the sound velocity perturbations using stochastic inverse method. For formulation of stochastic inverse, generation of data-data (Eq.9) and model-data (Eq.8) covariance matrices are needed for computing the modal amplitudes (Eq.10) to operate on observed time perturbations (δT^{M}) to estimate the sound velocity perturbations $\delta\hat{C}(x,z)$. For the present simulation, the dependence of travel time on the oceanic sound velocity has been calculated using ray paths traced by computer with mean velocity profile while vertical and horizontal velocity structures were modeled as a combination of Empirical Orthogonal Functions (EOFs) from in situ CTD data in the study area.

Table 11. Ray Identifier

Launch Angle(θ_0^0)	Total turning(N)	Upper-Lower turnings	Total Travel time T(s)
8.10	9	(5, -4)	181.4543
6.87	8	(4,-4)	181.2587
5.39	8	(4,-4)	181.2166
4.30	9	(5,-4)	181.2865
3.32	10	(5,-5)	181.3380
2.13	11	(6,-5)	181.3595
-1.56	12	(6,-6)	181.3763
-3.14	11	(6,-5)	181.3494
-5.30	9	(5,-4)	181.2357

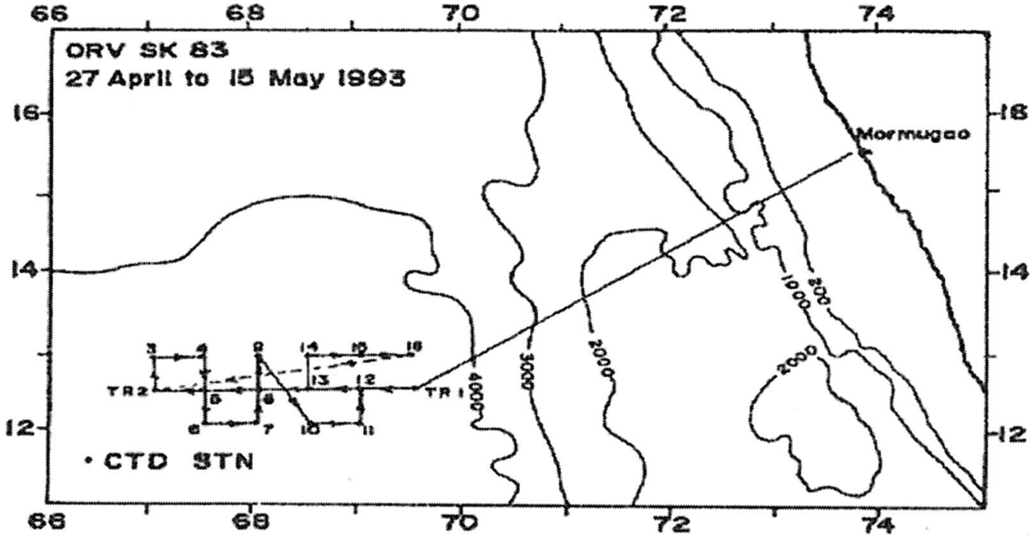

Figure 13. Location map showing transceiver (TR1 & TR2) moorings and CTD stations.

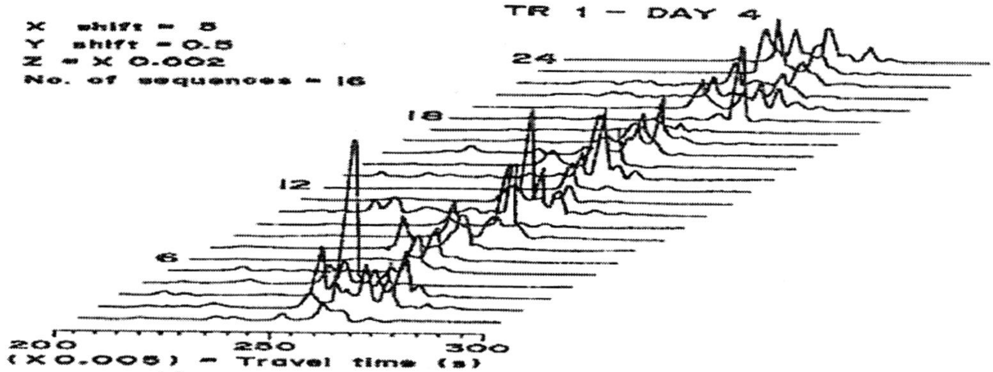

Figure 14. Waterfall plot showing measured signal arrivals at the receiver.

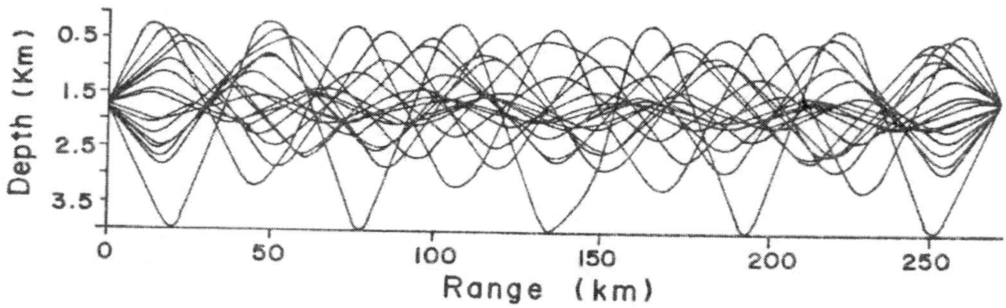

Figure 15. Eigen rays from TR2 to TR1.

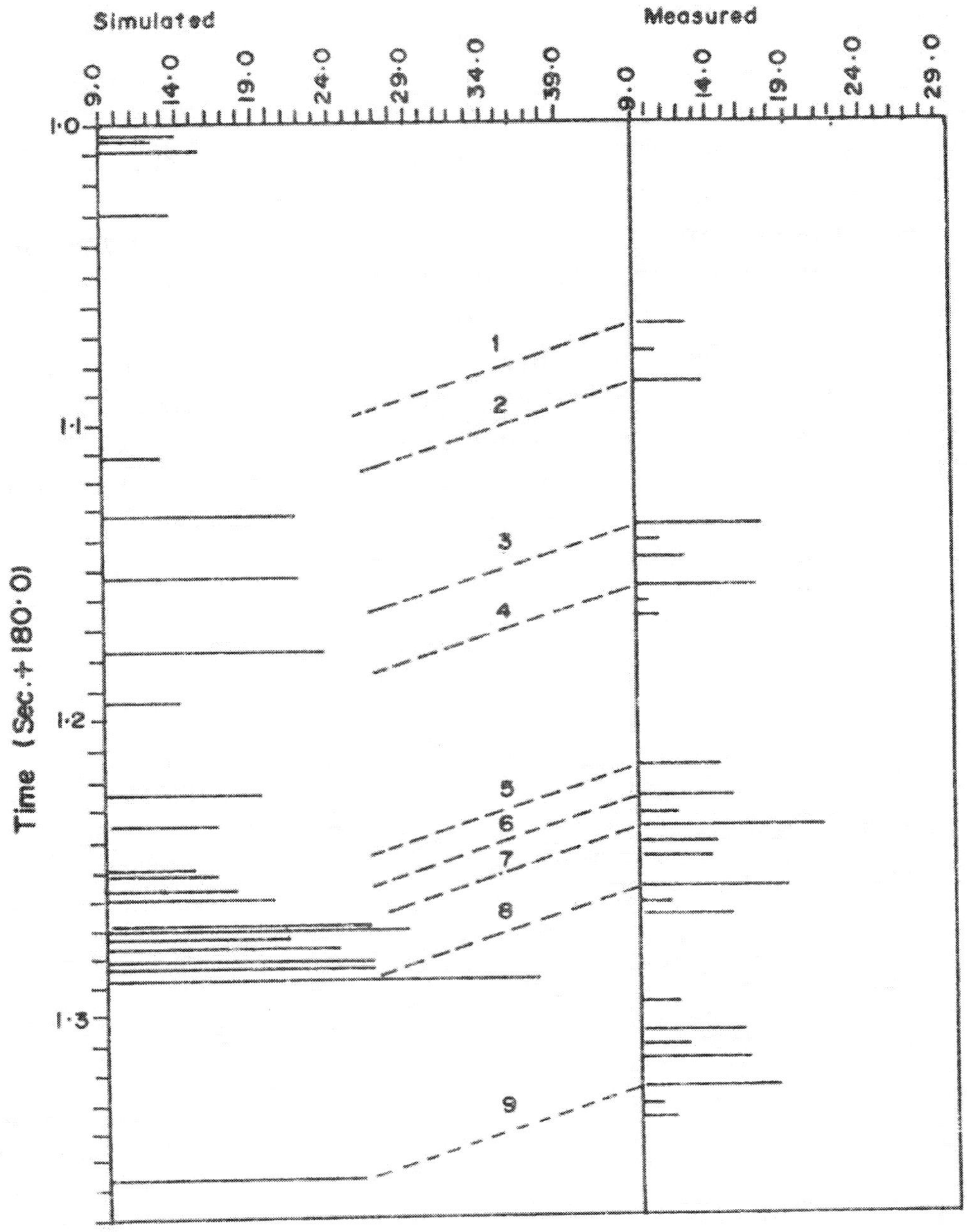

Figure 16. Identification of eigen rays predicted with measured arrivals.

A linear empirical relation $\delta\theta = \delta C/(aC_{oj})(a = \sim 3.2\text{x}10^{-3}/{}^{0}c)$ the coefficient of thermal expansion) was used to transform sound speed perturbation (δC) in the vertical plane to temperature perturbations ($\delta\theta$) (Fig.17), for the first four days of transmission, following Munk and Wunsch (1979). The 2-D temperature anomaly derived from the six hourly mean travel-time data showed a gradual warming of the top layers, indicating significant diurnal variability and intrusion of Red sea water.

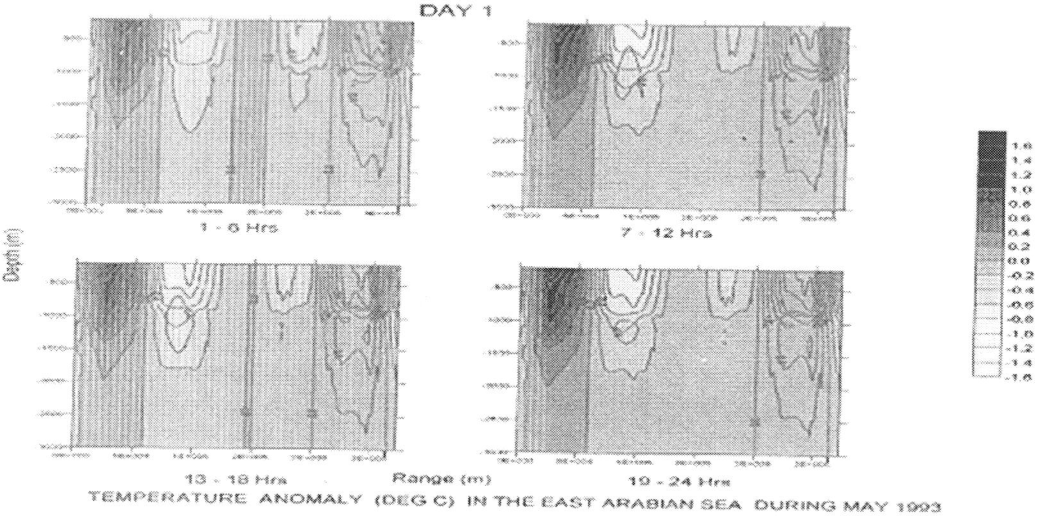

Figure 17. Objective mapping of temperature anomaly using OAT's principle applied to acoustic field transmission data.

ACKNOWLEDGMENTS

The authors are thankful to Dr. S.R. Shetye, Director, National Institute of Oceanography, Goa, for his interest and encouragement. The OAT group of NIO, Goa during 1982-1997 was also acknowledged for their long association with the first author of this chapter for the interaction and discussions to gain knowledge in marine acoustics. This is NIO contribution number 4982.

APPENDIX A

A1. Elements of Ocean Acoustics

A1.1. Wave Equation

Acoustic wave propagation is an unsteady fluid flow problem related to mechanical vibration of ocean water in the range of tens of hertz to few hundred kilohertz. These vibrations are assumed to cause small departures from what is otherwise a static state in the fluid. As these vibrations are very fast and small, we ignore heat transfer between fluid elements. We also ignore viscosity since fluid motions are small. Now the equations governing this flow field are the continuity equation (Ramana Murty & Mahadevan, 1994)

$$\frac{\partial \rho}{\partial t} + \nabla.\rho V = 0 \qquad \text{(A1)}$$

and the momentum equation,

$$\rho\left(\frac{\partial V}{\partial t}+V.\nabla V\right)=-\nabla P-\rho g\hat{k} \tag{A2}$$

where ρ is the density of the fluid, V is the velocity of a fluid element with components (u,v,w), ρ is the pressure, $\hat{k}$ is a unit vector in the vertical **(z)** direction and g is the acceleration due to gravity.

The flow variables associated with acoustic wave propagation problem are treated as small departures from the undisturbed fluid state, and they are represented in the form $\rho=\rho_0+\rho_1$, $V=V_0+V_1$ and $P=p_0+p_1$. Here, the subscripts 'o' denotes the undisturbed fluid state (i.e. the static state with $V_0=0$) and '1', the departures in field variables due to the presence of acoustic waves. Then, substituting these flow variables in Eqns. (A1) and (A2) and neglecting terms involving products of acoustic wave induced variables, we get the liberalized form of the governing equations as

$$\frac{\partial\rho_1}{\partial t}+\nabla.(\rho_0 V_1)=0 \tag{A3}$$

and

$$\frac{\partial}{\partial t}(\rho_0 V_1)=\nabla P_1 \tag{A4}$$

Eliminating V_1 between these equations, we get

$$\frac{\partial^2\rho_1}{\partial t^2}=\nabla^2 P_1 \tag{A5}$$

The frequency of acoustic oscillations is quite high and the associated pressure fluctuations in the fluid could be assumed as adiabatic. Then the equation of state is given by $P_1=C^2\rho_1$ and eqn. (A5) can be written as (subscript is omitted for convenience).

$$\frac{\partial^2 P}{\partial t^2}=C^2\nabla^2 P, \tag{A6}$$

where C is the sound velocity through the propagating medium. The above equation, referred to as the wave equation, defines the pressure field associated with acoustic waves propagating in ocean waters.

The solution to this equation (i.e. to the acoustic wave propagation problem in ocean waters) can be approached by acoustic ray theory applicable to high frequency waves or the normal mode theory. We will discuss here the ray theory, which has been successfully used in ocean acoustic tomography and other relevant sound propagation studies (Ramana Murty and Mahadevan, 1994).

A1.2. Ray Theory

Ray theory is widely used in underwater acoustics since it has many advantages. Some of them are 1) Rays are easily drawn 2) Real boundary conditions are inserted easily, e.g. a sloping bottom 3) It is independent of source.

This method is based on an equation called eikonel equation. This eikonel equation is obtained from the wave equation by assuming the solution in a particular form.

We initially seek a periodic solution in time to the wave equation in the form $P(x, y, z, t) = p(x, y, z)\exp(-i\omega t)$. This reduces the wave equation to the Helmholtz equation given by,

$$\nabla^2 p + k^2 p = 0 \qquad \text{(A7)}$$

where $k = \omega / C$, is the wave number. The variation in sound velocity in the ocean is very small and the solution to the Helmholtz equation can be assumed in the form

$$P(x, y, z) = A(x, y, z)\exp[i k_0 W(x, y, z)] \qquad \text{(A8)}$$

where A is a slowly varying function of position, k_0 is a constant reference wave number and W is called the eikonel function. The functional form of this solution can be easily recognized to be the same as that of a plane wave solution in a homogeneous medium. i.e.,

$$P(x, y, z) = A\exp\{iS(x, y, z, t)\} \qquad \text{(A9)}$$

where A is the constant amplitude of the wave and

$$S(x, y, z, t) = k_x x + k_y y + k_z z - \omega t$$

is called the phase function. Here k_x , k_y and k_z are the components of the wave number vector K of the plane wave. Note that, for a non-homogeneous medium, from eqn. (A8), the phase function can be defined as

$$S(x, y, z, t) = k_0\ W(x, y, z) - \omega t$$

Substituting the solution, eqn. (A8), in the Helmholtz equation, the real part gives

$$\nabla W . \nabla W - \omega^2 / (k_0^2 C^2) = \nabla^2 A / (k_0^2 A) \qquad \text{(A10)}$$

The RHS of the above equation can be shown to be small when the variation of **C** is small over one wavelength of an acoustic wave. This condition is satisfied in the case of high frequency sound propagation in ocean waters. Then we get

$$\nabla W.\nabla W = n^2 \tag{A11}$$

where the refractive index $n = C_0 / C$, is a function of position and $C_0 = \omega / k_0$. The eqn. (A11) is called the eikonel equation.

Now at any time t, $S(x, y, z, t)$ = constant, implies that the eikonel function is a constant. Hence W = constant defines a surface of constant phase and ∇W defines the normal to the surface of constant phase or the direction of ray propagation or ray path. In eq. (11), ∇W is a vector in the direction of ray path and n is the magnitude of this vector. Ray acoustics provides a computationally convenient and easy interpretation of acoustic propagation to a degree of accuracy. Ray tracing is a well-established technique for determining the character of sound in the ocean. The trajectory of a ray can be described by ray equations derived from the geometrical optics.

Let $dr = (dx, dy, dz)$ be the displacement vector along the ray path and

$ds = | dr |$. Then

$$n\frac{dr}{ds} = n\left(\frac{dx}{ds}, \frac{dy}{ds}, \frac{dz}{ds}\right) = \nabla W \tag{A12}$$

Now $\frac{d}{ds}\left(n\frac{dx}{ds}\right) = \frac{d}{ds}\left(\frac{\partial W}{\partial x}\right) = \frac{\partial}{\partial x}\left(\frac{dW}{ds}\right)$

$$= \frac{\partial}{\partial x}\left(\frac{\partial W}{\partial x}\frac{dx}{ds} + \frac{\partial W}{\partial y}\frac{dy}{ds} + \frac{\partial W}{\partial z}\frac{dz}{ds}\right)$$

$$= \frac{\partial}{\partial x}\left\{n\left[\left(\frac{dx}{ds}\right)^2 + \left(\frac{dy}{ds}\right)^2 + \left(\frac{dz}{ds}\right)^2\right]\right\} = \frac{\partial n}{\partial x}$$

Hence we get

$$\frac{d}{ds}\left(n\frac{dx}{ds}\right) = \frac{\partial n}{\partial x}, \frac{d}{ds}\left(n\frac{dy}{ds}\right) = \frac{\partial n}{\partial y} \text{ and } \frac{d}{ds}\left(n\frac{dz}{ds}\right) = \frac{\partial n}{\partial z} \tag{A13 a, b, c}$$

These equations state that the variation along the ray path of the product of the index of refraction and the direction cosine is equal to the space rate of variation of the index of refraction with respect to the appropriate coordinate. Eqns. (A13) are the generalized form of Snell's law.

Let us consider the sound propagation problem in two dimension in x-z plane (see Fig.1). Then along the ray path

$$\frac{dx}{ds} = \cos\theta \text{ and } \frac{dz}{ds} = \sin\theta \qquad \text{(A14 a,b)}$$

Substituting eq. (A14 a) in (A13 a) and $n = C_0 / C$, we get along the x-direction

$$\frac{d}{ds}\{(C_0 / C)\cos\theta\} = \frac{\partial}{\partial x}(C_0 / C)$$

$$\frac{\partial C}{\partial x} = \cos\theta \frac{dC}{ds} + CSin\theta \frac{d\theta}{ds} \qquad \text{(A15)}$$

Similarly along the **z**-direction

$$\frac{\partial C}{\partial z} = Sin\theta \frac{dC}{ds} - CCos\theta \frac{d\theta}{ds} \qquad \text{(A16)}$$

Eliminating dC/ds from eqs. (A5) and (A16), we get

$$\frac{d\theta}{ds} = \frac{1}{C}\left[Sin\theta \frac{\partial C}{\partial x} - Cos\theta \frac{\partial C}{\partial z}\right] \qquad \text{(A17)}$$

Now the eqs. (A14) and (A17) describe the ray paths and the numerical integration of these equations for a given sound speed distribution, provides the coordinates of the ray paths.

Intensity computations—The acoustic intensity at any point along the ray path is given by the expression (Krol, 1973).

$$Intensity = I(x)/ I_0 = (\sin\theta_0 / x\cos\theta).(\Delta\theta_0 / \Delta z) \qquad \text{(A18)}$$

where I_0 = reference intensity at the source; θ_0 = eigen ray angle at the source with reference ti the horizontal; θ = ray angle at the point of measurement with reference to the horizontal; X=distance from the source to the point of measurement; Δz = vertical distance at the point of measurement between eigen ray and ray in its immediate vicinity; and $\Delta\theta_0$ = angle between eigen ray and ray in its immediate vicinity at the source.

At the point where $\Delta z \rightarrow 0$, the above expression will not hold good (tends to infinity mathematically) and the previous value only gets substituted for computational continuity.

Considering the phase shift for any given frequency the intensity has been computed following (Moler & Soloman, 1970).

$$I/I_0 = \left|\sum P_\lambda e^{i2\Pi f t_k}\right|$$

$$= \sum_j \sum_k P_j P_k \cos\left[2\Pi f\left(t_j - t_k\right)\right] \quad \text{(A19)}$$

where $P_k = \left[I_k(x)/I_0\right]^{1/2}$ and t_k are the relative pressures and arrival times of the eigen rays. These have wider applications in signal detection.

APPENDIX-B

B1. Computational Procedures:

Following Lawson and Hanson (1974), 2D sound velocity perturbation $\delta C(x, z)$, (ignoring time dependency) is written as product of three matrices in component form:

$$\delta C(x, z) = \sum_{i=1}^{r} u_i(z) v_i(x) l_i^2 \quad (B1)$$

$x = 1, \ldots\ldots, n; z = 1, \ldots\ldots, m$; r = rank of D

where n is the number of vertical layers in the ocean model, m is the number of horizontal grid points, the $u_i(z)$ are the vertical modes and the $v_i(x)$ are the horizontal coefficients. Define D = $\delta C(x, z)(= C(x, z) - C_0 = D)$

Eqn (B1) is expressed in Singular Value Decomposition (SVD) form:

$$D = U \Gamma V^T \; (B2) \quad (m\, x n) \quad (m\, x\, r)(r\, x\, r)(r\, x\, n)$$

The dimensions of matrices D, U, Γ and V^T are (m,n), (m,r), (r,r), and (r,n) respectively. The matrices U and V are orthonormal i.e.; $U^T U = I_m$ and $V^T V = I_n$ where I_m and I_n are the identity matrices of order $(m\, x\, m)$ and $(n\, x n)$ respectively. In general, $U\, U^T \neq I_m$ and $V\, V^T \neq I_n$ and these matrices are called data and model resolution matrices. The matrix Γ is a diagonal matrix of nonzero "singular values" of D in descending order $(I_1 > I_2 > \ldots\ldots\ldots., I_r)$ and r $[r \leq \min(m, n)]$ is the rank of matrix D. Obviously, the rank of all the three component matrices will be $\boldsymbol{r}$, and hence all diagonal elements of Γ are squares of singular nonzero values $\boldsymbol{l_i^2}$. $\boldsymbol{U}$ and $\boldsymbol{V}$ are the respective coupled eigenvector matrices for the eigenvalue problem defined as

$$(D\, D^T)\, u = l^2 u \quad (B3)$$

and $(D^T\, D)\, v = l^2 v$ $\quad (B4)$

The matrices $D\,D^T$ and $D^T D$ are called the covariance matrices, of orders $(m\,x\,m)$ and $(n\,xn)$ respectively and , from which the EOF's are generated. The two sets, i.e depth-dependent $s(z)=\{u_1, u_2, ...u_p\}$ and range-dependent $s(z)=\{v_1, v_2, ...v_p\}$ are the orthogonal vectors (column vectors of u and v) of the matrix D respectively and $p(\leq r)$ are the number of factors to be considered. EOF depicts the underlying dynamics describe statistically to determine the direction of maximum variability. These directions called the "principal components" of EOFs of the second order statistics or "moments".

If $m<n$ then, first compute matrix U by solving eigenvalue problem (Eqn.B3) and matrix V is computed by relation $V = D^T U \Gamma^{-1}$

If $m>n$ then, compute matrix V by solving eigenvalue problem (Eqn.B4) and matrix U is computed by relation $U = DV\Gamma^{-1}$

Having computed the u_i, the modal amplitudes $a_i(x)$ at each horizontal grid points are easily computed by projecting the layered perturbation profiles on the layered EOF's, i.e;

$$a_i(x) = \sum_{K=1}^{r} \delta C(x, z_k) u_i(z_k),\ i, k = 1,.., r \qquad (B5)$$

where, i – indicates the mode index and n- the layer index. Here i can go to a number $p \leq r$ *(rank of D)*, if we truncate the mode sum. This is done by judiciously selecting the p eigenvectors or ranking the singular values of the Γ matrix in descending order. We then spline fit the modes, their coefficients, and the mean profiles to get a field with continuous derivatives at any point. The mode contribution of sound velocity perturbation field in terms of percentage is as follows:

B2. Closeness Ratio

The ratio between the sum of the factors considered and that of data matrix (Kernel) is the measure of closeness of the model data (Ramana Murty et. al., 1986)

$$\text{Measure of closeness} = \sum_{i=1}^{p} l_i^2 \Big/ \sum_{i=1}^{r} l_i^2 \qquad (B6)$$

where r is the rank of the data matrix $\boldsymbol{D}$ (kernel) and $p \leq r$. The first eigen function, u_1 in depth dependent vector space $\boldsymbol{s}(z)$, associated with the largest eigenvalue l_1^2 represents the gross features (i.e. $\| u_1^T D \|$) in the data in the least square sense, while the second function u_2 in the space $\boldsymbol{s}(z)$ associated with second largest eigenvalue l_2^2 describes the residual mean square data [i.e. $\| u_2^T (D - (u_1 u_1^T D)) \|$] in the least square sense and so on. The similar explanation can be given for range dependent vectors $s(r)$. The closeness ratio is expressed in percentage to judge the contribution of different parameters. $s(z)$ and $s(r)$ distribution of individual percentage contribution of jth element of, k^{th} eigen function for present analysis are given by (Ramana Murty et.al., 2007)

$$\left[\frac{u_{jk}}{\sum_{i=1}^{m} |u_{ik}|}\right]\left[\frac{l_k^2}{\sum_{i=1}^{r} l_i^2}\right] * 100 \qquad (B7)$$

$$\left[\frac{v_{jk}}{\sum_{i=1}^{n} |v_{ik}|}\right]\left[\frac{l_k^2}{\sum_{i=1}^{r} l_i^2}\right] * 100 \qquad (B8)$$

REFERENCES

Acoustic Group," *Acoustic Transmission Experiment – 93*", Report of National Institute of Oceanography, Dona Paula, Goa., 1993.

Aki, K., and Richards, P.G., "*Quantitative Seismology, theory and methods, Freeman, W.H., San Francisco*, 1980.

Aubray, D., "*Statistical and dynamical prediction of changes in natural sand beaches*", *Ph.D. dissertation* (Scripps Institute of Oceanogr., San Diego, California) 1978.

Babu, M.T., Prasanna Kumar, S., and Rao, D.P., "A subsurface cyclonic eddy in the Bay of Bengal", *J. Mar. Res*., 49, 404-410, 1991.

Babu, M.T., Sarma, Y.V.B., Murty, V.S.N., and Vethamony, P.; On the circulation in the Bay of Bengal during northern spring inter-monsoon (March-April) 1987. *Deep sea Res. II* 50, 855-865, 2003.

Chen, C.T., and Millero, F.J., "Speed of sound in sea water at higher pressure"; *J. Acoust. Soc. America*, 60, 129-135, 1977.

Chelton, D.B., DeSzoeke, R.A., Schlax, M.G., E1Naggar, K., and Siwertz, N., "Geophysical variability of the first baroclinic Rossby radius of deformation", *J. Phys. Oceanogr*., 28, 433-460, 1998.

Cornuelle, B.D., *"Inverse methods and results from the 1981 ocean acoustic tomography experiment"*, *Ph.D.Thesis*, Woods Hole Oceanographic Institution, Mass., 1983.

Denman, K.L., and Dower J.F., "*Patch dynamics in encyclopedia of ocean sciences, (Ed by J.H. Steele et al)*", Academic Press, New York, 2107-2114, 2001.

DeSanto, J.A., "Theoretical Methods in Ocean Acoustics", *In Ocean Acoustics*, edited by DeSanto J.A., 7-77, 1979.

Ewing, W.M., and Worzel, J.L. "Long range sound transmission", *Geol. Soc. Am. Mem.* 27, part III, 1-35, 1948.

Flatte, M., Dashen, R., Munk, W.H., Watson, W., Zachariasen, F., *"Sound Transmission Through a Fluctuating Ocean"*, Cambridge University Press, England, 56-57, 1979.

Fukumori, I., *Efficient representation of the hydrographic structure of the North Atlantic Ocean and aspects of the circulation from objective methods,* Ph.D. dissertation (Joint program in Oceanography, Massachusetts Institute of Technology – Woods Hole Oceanographic Institution), 236, 1989.

Gregg, M.C., and Briscoe, M.G., "Internal waves, finestructure, and mixing in the ocean," *Rev. Geophys. Space Phys.* 17, 1524-1547, 1979.

Hamilton, K.G., Siegmann, W.L., and Jacobson,M.J., "Simplified calculation of ray phase perturbations due to ocean environmental variations" *J.Acoust.Soc.Am.,* 67, 1193-1206, 1980.

Hua, B.L., McWilliams, J.C., and Owens W.B., "An objective analysis of the POLYMODE local dynamics experiment. Part II: streamfunction and potential vorticity fields during the intensive period" *J. Phys. Oceanogr.*, 16, 1985.

Jackson, D.D., Interpretation of inaccurate, insufficient and inconsistent data, *Geophys. J.Astron.Soc.*, 28, 97-109, 1972.

Krol H R., "Intensity calculations along a single ray", *J Acoust Soc Am.*, 53, 864-868, 1973.

Levitus, S., "*Climatological Atlas of the World Ocean*", Professional Paper 13, NOAA, Rockville, MD, 1982.

Liebelt, P.B., *"An Introduction to Optimal Estimation",* Addision-Wesley Publishing Company, London, pp 272, 1967.

Lawson, C.L., Hanson, R.J. "Solving least squares problem", *Prentice Hall,* New Jersy, 1974.

Legeckis, R., "Satellite observations of a western boundary current in the Bay of Bengal" *J.Geophys.Res.*, 92, 12974-12978, 1987.

Madhusoodhanan, P., and James, V.V., "Thermohaline features of the subsurface cyclonic eddy in the south central Bay of Bengal during August 1999". *Proc.Ind.Acad.Sci (Earth and Planet Sci.),* pp 112, 2003.

Menk, W., "*Geophysical data analysis: Discrete Inverse Theory*", 2nd *Academic Press*, San Diego, 1989.

Moler, C.B., Solomon, L.P., "Use of splines and numerical integration in geometrical acoustics. J. of Acou. Soci. of Amer. 48, 739, 1970.

MODE group, Mid-ocean dynamics experiment, *Deep Sea Res*, 25, 859-910, 1978.

Munk, W.H., and Wunsch, C., "Ocean acoustic Tomography: A scheme for large scale monitoring"; *Deep Sea Res.*, 26A, 161, 1979.

Munk, W.H., and Wunsch, C., "Observing the oceans in the 1990s", *Phil. Trans. Royal Soc.* London. A307, 439-464, 1982.

Munk, W.H., and Forbes, A.M.G., "Global Ocean Warming: An acoustic measure?" *J. Phys. Oceanogr.* 19, 1765-1778, 1989.

Park, J.H., "Study on the analyses of the Coastal Acoustic Tomography Data", *Doctor Dissertation, Graduate School of Engineering, Hiroshima Unversity*, 2001.

Pedlosky, J., "Geophysical Fluid Dynamics", *Springer-Verlag, New York*, pp.625, 1980.

Prasanna kumar, S., Ramana Murty, T.V., Somayajulu, Y.K., Chodankar, P.V., and Murty, C.S., "Reference sound speed profile and Related Ray Acoustics of Bay of Bengal for Tomographic Studies", *Acustica,* 80, 127-137, 1994.

Prasanna Kumar, S., Ramana Murty, T.V., Somayajulu, Y.K., Saran, A.K., Navelkar, G.S., Almeida, A.M., Fernando, and Murty, C.S., "Acoustic Tomography Experiment in the Eastern Arabian Sea", *ACUSTICA Acta acustica*, 85, 31-38, 1999.

Prasanna Kumar, S., Nuncio M., Jayu Narvekar, Ajay Kumar, Sardesai S., Desouza, S.N., Mangesh Gauns, Ramaiah N., and Madhupratap, M.; "Are eddies nature's trigger to enhance biological productivity in the Bay of Bengal?", *Geophys. Res. Lett.*, 31, L07309, doi: 10.1029/2003G1019274, 2004.

Ramasastry, A.A., and Balaramamurty, C., "Thermal fields and oceanic circulation along the east coast of India", *Proceedings of Indian Academy of Science*, 46, 293-323, 1957.

Ramana murty, T.V., Veerayya, M., and Murty C.S., "Sediment size distribution of the beach and near shore environs along central west coast of India: An analysis using EOF", *J. Geophys. Res.*, 91, 8523-8526, 1986.

Ramana Murty, T.V., Prasanna Kumar, S., Somayajulu, Y.K., Sastry, J.S., De Figueiredo, R.J.P., " Canonical sound speed profile for the central Bay of Bengal", *J. Eart. Sys. Science*, 253-263, 1989.

Ramana Murty, T.V., Somayajulu, Y.K., and Sastry, J.S., "Computation of some acoustic ray parameters in the Bay of Bengal", *Indian J.Mar.Sci.* 90, 235-245, 1990.

Ramana Murty, T.V., Somayajulu, Y.K., Mahadevan, R., Murty C.S., and Sastry J.S., "A solution to the inverse problem in Ocean acoustics", *Defence Science Journal*, 42, 89-101, 1992.

Ramana Murty, T.V., Mahadevan, R., *Stochastic Inverse Method for Ocean acoustic tomography studies-A simulation experiment, Technical report No. 31*, Indian Institute of Technology, Madras, India, 1994.

Ramana Murty, T.V., and Mahadevan, R., "The stochastic inverse method for Ocean acoustic tomography studies – a simulation experiment", *Acust. Lett., 19, 15-19, 1995.*

Ramana Murty, T.V., Malleswara Rao, M.M., Surya Prakash, S., Chandramouli, P., Murty, K.S.R., *Algorithams and interface for ocean acoustic ray tracing, Technical report No. NIO/TR-09/2005,* National Institute of Oceanography, Dona Paula, Goa, India, 2005.

Ramana Murty, T.V., Mohan Rao, K., Malleswara Rao, M.M., Lakshmi Narayana, S., and Murty, K.S.R., "Sediment – size distribution of innershelf off Gopalpur, Orissa Coast using EOF analysis", *Journal Geological Society of India,* 69, 133-138, 2007.

Ramana Murty, T.V., Malleswara Rao, M.M., and Sadhuram, Y., Sujith Kumar, S., Sai Sandhya, K., Maneesha, K., and Murthy, K.S.R., "*Identification of internal waves off Visakhapatnam from Thermister Chain", Proceedings*, National Symposium on "Emerging trends in Meterology & Oceanography, METOC, Kochi, 2007a.

Ramana Murty, T.V., Malleswara Rao, M.M., and Sadhuram, Y., " *Objective mapping of temperature field by stochastic inverse method using acoustic tomography experimental data of eastern Arabian sea", Proceedings,* National Symposium on "Emerging trends in Meterology & Oceanography, METOC, Kochi, 2007b.

Rao, D.P., and Sastry, J.S., "Circulation and distribution of some hydrographical properties during the late winter in the Bay of Bengal" , *Mahasagar-Bull. Natl.Inst. Oceanogr.*, 14, 1-16, 1981.

Richman, J.G., Wunsch C., and Hogg N.G.,"Space and time scales of meso-scale motion in the western North Atlantic", *Rev. Geophys. and Sp. Phys.* ,L5, 385, 1977.

Robinson, A. R, "*Eddies in marine sciences", Springer Verlag* , New York, 1983.

Robinson, A.R, and Ding Lee, "*Oceanography and Acoustics*", *AIP Press*, 257, 1994.

Sastry, J.S., Somayajulu, Y.K., Prasanna Kumar, S., Ramana Murty, T.V., and Raghavan, K., " *Forward problem related to Ocean acoustic tomography*", *Technical report No. 3/84,* National institute of Oceanography, Dona Paula, Goa, 1984.

Simmen, J., Flatte, S., & Wang, G., "Wavefront folding, chaos, and diffraction for sound propagatin through ocean internal waves", *Jouranl of the Acoustical Society of America,* 102(1), 235-255, 1997.

Somayajulu, Y.K., "*Some aspects of simulation of sound propagation in the Arabian Sea* ", Ph.D Thesis, Goa University, India, 25, 1993.

Somayajulu, Y.K., Ramana Murty, T.V., Prasanna Kumar, S., and Murty, C.S., "Some studies related to acoustic propagation in the Arabian Sea", *Acoust.Lett.*, 17, 173-184, 1994.

Spiesberger, J.L., Spindel, R.C., and Metzger, L., "Stability and identification of ocean acoustic multipaths", *J. geophys. Res.* 96, 4869-4889, 1980.

Spiesberger, J.L., and Metzger, K.M., "Basin-scale tomography: A new tool for studying weather and climate," *J. Geophys. Res.* 96(C3), 4869-4889, 1991.

Sridevi, B., Ramana Murty, T.V., Sadhuram, Y., Rao, M.M.M., Maneesha K., Sujith Kumar, S., Prasanna P.L., "Impact of Internal waves on sound propagation off Bhimilipatnam, east coast of India", *Eastuarine Coastal and Shelf Science*, 88, 249-259, 2010.

Sridevi B., Ramana Murty T.V., Sadhuram Y., Murty V.S.N., "Impact of Internal waves on the acoustic field at a coastal station off Paradeep, east coast of India", *Natural Hazards,* 57, 563-576, 2011.

Stefanie E.W., "*Effects of internal waves and turbulent fluctuations on underwater acoustic propagationt*", *Ph.D.Thesis*, Worcester Polytechnic Institution, Worcester, USA., 2006.

Stidd C.K., "The use of eigenvectors for climate estimates", *J. Appl Meteor.*, 6, 255-271, 1967.

Swallow J.C., "*Eddies in the Indian Ocean*" (Ed. By A.R.Robinson), *Springer_Verlag Berlin Heidelberg*, 1983.

Wiggins, R.A., "The general linear inverse problem : Implication of surface waves and free oscillations for earth structure", *Rev. Geophys. Space phys*, 10, pp.251-285, 1972.

Winant, C.D., Inman, D.L., Nordstorm, C.E., "Description of seasonal beach changes using empirical eigen functions", *J. Geophys. Res.*, 80, 1979-1986, 1975.

Worcester, P.F., Spindel, R.C., and Howe, B.M., "Acoustic transmission: Instrumentation for meso-scale monitoring for ocean currents", *IEEE J. Ocean Engg.* OE-10, 123-137, 1985.

Worcester, P.F. and Cornuelle, B.D., "A review of Ocean Acoustic Tomography: 1987-90", *Rev. Geophys., Suppl.*, 855-865, 1991.

INDEX

A

B

C

D

E

F

G

O

P

Q

R

S

T

U

V

W

Y